Analysis of Rubber and Rubber-like Polymers

THIRD EDITION

Analysis of Rubber and Rubber-like Polymers

THIRD EDITION

W. C. WAKE

Ph.D., D.Sc.(Lond.), F.R.S.C., F.P.R.I., Hon.D.Sc.(The City University)

Hon. Fellow, The City University, London, UK

B. K. TIDD

Ph.D., M.R.S.C.

Head of the Analytical Group, Malaysian Rubber Producers' Research Association, Brickendonbury, Hertfordshire, UK
Chairman of BSI Committee RUC/37 (Chemical Analysis of Rubbers)

and

M. J. R. LOADMAN

M.Sc.

Deputy Head of the Analytical Group, Malaysian Rubber Producers' Research Association, Brickendonbury, Hertfordshire, UK

APPLIED SCIENCE PUBLISHERS
LONDON and NEW YORK

APPLIED SCIENCE PUBLISHERS LTD
Ripple Road, Barking, Essex, England

Sole Distributor in the USA and Canada
ELSEVIER SCIENCE PUBLISHING CO., INC.
52 Vanderbilt Avenue, New York, NY 10017, USA

First edition 1958
Second edition 1969
Third edition 1983

British Library Cataloguing in Publication Data

Wake, William C.
Analysis of rubber and rubber-like polymers.—3rd ed.
1. Rubber chemistry
I. Title II. Tidd, B. K. III. Loadman, M. J. R.
678′.4 TS1892

ISBN 0-85334-215-6

WITH 60 TABLES AND 71 ILLUSTRATIONS

Photoset in Great Britain by Keyset Composition, Colchester
Printed in Great Britain by Galliard (Printers) Ltd, Great Yarmouth

Preface

The first edition of this book described the analytical situation which had existed for a number of years for maintaining quality control on vulcanizates of natural rubber. This situation had recently been disturbed by the introduction of a range of synthetic rubbers which required qualitative identification and needed to be estimated quantitatively. For the former purpose 'wet' chemistry based on various imperfectly understood organic reactions was pressed into service. Alongside this was the first introduction of instrumental analysis by 'fingerprinting' the infrared spectra either of the polymers or, more usually, their pyrolytic products. The identification of materials from the range of organic accelerators and antioxidants introduced during the 1930s was, in the first edition, dealt with by a combination of column chromatography and IR spectroscopy or by paper chromatography. Quantitative procedures were, however, still classical in the tradition of gravimetric or volumetric analysis with an initially weighed sample yielding, after chemical manipulation, a carefully precipitated or dried end product or a solution of known simple composition whose weight or titre, as a percentage of the initial sample, defined the function being determined. An example of such was the determination of 'total' and 'free' sulphur which was the major analytical operation connected with the rubber industry before 1940.

The second edition of this work consolidated the newer techniques introduced in the first without adding to them although, in other applications of analytical chemistry, instrumental analysis had already brought about a transformation in analytical practice.

The third edition gives full credit to modern instrumentation in all aspects of the analysis of rubber and rubber-like polymers, describing techniques and illustrating applications where equipment still at the 'research stage' could add to the strength of the analyst's armoury of the future. Nevertheless the financial limitations in force today are appreciated, and within each area of analysis is covered a variety of techniques from the 'burn test', costing nothing, to those using instrumentation costing many thousands of pounds. Whatever equipment the analyst or technologist has, it is intended to indicate how best it can be used, and where the 'next instrumental step forward' could lie.

Because most of the queries the analyst receives are from technologists the policy has been to show how specific analytical data can be intercorrelated and how this can then be expressed in the technological language of 'formulation'. At the same time the validity of these results is discussed in detail in an attempt to prevent the biggest problem the analyst faces—unintentional misquotation!

Two of us (B.K.T. and M.J.R.L.) acknowledge with gratitude the assistance given by our colleagues, both in discussion and in the mass of previously unpublished data which they have made available to us. We trust that their contribution will be apparent through references within the text, as indeed will those of all whose work has been quoted.

Acknowledgement is also due to the Board of the Malaysian Rubber Producers' Research Association (MRPRA) for permission to undertake this project and for the facilities made available to two of us (B.K.T. and M.J.R.L.); without the agreement of the Director of Research (Dr L. Mullins, CMG, JMN) this edition would not have been possible. Finally, our most sincere thanks are due to Mrs Joyce Davies who managed to translate many pages of manuscript into an intelligible typescript.

W. C. WAKE
B. K. TIDD
M. J. R. LOADMAN

Acknowledgements

In a book of this nature it is inevitable, and indeed desirable, that a wide range of publications be consulted to afford as balanced a picture as possible of the current position in the analysis of rubber and rubber-like polymers.

From these publications many tables and figures have been culled to illustrate relevant points throughout the text and it is with much gratitude that we, and the publisher, thank the copyright holders for permission to use their data.

The very number of these necessitates an almost tabular presentation, but the brevity in no way reduces the sincerity of our appreciation to:

The British Standards Institution for Figs. 6.3 and 6.4 (BS 903 Pts 6–10), and the American Society for Testing and Materials, together with A. Krishen (1974) for Figs. 8.6, 8.7 and 8.8. Full copies of these documents may be obtained from the appropriate organization at 2 Park Street, London, and 19 Race Street, Philadelphia, PA 19103, respectively.

The National Bureau of Standards, with J. J. Maurier (1970) for Table 11.7.

John Wiley & Sons Inc., together with: I. R. Gelling, M. J. R. Loadman and B. D. Sidek (*J. Polym. Sci. Polym. Chem. Edn.*, 1979) for Figs. 8.12, 8.13 and 8.14; H. G. Kim and L. Mendelkern (*J. Polym. Sci. Pt. A2*, 1972) for Fig. 8.16; M. B. Evans, G. M. C. Higgins, D. F. Lee and W. F. Watson (*J. Appl. Polym. Sci.*, 1960) for Fig. 7.2 and Table 7.4, and B. Lee and C. Singleton (*J. Appl. Polym. Sci.*, 1979) for Fig. 8.17. Also, from the *Textbook of Polymer Science* (2nd ed., 1971, F. W. Billmeyer Jr.), Fig. 9.3 and from the *Encyclopedia of Industrial and Chemical Analysis* (Vol. 8, 1969), the method for iodine adsorption on carbon black as described by B. Schubert, E. P. Ford and F. Lyon, and, from Volume 12 (1971), Fig. 4.1 and Table 4.6 (W. Cooper, F. C. J. Poulton and P. R. Sewell).

The American Chemical Society, with C. J. Carman (*Macromolecules*, 1973) for Fig. 9.12 and with A. Krishen (*Analyt. Chem.*, 1972) for Fig. 8.5.

The Editor of *Rubber Chem. Technol.*, with: S. J. Swarin and A. M. Wims (1974) for Figs. 12.2, 12.3 and 12.4 together with Tables 12.6 and 12.7; D. W. Brazier and G. H. Nickel (1975) for Table 12.5; R. Pautrat, B. Metivier and J. Marteau (1976) for Fig. 11.3; A. K. Sircar and T. G. Lamond (1978) for Fig. 12.5; D. W. Brazier (1980) for Table 12.3, and E. L. Charsley and J. G. Dunn (1982) for Fig. 11.4 and their subsequent conclusions therefrom, quoted verbatim.

The Editor of *Rubber Age*, with J. J. Maurier (1970) for Figs. 11.1, 11.2 and Table 11.2.

The Plastics and Rubber Institute with: J. R. Davies and F. W. Kam (1967) for Table 11.3; E. A. Ney and A. B. Heath (1968) for Figs. 8.3 and 8.4; B. K. Tidd (1977) for Figs. 3.5 and 6.6 together with Tables 12.8 and 12.9, and J. E. Davey, A. D. Edwards and G. M. C. Higgins (1978) for Fig. 6.5.

The Editor of *Rubber J.*, with T. G. Lamond and C. R. Gillingham (1970) for Tables 11.5 and 11.6, and with T. G. Lamond and C. R. Price (1970) for permission to reproduce the 'OT method' of carbon black analysis.

The Editor of *Rev. Gen. Caoutch. Plast.* with R. Pautrat, B. Metivier and J. Marteau (1975) for Table 12.4, and *Thermochim. Acta*, with S. H. Goh (1980) for Fig. 8.19.

The Literary Executor of the late Sir Ronald A. Fisher, FRS, Dr Frank Yates, FRS, and the Longman Group Ltd, London, for permission to use data from their book *Statistical Tables for Biological, Agricultural and Medical Research* (6th ed, 1974) (previously published by Oliver & Boyd Ltd, Edinburgh) in the construction of Tables 14.2 and 14.3, and Pergamon Press, with T. T. Gorsuch, for Fig. 10.1, from *The Destruction of Organic Matter* (1970).

H. A. Willis, for Fig. 7.1, taken from *The Identification of Plastics* (1965); Professor G. Lombardi of the City University, Rome, for Table 8.3, taken from the ICTA booklet *For Better Thermal Analysis* (1980), and W. A. Brown, D. E. Schleifer and A. C. Patel, at Ashland Chemical Co., Columbus, Ohio, for the data incorporated into Table 11.4.

Our thanks are also due to Dionex (UK) Ltd (Frimley, Camberley, Surrey), Stanton Redcroft Ltd (Copper Mill Lane, London), Hewlett Packard Ltd (Winnersh, Wokingham, Bucks) and the Parr Instrument Co. (Moline, IL 61265, USA) for permission to reproduce, from their brochures, Figs. 6.7 and 6.8; 8.9; 9.2, and 10.2 respectively.

Finally we record our thanks to the Director of the Rubber Research Institute of Malaysia, for allowing the use of the data in Table 14.6 and to the Board of the Malaysian Rubber Producers' Research Association for Figs. 3.4 and 8.2, taken from the 'house publication' *NR Technol.* and credited to G. M. C. Higgins (1978) and to G. M. C. Higgins and M. J. R. Loadman (1970) respectively.

Contents

Chapter 1

Introduction

THE NATURE OF THE MATERIALS

Most readers will have a clear idea of the sort of materials that are covered by the title of this book. It is not easy, however, to define them unambiguously so that it is immediately apparent whether or not a given material is, or is not, included. The difficulty has been in part avoided by giving an appendix of materials which have been considered in drawing up the methods and schemes of this book but, even so, some attempt must be made at definition, or at least at description.

Natural rubber derived from the tree *Hevea brasiliensis* is the prototype of the modern range of materials which have a high extensibility combined with an ability to recover from extension. These properties have been found to be dependent on a certain type of molecular structure, and rubber-like materials have physical properties similar to natural rubber because they, too, have the same molecular pattern in their structure. It has been usual to refer to the rubber-like materials as *highly elastic*, and to group their properties as *high elasticity*. The terms *highly elastic* and *high elasticity* refer to the degree of recoverable extension and not, as in classical physics, to the coefficient of restitution. The essential features of molecular structure that give rise to high elasticity are the ability of certain atoms to unite forming long, very flexible, chains and the existence of a range of molecular attractions between these chains which modifies the degree of flexibility. If the chains are perfectly uniform and symmetrical, the molecular attraction between them will reduce flexibility and even lead to crystal formation. If they are completely irregular the material will possess little strength and will break at a comparatively low extension. For dimensional stability over long periods of time it is further necessary that the molecular chains should be linked together by occasional crosslinks to form a three-dimensional network. All the materials in the appendix conform in structure to the first two of the above requirements but those which also conform to the last condition, that of possessing a crosslinked three-dimensional structure, are those referred to as *rubbers* whereas the remainder are *rubber-like.* In general one may say that the rubber-like materials are, in popular speech, the *flexible plastics* or *thermoplastics.* Rubber and rubber-like materials are then materials composed of long-chain molecules, which show high elasticity.

The term *polymer* has not yet been introduced in this discussion although the reader familiar with these materials will probably know them under the generic title of 'high polymers'. When the long molecular chain of plastic sulphur is considered, the sulphur atom forms the simple unit from which it is built, but most other long-chain molecules are formed by the repetition of a rather more complex unit consisting of several atoms to which other atoms or atomic groupings are appended. The simplest unit, repetition of which builds the long molecular chain, is the *monomer*, and structural formulae familiar to the organic chemist can be built up on a pattern $(\text{monomer})_n \rightarrow$ polymer. Details of the methods whereby polymers can be built from monomers are outside the scope of this book but it will be realized that more than one type of monomer can be used to build the polymer and, in fact, many important synthetic rubbers are *copolymers* derived from two monomers. Where two or more monomers are used, the chain molecule does not necessarily contain a regular and uniform alternation of the units; nor are the monomer units necessarily distributed randomly along the length of the chain. Indeed the extent to which monomers exist in blocks, and the length of these blocks, play an important part in the properties of a particular copolymer. The use of only one monomer naturally leads to rather greater regularity if only because the possible irregularities are fewer. However, even a relatively simple monomer such as 1,3-butadiene can give rise to polymeric irregularity due to combination in the 1,2 instead of the 1,4 position and due to the possibility of the spatial configuration around the central double bond in each unit of the polymer being *trans* as well as *cis*. It will have been appreciated that the polymerization of butadiene involves a loss of unsaturation since $CH_2{=}CH{-}CH{=}CH_2$ becomes in the polymer $({-}CH_2{-}CH{=}CH{-}CH_2{-})_n$, the monomer units connecting up end to end during the polymerization process. This method of polymerization is the *addition* method and the resulting polymers are *addition polymers.* Simple addition polymerization of monomers can thus lead to quite complex irregular structures which are, nevertheless, still chain molecules.

An alternative approach to the formation of a long molecular chain is that of the condensation of two or more types of molecule each possessing two, and no more than two, functional groups and accompanied by the elimination of some simple molecule such as hydrogen chloride or water. It is by this method that the familiar alkyd resins used in paint technology are made from dibasic acids and glycols with the elimination of water. There is, however, a limit to the size of the molecule that can be made in this way and this limit is below the size where useful rubber-like properties are developed. In order to achieve sufficient polymerization a further stage is added whereby the molecules obtained from the condensation process are linked together, usually by very highly reactive diisocyanates. This process is the basis of important synthetic rubber-like materials used as rubbers, such as highly elastic lacquers and foams.

Unlike addition polymers, polycondensates must, from their chemical nature, be completely regular although not all molecular chains will be of the same size. Irregularity is immediately introduced by the use of isocyanates in the second stage since molecules of differing sizes will be joined together, not necessarily end to end, in forming the longer chain.

So far, only the formation of chain molecules has been discussed but rubbers possess a crosslinked three-dimensional structure formed by joins from one chain to another. These crosslinks must be sufficiently frequent to prevent chains from sliding over each other but not frequent enough to destroy their essential flexibility; in practice about 1% of monomer units take part in the crosslinking process. Convenient crosslinking agents are chemically dissimilar from the chains they link and, as is well known, sulphur is the most common, its use being termed the *vulcanization* of rubber. More recently, the term *curing* has tended to replace *vulcanization*, and both are now used almost synonymously for any crosslinking of a long-chain substance to give a rubber.

Given suitable conditions of solvent and temperature, most high polymers will dissolve, but ease of solution decreases with increasing molecular size, and crosslinking leads, of course, to an effectively infinite molecular size. At such an extreme size, solution of the polymer in solvent cannot occur but mixing of solvent molecules with the network of the polymer is possible and this is the mechanism of solvent swelling. The swollen polymer still possesses the crosslinked structure of its unswollen state but is rendered softer, more flexible and weaker because the strength and rigidity of the network structure are no longer enhanced by the attraction of the polymer chains for each other. A limited mixing of polymer and a suitable solvent is sometimes advantageous and is the basis of the use of polyvinylchloride as a flexible plastic. Pure polyvinylchloride is a hard, rigid, horn-like material which can be mixed with a solvent such as dibutylphthalate to give the familiar plastic material which is sufficiently soft and flexible for garments to be made from it. The solvent material is referred to as a plasticizer and it functions by separating the molecular chains one from another. Popular technical usage seems to have restricted the term *plasticizer* to solvents used with synthetic materials such as polyvinylchloride and nitrile rubbers, preferring the term *softener* or *extender*, depending upon the level present, for materials performing a similar function in natural and general-purpose rubbers.

Rubber-like materials, or thermoplastics, can conveniently be divided into five types as illustrated in Table 1.1.

TABLE 1.1

CLASSIFICATION OF THERMOPLASTIC MATERIALS

Type	*Soft component* *A*	*Hard component* *B*	*Structure*
Polyvinyl rigid plastics	Plasticizer	Polyvinylchloride or ethylene–vinylacetate copolymer	Swollen polymer
Polyurethanes	Polyether glycol or polyester glycol	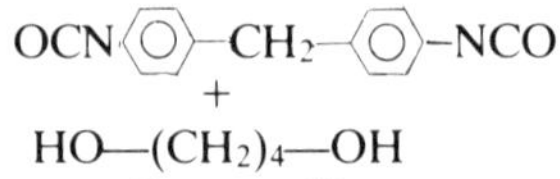	$(AB)_n$
Polyether esters	Polyether glycol	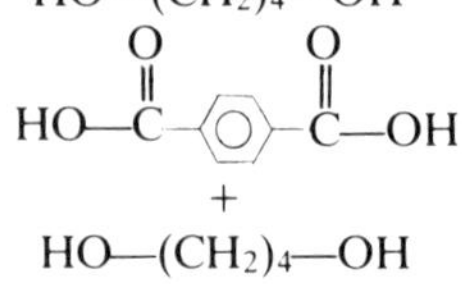	$(AB)_n$
Styrene block/star copolymers	Polyolefin	Polystyrene	ABA, ABA A
Thermoplastic elastomer	EP rubber or NR	Polypropylene or polyethylene	Blend

Yet a further type of elastomer is currently emerging from the interface of rubber and plastics technology. Dissolution of a rubber such as natural rubber, ethylene propylene terpolymer, or dihydrochlorinated chlorobutyl rubber in a monomer such as styrene or acrylonitrile, followed by peroxide-initiated polymerization and vulcanization, results in the plastic, which is incompatible with the rubber in that they have a negligible mutual solubility, precipitating as small reinforcing particles within the crosslinked elastomeric phase. The rubber can retain its elastic properties with up to 75% loading of polymerized monomer.

In addition to crosslinking and plasticizing, it is frequently desirable to incorporate powders into the materials to increase their bulk, alter their density, reduce their resilience, cheapen their cost by dilution, or to modify some special property. These powders are incorporated before crosslinking and are dispersed in the polymer, which then forms a continuous matrix. In the case of rubbers, the complexity does not end at this point because the crosslinking agent is rarely sulphur alone. Accelerators are added to speed the rubber–sulphur reaction, zinc oxide to 'activate' the accelerator, and some fatty acid, usually stearic acid, to assist the activation.

In general, organic materials are not as stable as inorganic materials. Just as wood or leather has a shorter life than stone, iron or copper, rubbers and rubber-like materials, although they may be synthetic, are organic in nature and age in the presence of air and light. This ageing is partially counteracted or deferred by small amounts of stabilizing agents which are added either during the manufacture of synthetic material or when mixing the polymer with plasticizer and powders. In rubbers possessing some degree of unsaturation the stabilization is with antioxidants or antiozonants, to protect against oxidative damage, whilst in PVC it is with a metal oxide to protect against loss of hydrogen chloride.

THE ANALYTICAL PROBLEM

The rubber or rubber-like material is in essence a mechanical mixture of polymer and inert powders, or fillers as they are usually called, together with a number of other ingredients which may be regarded as being dissolved in the polymer. The temperature of vulcanization and the presence of reactive species cause changes in the nature of these other ingredients so that they frequently no longer exist in their original form in the vulcanizate and it is necessary to identify the products derived from them to deduce their original presence.

The qualitative and quantitative separation and identification of any, or all, of these chemicals can be a complex and time-consuming process and it is thus important to consider the purpose for which an analysis is required, what degree of accuracy is needed, and which of a variety of available methods, if any, will enable it to be achieved. It is worth remembering that Parkinson's Law applies as much to the analytical laboratory as elsewhere, and may be stated as 'Whenever new equipment, techniques or automation are introduced, demand will increase proportionately'. Nothing is as cost effective as a sceptical approach to the question of the need for a particular analysis.

Analysis of rubbers or rubber-like materials seems to be carried out under the headings given below, although any order of priority must depend upon the type of laboratory carrying out the work and the purpose for which the work is required.

(1) Complete analysis of a competitor's product.
(2) Partial analysis, e.g. fillers only or nature and percentage of polymer, under similar circumstances.
(3) Reasonably complete analysis of representative samples purchased, e.g. by a Government Department, to a rigid specification.
(4) Specific analysis, e.g. combined and free sulphur, type of antioxidant etc.
(5) Analysis as a means of checking product behaviour, e.g. pH of aqueous extract of a gasket intended for use in contact with metal.
(6) Analysis of deteriorated or faulty products to determine, if possible, the cause.
(7) Analysis to detect factory errors.

The reasons for desiring to know the exact make-up of a competing product may or may not be regarded as an ethical problem by the reader but, fortunately, this has but little bearing on the analytical problem. If exact repetition is the reason for requiring the analysis then a full analysis will necessarily include sulphur, accelerators and antioxidants in the case of rubber, and plasticizers and stabilizers if the material is a flexible plastic, though quantitative assessment of accelerator or stabilizer will rarely be essential.

Sometimes, however, it is interest in the cost of materials that prompts analysis, and then chemical differentiation between various antioxidants or stabilizers would be unnecessary, for the cost difference, if one were substituted for another, would be insignificant relative to small bulk ingredient changes. In general, analysis for costing purposes only requires the identification and estimation of the polymer and bulk fillers.

The examination of a certain percentage of products, purchased to a specification which lays down their composition, is rarely carried out nowadays. Many authorities regard a specification as laying down the performance required, leaving the manufacturer to achieve this in his own way. This largely abolishes the need for lengthy chemical analysis and substitutes the easier and cheaper methods of physical testing, before and after accelerated ageing if this is required. This can, however, cause problems for the analyst if he is asked by a user to comment on the reasons for a particular product failing to meet its required performance specification. With no knowledge of the polymer, fillers or other chemicals present, a complete analysis will be necessary in order to see whether or not it would be expected to meet the specification, even if correctly mixed and cured, before considering possible errors in manufacture.

On the other hand, certain contracts contain a clause requiring disclosure of the materials of manufacture and some of the reasons for this are not sufficiently appreciated. Where the rubber or rubber-like material is in contact with complex materials, e.g. explosives, living tissue, food or medical supplies, the manufacturer cannot be expected to foresee all possible effects of the chemicals incorporated into the finished article which he supplies. Only the user can have anything approaching sufficient knowledge of which ingredients are, or are not, permissible. Disclosure by the manufacturer ensures that only suitable materials are used but disclosure without the possibility that subsequent departures from the disclosed formula will be detected is no safeguard. Where health or safety is at stake check analyses may be necessary on every batch. However, disclosure considerably lightens the analyst's task, since the analysis can be designed around the known formula with the omission of many steps that would be essential were the product unknown. Care must be taken, however, that in designing a specific analytical sequence it does not become so specific as to exclude the observation of extra-specification materials, i.e. in making sure that one particular antioxidant is present, one should also ascertain that all others are absent.

Further uses for the analysis of rubber and rubber-like materials are those of examining deteriorated or faulty products. These may be rejected by the purchaser, or be returned after use or misuse. In all cases it will be necessary to carry out an investigation to a greater or lesser extent and, almost inevitably, in the early stages of such an investigation the faulty article will be examined by analysis. It may be, for example, that a bloom has developed in the warehouse; the analyst is consulted on the nature of the bloom and once this is unambiguously known the problem is usually more than half solved. This leads to the consideration of analysis as an instrument of control within a factory during production. The improvement in physical testing of materials has led manufacturers increasingly to depend upon physical properties as a criterion of correct manufacture, but advances in instrumental techniques over recent years should make each factory manager consider whether any particular one could be of use to him. Typically, a vulcanized product can be analysed non-destructively for sulphur content in under two minutes whilst an 'oil', polymer, black, inorganic filler analysis can be obtained on a few milligrams of sample in under ten minutes. Perhaps more important still, batches of uncured compound can be checked and adjusted if necessary before cure, thus preventing possible wastage and reducing product variability.

SCOPE OF THE BOOK

There are available to readers of English three comprehensive sources of analytical methods for rubber: the publications of the International Organization for Standardization (ISO), British Standards Institution (BS) and the American Society for Testing and Materials (ASTM). Each gives standard methods for performing a range of analyses, the details given being precise and comprehensive, covering everything from the design of suitable apparatus and the quality of reagents to the manipulative details for each step. For many estimations alternative methods are given. It is not proposed that this book should supersede the published works of the standardizing bodies but rather that it should supplement them and for reference a list of current ISO, BS, ASTM and DIN (Deutsches Institut Für Normung. e.V.) documents relevant to the analysis of rubber and rubber-like materials is given in Appendix I.

There are two ways in which supplementation is needed and should be useful. In the first place, the standard methods often give no indication of when they should be used or why one method is preferable to another. In the second place, there is no attempt to incorporate the discrete methods into an analytical scheme designed for this or that purpose. When, in the course of general analytical practice, an analyst is called upon to estimate, say, the total sulphur content of sample A, he rightly turns to BS 903: Part B6-1958, looks up the method and meticulously follows the instructions. He there finds all the information he needs and, if he is competent, the right answer follows automatically. Suppose, however, that he is asked to investigate a faulty product, or to advise on suitable procedure for factory control; he then needs a conspectus of methods available together with information illustrative of their use. He will need to know what has been attempted before and what suggestions have been made for dealing with the sort of problem with which he is faced. He will need, in short, a critical assessment of analytical practice in the field of the material in question, and it is for such a person that the present work is intended.

In the following chapters an attempt is made to assess critically the tools and practice of analysis applied in the field of rubber and rubber-like materials. Although the major concern of this book is the identification and estimation of the components of the complex material of a manufactured product, this includes, of necessity, certain aspects of raw rubber analysis. Standard methods, already published, are not in general repeated here and only where a method is not in a British or International Standards publication are procedural details fully set out.

The advent of new instrumental techniques since the first edition of this book was published, together with the better understanding of the significance of such factors as microstructure and molar mass distribution, has led to the introduction of a chapter on basic aspects of polymer characterization.

At the time of publication of the first edition, books devoted solely or even principally to the analysis of rubber and rubber-like materials had been rare although most textbooks on rubber chemistry and, more recently, on high polymers, devote some space to the topic. The first textbook of analysis was either that of Ditmar or that of Pontio, since both were published in 1909. *Die Analyse des Kautschuk der Guttapercha Balata und ihrer Zusatze* is the title of Ditmar's work. It contains much discussion on the theory of the constitution of rubber, the preparation of chemical derivatives such as brominated rubber, and the alleged structures of these. Pontio's *Analyse du Caoutchouc et de la Gutta-Percha* is altogether lighter but nevertheless contains the essential processes for the examination of rubber from various botanical sources as well as alternative methods for analysing vulcanized rubber. The first work in English seems to have been that of Caspari (1914) which, in spite of its title, *India-Rubber Laboratory Practice*, was concerned almost exclusively with analysis. Of course, analysis had been dealt with extensively in Weber's much earlier book *The Chemistry of Rubber* (1902), and the popularity of this work may have accounted for the lack of a book specifically on the subject. Tuttle followed in 1922 with the first American book, *The Analysis of Rubber*, and after this there was a gap until the Ministry of Supply published, at first for limited circulation only, its *Users' Memorandum No U.9, Identification and Estimation of Natural and Synthetic Rubbers*, in 1944, and a revised edition in 1946. This was actually a pamphlet rather than a book, and the first real textbook of analysis dealing with synthetic as well as natural rubber was that of H. E. Frey, *Methoden zur Chemischen Analyse von Gummimischungen*, published by Springer in 1953. Roff in 1956 dealt extensively with analytical matters in his reference book *Fibres, Plastics and Rubbers* which has the advantage of giving the salient features in a concise form and setting them out in relation to other high polymers covering a greater range of properties than are dealt with here.

The journals *Analytical Chemistry* and *Rubber Chemistry and Technology* have published critical reviews in the field, such as *Analysis, Composition and Structure of Rubber and Rubber Products* (Tyler, 1967). Full textbooks of methods and critical discussions have also been published in the United States by Tryon and Horowitz (1963), Tyler and Tryon (1963) and the very extensive study in three volumes edited by Kline (1959, 1962). In England, the publication by Haslam and Willis (1965) entitled *Identification and Analysis of Plastics*, now in its second edition with Squirrell as co-author (1972), includes many data on rubbers as well as plastics.

The two atlases, *Infra Red Analysis of Polymers, Resins and Additives* Volumes I and II by Hummel and Scholl (1969, 1973), recently revised in three volumes, *Atlas of Polymer and Plastics Analysis* (Hummel, 1981*a*, *b*; Scholl, 1981), provide many thousands of reference spectra as well as much practical analytical advice.

The more recent publication of new books in this field has been limited; although the *Handbook of Analysis of Synthetic Polymers and Plastics*, by Urbanski *et al.*, appeared in 1977 it is a reprint of a Polish publication of 1972.

There is thus now a wide range of books dealing with the analysis of rubber and although the first edition of this work could claim priority in its field this was neither the case for the second nor is it for this third edition. The justification for the second edition still holds true for the third; no other work seems to deal with the problems of the general analyst or technologist, nor do other books discuss the significance of each individual analysis in the total concept of the vulcanizate formulation, the relevance of state of cure or of blooming, or the analysis of degraded materials to provide data on the reason for, or mechanism of, degradation. The opportunity has been taken to include more details than previously on a number of instrumental techniques such as thermogravimetric analysis, gas and liquid chromatography, and nuclear magnetic resonance which, although expensive, provide substantial amounts of data much more rapidly than the more classical chemical and manual methods. It is appreciated, however, that many laboratories and factories will not have these instrumental facilities and thus the original pre-instrumental methods are still covered, providing as wide a range as possible for each type of analysis.

The practising technologist, or rubber-chemist, who provides an analytical service will soon find that as well as analysing vulcanizates, he will be asked to study thermoplastics, compounds prior to vulcanization, raw rubbers, and possibly latex. At each stage throughout this edition, variations in experimental technique which will broaden the scope of the analytical procedure are described and discussed.

GENERAL PRINCIPLES

A vulcanizate can be considered to consist of five major classes:

(1) polymers,
(2) plasticizers/oils,
(3) solid fillers,
(4) ancillary chemicals and their residues,
(5) adventitious materials.

In general the first stage of a routine comprehensive analysis is extraction of the total material, thus separating the oils or plasticizers, together with the ancillary chemicals and some of the adventitious materials, from the bulk of the sample in a state suitable for further examination, at the same time leaving a much cleaner bulk residue, containing only classes (1), (3) and some of (4) and (5).

(1) Polymers

Polymer analysis can be considered in two distinct parts; what is the polymer and how much is there? If a second polymer is present the situation becomes more complex; are the polymers blends, copolymers (block or random), vulcanized or not, and what is their ratio one to the other? It is not surprising, therefore, that this topic accounts for more chapters than any other in this edition.

(3) Solid Fillers

The bulk filler may consist of a single material or may be a mixture of quite a number of components. Apart from obvious error due, for example, to failure to retain some of the filler on the filter pad—a trouble largely confined to carbon black though not unknown with precipitated calcium carbonate—if a 'wet' separation is adopted, or the loss of carbon dioxide from calcium carbonate in a pyrolytic method, error in determination of total filler can arise from the nature of the fillers themselves. Thus, precipitated calcium carbonate may contain up to 5% of stearic acid and, since calcium stearate is soluble, the material remaining will differ from that originally added to the polymer by amounts up to 5·5%. In most rubber compounds up to 5 parts of stearic acid in 100 parts of the rubber are added and some of this is likely to react with calcium carbonate although the amount is probably small. Some clays used in rubbers and PVC compounds contain added organic materials, 2–3% of which is extractable and which leads to analytical figures differing from recipe figures, part of the bulk filler being transferred to the figure for extractable materials.

Also included in this class would be zinc oxide although, at its usual level of 2–5%, it would be wrong to class it simply as a filler.

In describing the analysis of fillers, a distinction is made between carbon black and inorganic fillers as their identification requires completely different techniques.

(5) Insoluble Adventitious Materials

This would normally include dirt contamination present in either the polymer or compounding ingredients, together with protein and other insoluble non-rubbers from natural rubber, or catalyst residues from synthetic polymers, if present. The analysis of any of these could be significant in terms of both problem solving and polymer identification.

(2), (3), (4), (5) Extractable Materials

When simple division into broad groups suffices for an analysis, the extraction procedure is carried out quantitatively as the first stage of the determination of the polymer loading, and to give an indication of added softener, extender, or plasticizer. If the sample is a rubber, knowledge of the nature of the polymer and its loading will enable an estimate to be made of the associated extractable material, and if the extract exceeds this, oil may be presumed present. The extract from PVC compounds will be plasticizer since the various chemicals added to rubber have no analogues in PVC compounding and a simpler system results.

Complete analysis of the extract from a rubber vulcanizate is a lengthy and expensive undertaking. The extract is a complex mixture of materials initially present in the rubber, vulcanizing and other ingredients added during compounding, together with reaction and breakdown products of these.

In contrast to rubbers, flexible polymer compounds yield simple extracts consisting almost entirely of plasticizer, although small amounts of stabilizers may be, and usually are, present.

No discussion of rubber analysis is complete without intelligent anticipation of the errors expected, and their significance in the interpretation of the results. In some areas of chemical analysis it is quite possible, and reasonable, to quote percentages to two places of decimals, with equivalent implied precision for those components present at much lower levels. In the field of rubber analysis these levels of accuracy are neither sought nor usually obtained and typically one would expect an accuracy of no better than 1–2% of the measured value. The meaning of the term 'accuracy' itself is discussed at length in the final chapter of this book, but it should be borne in mind at this stage that there is little to be gained by analysing components to a much greater accuracy than that with which they can be added to the mix, whilst arguably the accuracy need only be sufficient to indicate variations from a norm great enough to affect the properties, either during manufacture or use, of the product. It should also be remembered that virtually none of the materials being considered herein is 'pure' as one would normally define the term, and thus, however good the analysis, it will not enable a more 'accurate' estimation of the actual added material to be made.

REFERENCES

Caspari, W. A. (1914) *India-Rubber Laboratory Practice*, Macmillan, London.

Ditmar, R. (1909) *Die Analyse des Kautschuks der Guttapercha Balata und ihrer Zusatze*, Hartleben, Vienna and Leipzig.

Frey, H. E. (1953) *Methoden zur Chemischen Analyse von Gummimischungen*, Springer Verlag, Berlin.

Haslam, J. and Willis, H. A. (1965) *Identification and Analysis of Plastics*, Iliffe, London.

Haslam, J., Willis, H. A. and Squirrell, D. C. M. (1972) *Identification and Analysis of Plastics*, 2nd ed., Iliffe, London.

Hummel, D. O. (1981*a*) *Atlas of Polymer and Plastics Analysis*, Volume I, *Polymers, Structures and Spectra*, Carl Hanser Verlag, Munich.

Hummel, D. O. (1981*b*) *Atlas of Polymer and Plastics Analysis*, Volume II, *Plastics, Fibres, Rubbers, Resins*, Carl Hanser Verlag, Munich.

Hummel, D. O. and Scholl, F. K. (1969) *Infra Red Analysis of Polymers, Resins and Additives, an Atlas*: Volume I, *Plastics, Elastomers, Fibres and Resins*, Carl

Hanser Verlag, Munich.

Hummel, D. O. and Scholl, F. K. (1973) *Infra Red Analysis of Polymers, Resins and Additives, an Atlas*: Volume II, *Additives and Processing Aids*, Carl Hanser Verlag, Munich.

Kline, G. M. (1959) *Analytical Chemistry of Polymers* **I**, Interscience, New York. Idem. **II** and **III** (1962).

Ministry of Supply (1944) *Identification and Estimation of Natural and Synthetic Rubbers, Users' Memorandum U.9*, London, and (1946) *Users' Memorandum U.9A*, London.

Pontio, M. (1909) *Analyse du Caoutchouc et de la Gutta-Percha*, Gauthier-Villars, Paris.

Roff, W. J. (1956) *Fibres, Plastics and Rubbers*, Butterworth, London.

Scholl, F. K. (1981) *Atlas of Polymer and Plastics Analysis*, Volume III, *Additives and Processing Aids*, Carl Hanser Verlag, Munich.

Tryon, M. and Horowitz, E. (1963) Methods for the analysis of rubber and related products, in *Handbook of Analytical Chemistry*, Meites, L., Ed., McGraw-Hill, New York.

Tuttle, J. B. (1922) *The Analysis of Rubber*, Chemical Catalog Co., New York.

Tyler, W. P (1967) *Rubber Chem. Technol.* **40**, 238.

Tyler, W. P. and Tryon, M. (1963) in *Standard Methods of Chemical Analysis*, 6th ed., Welcher, F. J., Ed., **2B**, 43, Van Nostrand, Princeton.

Urbanski, J., Czerwinski, N., Janicka, K., Majewska, F. and Zowall, H. (1977) *Handbook of Analysis of Synthetic Polymers and Plastics*, Halsted Press, New York.

Weber, C. O. (1902) *The Chemistry of Rubber*, Griffin, London.

Chapter 2

Sampling and Sample Preparation

It is essential that the material actually analysed either be representative of the material available, or be that most appropriate for solving the particular problem presented.

International and British Standards distinguish between the sample (that which one is given to analyse) and the test portion (that which one separates from the sample and uses entire for a given investigation). Using this differentiation, our discussion centres upon the choice of a test portion and its subsequent treatment so that the maximum relevant information is obtained.

Procedures for the sampling of both natural and synthetic latices are fully detailed in ISO 123-1974 whilst for raw rubber ISO 1795-1974 should be consulted. It is not intended to cover these here as they are extremely detailed and specific. Our more immediate concern is to indicate the problems confronting an analyst when faced with a compounded material.

In mixing and processing rubber and rubber-like polymers, powders are added and form a disperse phase in a matrix or continuous phase of polymer. The degree of dispersion may vary considerably both over short distances and long distances. The analyst is usually given a sample on which an analysis is needed, and before taking the test portion required for a given determination he ensures that the test portion is appropriate for that particular analysis.

ANALYSIS OF AVERAGE COMPOSITION

In those cases where one or more aspects of the overall formulation are to be determined, it is necessary that the test portion is taken from a large enough volume to ensure that inhomogeneities arising either from mixing or from peculiarities of the particular manufacturing process can be averaged out by homogenization prior to the relevant analysis. The following examples, though not exhaustive, are illustrative of the general principles to be observed.

Thin Calendered Sheets or Proofings

Economic considerations will usually prevent sampling from the centre of the length of a roll but the extreme ends should be avoided and samples taken near to both ends rather than one only. The samples should be from the entire width, preferably cut diagonally. Whether separate test portions are cut from a sample from each end or, alternatively, whether the samples from the two ends are blended and homogenized before taking the test portion, will depend on the nature of and reason for the analysis.

Small Moulded Articles

The quantity of material required for a particular set of analyses will largely determine the number of mouldings required. A sufficient number of mouldings to allow all necessary analyses to be carried out should be homogenized together before taking the test portion(s).

Large Manufactured Articles

In this case the nature of, and reason for, the analysis will influence profoundly the procedure to be followed. Composite articles such as tyres must be sectioned and dismantled, the various components being separated and handled separately. Homogenization of the separated components will usually but not necessarily be carried out.

Rubberized Fabrics

Thick rubberized fabrics may sometimes be separated by cutting with a razor blade, but in cases where this is not feasible it is often possible to separate rubber from fabric after swelling the rubber with vapours of a suitable solvent such as chloroform or dichloromethane. The rubber is freed of solvent by evaporation in air or vacuum at room temperature, and then homogenized.

If rubber cannot be separated from the fabric, then the material must be analysed as a whole, after cutting into small pieces (ISO, to pass a 2 mm sieve; ASTM, 1·5 mm square).

HOMOGENIZATION OF SAMPLE

Two methods are available for rubbers and rubber-like polymers: (i) the sample may be finely divided by cutting or grinding and the cuttings well mixed before taking the test portion, or (ii) the mixing may be carried out by passing through the tightly closed nip of a roll mill, the cutting of the test portion being delayed until after sheeting the sample. In some cases a piece of the homogenized sheet can itself form the test portion without the necessity for finely dividing by cutting. In all cases it is essential to ensure that any extraneous foreign matter is excluded from the sample prior to homogenization.

International, British and American standardizing bodies prefer comminution of the material by passing through the cold, tightly closed rolls of a two-roll rubber-mill. If this machinery is available it is undoubtedly the best way to prepare the material for most analytical procedures and anyone regularly analysing rubber-like samples would be well advised to install one. The rolls need not be machine driven as perfectly satisfactory results can be obtained with a long handle on each roll and human effort to turn them. Failing this, a rotating rasp may be used but is not favoured. Rasping causes a considerable local temperature rise which can lead to chemical reactions such as 'maturing' processes, and reaction of any residual sulphur, whilst oxidation occurs with consequent increase in extractable material. Also, it is unsuitable for unvulcanized rubber and the rubber-like plastics. The obvious alternative of grinding or buffing is not acceptable; the powder obtained is oxidized and is contaminated with material from the grinding wheel. Cutting with scissors or knife (razor blade) is laborious but is essential if the roll mill is not available. The International Standard ISO 4661 Part II, at present at the draft stage, allows cutting and specifies that material 'shall be comminuted to pass a 1·7 mm aperture sieve'. The ASTM Standard on rubber products, ASTM D 297-1981, also allows cutting but requires the sample to be rubbed or passed through a 14-mesh sieve (this sieve has an opening of 1·40 mm). BS 903: Part B1-1958 also allows cutting, the cut-up material to pass a BS 10-mesh sieve which has an opening of 1·676 mm. All three specifications require the sheeting, if this is the method of preparation used, to be to 0·5 mm or less in thickness.

ANALYSIS OF LOCALIZED COMPOSITION

There are many occasions when homogenization of a sample destroys the very features which are important in a particular investigation. The situations where homogenization is inappropriate are too varied for a comprehensive discussion to be presented; nevertheless, the following examples are illustrative and highlight the need for closely defining the analytical problem and designing both the sampling and the analysis procedures appropriately.

Vulcanization state of thick articles. An example where homogenization might not be desirable is the determination of the 'free' sulphur content of a truck tyre. Assuming that this is required because inadequate vulcanization is suspected, it would be reasonable to take the test portion only from the inner face of the tread rubber rather than from a homogenized cross-section of the tread. Similarly with large blocks of rubber for mounting engines or for bridge bearings, where the state of cure may well vary with the distance from the surface, free sulphur determinations carried out on test portions taken separately from the centre and outer parts of the block would be more informative than those carried out on a test portion taken from a homogenized cross-section of the block.

Analysis of blooms (see Chapter 13). Where blooms form on the surface of a rubber mix or rubber article, it is clearly inappropriate to homogenize the bulk material prior to identification of the bloom.

Bond failure problems. Bonds between rubber and metal are sensitive to the state of cure of the rubber. It is, however, the state of cure in the immediate vicinity of the metal which is important, and so the test portion must be taken from this area rather than from a homogenized cross-section.

Inhomogeneity and poor dispersion. These can cause a variety of problems, such as variable physical properties, article-to-article variation, unevenness of colour etc. Such problems can be investigated by the reverse of the normal procedure. By cutting down on the size of the test portion, and with no homogenization stage, an idea may be obtained of the degree of inhomogeneity existing, provided that the analytical technique employed is sufficiently sensitive to cope with the small sample size.

SIZE OF TEST PORTION

The size of the test portion must be chosen with several factors borne in mind.

(1) It must be sufficiently large to allow the carrying out of all the analytical techniques which might be required.

(2) It must be sufficiently large to give adequate sensitivity for each technique being employed.
(3) It must be sufficiently large to average out any irrelevant inhomogeneities.
(4) It must be within reasonable limits such that handling during subsequent analysis is not an insuperable problem.
(5) It must be sufficiently small that reagent volumes and apparatus are not impractically large.
(6) It must be sufficiently small that relevant inhomogeneities or adventitious contaminants are not swamped out.

SAMPLE PREPARATION

Although International and other standards organizations define the material actually being analysed as the *test portion*, the general analyst, and indeed the scientist, in the English-speaking world, uses the word *sample* to refer to the material he is actually analysing. Indeed this is implicit both in the heading of this section, and in the heading of the corresponding parts of International Standards, which also use the word *sample.*

In conformity with common convention, therefore, the word *sample* is used from this point onwards throughout the book to refer to that piece of material which the analyst is actually using. The *sample* may be one section of the *test portion*, taken to carry out one of a series of interrelated analyses, or it may be a discrete microportion on which one specific analysis will be carried out, and which will not necessarily be representative either of the bulk material or of another microsample.

Having selected a *test portion* which is most appropriate for a given analytical problem, the analyst must then decide on the most appropriate procedure for preparation of the sample to be analysed.

In general, sample preparation techniques other than the initial homogenization procedure are specific to the analytical technique being employed. An exception to this is solvent extraction, partial or exhaustive, which is considered in some detail in the next chapter.

Other preparation techniques range in complexity from cutting into thin strips with scissors, through hot pressing, microtoming, or ashing, to sophisticated total or selective degradative procedures used primarily for infrared or nuclear magnetic resonance spectroscopic investigation. These preparation techniques are considered during the discussions of the particular analytical technique, in subsequent chapters.

Chapter 3

Extraction

PRELIMINARY REMARKS

Although the concept of extraction is thoroughly understood by most analysts, its applications to the analysis of rubbers and rubber-like materials are diverse and complicated. As a first step it is advisable to differentiate between *extraction*, *solution* and *dissolution*. Extraction is here defined as the procedure for removing organic additives from the polymer/black/inorganic components without simultaneously removing significant amounts of the polymeric phase, whilst solution and dissolution involve the removal of polymer from the remaining components. It must, however, be borne in mind that most polymers, even in the uncompounded state, contain non-rubbers which will be extracted by these techniques, and in any quantitative extraction due correction must be made for them. In general those extracted organic materials are of low molar mass, but also included are polymeric plasticizers, factice and mineral rubber, more realistically considered as plasticizers than polymers.

Extractions do not necessarily require solvents; useful information may be provided by a thermal extraction whilst extraction using a solvent may be carried out in the cold, or heated, for periods of time ranging from seconds to days, and be either quantitative, qualitative, or selective depending upon the exact nature of the experiment.

It would be realistic to say that in the vast majority of cases the purpose of an 'extraction' is to use an appropriate solvent to provide essentially complete separation of the extractable materials from the polymer so that each can be examined without interference from the other; for this reason the basic theory of extraction merits a detailed consideration.

The choice of an 'appropriate solvent' could cause the first difficulty.

Until a completely extracted sample is available identification of an unknown polymer may not be possible but, paradoxically, until the polymer has been identified, one does not know the correct solvent to use for extraction. This difficulty is more apparent than real since the analyst usually obtains some information from the appearance of the sample, its use if known, and its colour or smell. If a wrong solvent is mistakenly used for extraction, the fault will be detected at an early stage of subsequent work and remedial action taken. All the solvents commonly employed for hot extractions and certainly those recommended in these pages, will usually extract all extractable material; where they fail is in extracting some polymer as well as the non-polymeric material. In such a case, the polymer is available for identification but the extract will be too great and may mislead the analyst as to its composition and nature. As soon as the polymer is identified, however, the analyst will realize his mistake and will take steps to correct it if information on its amount or composition is required.

Natural rubber, being the oldest of the class of materials we are considering, serves as the prototype for extraction procedures. Henriques (1892) introduced extraction with alcoholic 'potash' to remove factice and an abstract by Weber in 1894 records the extraction of 'asphaltum' by cold nitrobenzene. Henriques also used carbon disulphide in 1894 to extract vulcanized rubber, and Holde, about the same time, used ether–alcohol mixtures. Acetone seems to have been used at the turn of the century and Weber (1902) argues in favour of its use, while Caspari (1914) records it as the standard procedure. Acetone is very suitable for the extraction of natural rubber but is not the best solvent for use with all synthetic rubbers. It is not suitable for extracting unvulcanized synthetic rubbers or thermoplastic materials such as polyvinylchloride; these remarks are amplified later in this chapter.

NATURE OF THE EXTRACTION PROCESSES

The extraction of soluble substances from a rubber by a solvent is a diffusion phenomenon. In the case of a vulcanized rubber, the substance on which most extractions are done, the rubber acts as a semi-permeable membrane. A short time after the start of the extraction, the rubber is swollen to its maximum extent by imbibition of the extracting liquid which forms a relatively concentrated solution inside the rubber of the substances to be extracted. The solvent outside the rubber is continually renewed so that the concentration outside the rubber is virtually zero and diffusion of the soluble substances naturally occurs in the direction of the concentration gradient. The rubber acts as a semi-permeable membrane by reason of its crosslinked nature giving a mesh the size of which limits absolutely the size of the molecules that can diffuse out. Since the process is taking place in a (relatively) non-ionizing solvent with largely neutral molecules and the mesh itself is non-ionizing, many of the considerations normally important with semi-permeable membranes can be ignored. Instead of a single mesh size there will be, of course, a size-range depending on the distribution of the crosslinks forming the structure, and thermal agitation will lead to variation in the space through which a molecule could move. However, the occurrence of the maximum space given by the fully extended chains forming the sides of any mesh will have a finite probability and therefore molecules corresponding to this fully stretched mesh size, or smaller, will be extractable. For a soft vulcanized natural rubber this mesh will have sides of about 20 nm so that molecules whose smallest dimension, considered as a radius, exceeds 10 nm will be inextractable. This measurement could correspond to a very large molecule though in actual fact we know the proteins in natural rubber to be inextractable whilst the naturally occurring resins with molar masses of a few thousand come out very easily. The more the rubber is swollen by solvent, the more rapidly will extraction occur but the absolute limit to the size of molecules which can be extracted will not be affected. When the molecular size is near the limit of extraction the effect of swelling on the rate of extraction can make all the difference between extraction in a few hours and extraction necessitating weeks. This may be illustrated by reference to bitumen. The higher molar mass portion of this can be extracted from vulcanized natural rubber in a few hours if chloroform is used but continuous extraction is obtained with acetone, a constant weight never being achieved.

The fundamental law in the study of extraction processes as examples of diffusion phenomena is Fick's first law (see Eqn (3.1) below).

If $\mathrm{d}m$ g of the substance diffuses in time $\mathrm{d}t$ across an area A under a concentration gradient $\mathrm{d}c/\mathrm{d}x$, then

$$\mathrm{d}m = -DA\frac{\mathrm{d}c}{\mathrm{d}x}\mathrm{d}t \tag{3.1}$$

where D is the diffusion coefficient, which is defined by Fick's law. The elimination of m from Fick's first law gives the general differential equation of diffusion sometimes known as Fick's second law. It is

$$\frac{\partial c}{\partial t} = D\frac{\partial^2 c}{\partial x^2} \tag{3.2}$$

The various solutions to this which can be obtained after the imposition of certain boundary conditions are conveniently discussed by Barrer (1941).

The case with which we are concerned may be visualized as diffusion from a thin membrane of, say, 0·5 mm thickness, the concentration of soluble material in the membrane being given by the ratio of the extractable solids to the solvent imbibed, and the concentration falling discontinuously to zero at the interface between membrane and liquid. This is, of course, an idealized approximation, but as the solvent surrounding the sample is continuously agitated by the arrival of freshly condensed solvent and is also completely drained at frequent intervals, it suffices as a model. For this model, the amount of extractable material (m) left in the membrane at any time t is given by

$$m = 0{\cdot}0405C_0\sum_{n=0}^{\infty}\frac{1}{(2n+1)^2}\exp[-(3{\cdot}95\times10^3)(2n+1)^2Dt] \tag{3.3}$$

In this equation C_0 is the initial concentration of extractable material in the volume of the membrane, and provided consistent units are used for C_0 and m (grams or moles) the diffusion coefficient (D) is given in $cm^2\ s^{-1}$.

The magnitude of D is a function of molecular size and some idea of its variation can be obtained from Fig. 3.1, which gives D plotted against log (molar mass) for an aqueous system and is derived from data given by Alexander and Johnson (1949). It will be observed that the diffusion coefficient for oxygen in water is about 2×10^{-5} (at 18 °C) whereas values given in Table 3.1 for nitrogen, which has about the same molecular size as oxygen, show that for normal vulcanized rubbers it is of the order of 2×10^{-7}.

Increase in the value of D due to the higher temperature and to the presence of the solvent needs to be set against its decrease with increasing molar mass, but with the smaller molecules the presence of solvent would bring D up to the same order as exhibited by molecules of similar size in water. It is clear from Fig. 3.1 that, in the absence of the restraining influence of a semi-permeable membrane, the diffusion coefficient decreases relatively slowly with increasing molar mass but once the diffusing molecule approaches the size of the membrane mesh, it will cause

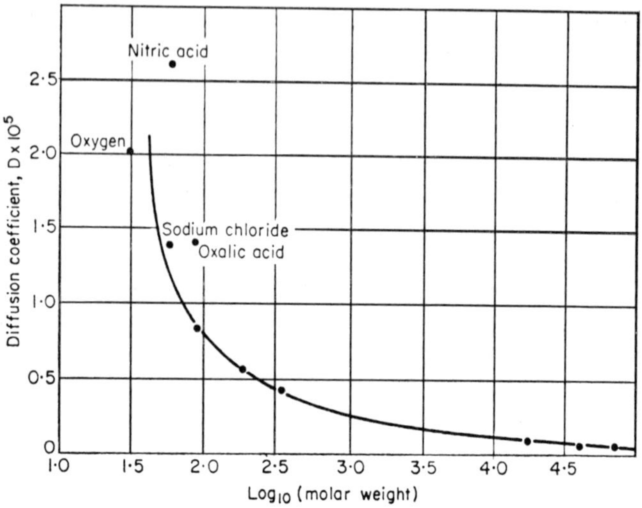

FIG. 3.1. Coefficient of diffusion in water as a function of molecular size.

D to drop rapidly to zero. It seems likely that for the resins and plasticizers normally extracted from rubbers, D is about 10^{-6}–10^{-7}.

We are now able to discuss the question of completeness of extraction. If we return to Eqn (3.3) it will be seen that it converges very rapidly indeed and a reasonable approximation can be obtained by expanding for two terms only and rearranging to give Eqn (3.4):

$$\frac{m}{0{\cdot}0405C_0} = \exp(-3{\cdot}95 \times 10^3 Dt) + \frac{1}{9}\exp(-35{\cdot}55 \times 10^3 Dt) \qquad (3.4)$$

Initially, when $t = 0$, $m/0{\cdot}0405C_0 = 1{\cdot}111$, whereas if the summation had been carried to infinity this expression would be equal to 1.234 since,

TABLE 3.1

DIFFUSION COEFFICIENTS FOR NITROGEN IN POLYMERS

Polymer	*Temperature*, °C	D, cm^2 s^{-1}
Vulcanized polychloroprene	27·1	$1{\cdot}9 \times 10^{-7}$
Butadiene copolymers:		
(a) acrylonitrile	17	$0{\cdot}66 \times 10^{-7}$
(b) methacrylate	20	$3{\cdot}4 \times 10^{-7}$
(c) styrene	20	$2{\cdot}4 \times 10^{-7}$

After Barrer (1941).

by definition and for the thickness of membrane taken, $m/C_0 = 0{\cdot}05$. Figure 3.2 shows a plot on double logarithmic scale of $m/0{\cdot}0405C_0$ obtained from the right-hand side of Eqn (3.4) against $\log t$ for a time scale of 10–10^5 s (about 28 h) and for two values of D. Whilst it is true that an infinite time is required to complete the extractions, the amount unextracted after even a short time can be below that detectable by the analytical operation involved. When $D = 10^{-7}$ the unextracted material is reduced to 1% of its initial value in 3 h and to 0·1% in 5 h after which the amount remaining becomes too small to have any analytical meaning.

The diffusion equation with the awkward summation of exponentials in its integrated form has been avoided by experimentalists in discussing the

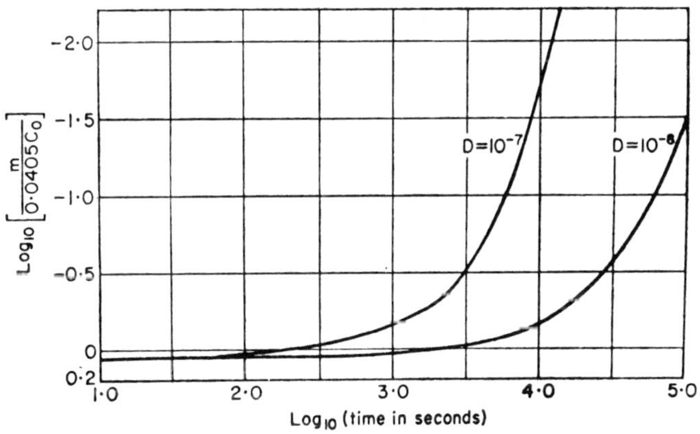

FIG. 3.2. Influence of diffusion coefficient on extraction time.

effect of extraction time and of other variables. In addition to this it is rare in analytical practice to be extracting from a polymer a single molecular species of definite molar mass. In the past, extraction was always from raw or vulcanized natural rubber where the mixture of extractable substances certainly defied any attempt to ascribe a definite, even if average, value of D, because the range of molar masses of the non-rubber constituents increases smoothly from that of quebrachitol and those of the fatty acids to molar masses of tens of thousands. However, the shape of the curves in Fig. 3.2 indicates that definite extract levels should be obtained if the polymer is a true membrane, i.e. if it is either crosslinked or, if free from crosslinks, then also free of polymer material of low molar mass that might be soluble in the extracting solvent. Failure to extract to constant weight is most likely to be a chemical phenomenon due to slow oxidative scission giving a constant supply of material of low molar mass derived from the polymer. The simple application of the diffusion equation assumes the concentration to fall to zero at the surface of the rubber. This is not true, for the rubber is always wetted by liquid and a layer remains even after syphoning has removed the bulk of the solvent. The effect of this will be to increase the apparent distance through which the substances must diffuse and will decrease the concentration gradient thus depressing the value of the diffusion coefficient.

The diffusion theory expounded above gives a reasonable physical picture of the extraction process and when applied quantitatively gives values for the parameters of the equation used which are of the right order. The corollary, that extraction can never be complete, is not of analytical significance for the amount remaining unextracted when D is of the order of 10^{-6} can be reduced below the limits of analytical sensitivity within reasonable periods of extraction. This is no longer so when high molar mass polymers are to be extracted from another, crosslinked polymer. Polymethylmethacrylate of high molar mass (intrinsic viscosity 6·6) was extracted with acetone from a natural rubber vulcanizate only to the extent of 14·6% of the amount there, after 27 days (Cooper and Smith, 1962). Their data do not allow calculation of the diffusion coefficient but an approximate treatment suggests that it is smaller than 10^{-10}.

STANDARD APPARATUS FOR DETERMINATION OF EXTRACT

The apparatus used for extraction should preferably be of all-glass construction and two forms preferred by the authors are shown in Figs. 3.3(a) and (b). These two are also recommended in the International Standards Organization document ISO 1407-1976 which also allows a form of extraction apparatus in which a coiled metal condenser is inserted into the neck of a conical flask, which is closed by a metal disc through which the condenser tube passes, and is integral with it. This form of apparatus, usually known as the 'Underwriters' and illustrated in Fig. 3.3(c), was also specified in BS 903-1940. The objections to it are that when several are connected in series the rubber tubing used for the purpose tends to prevent the closing plate sitting squarely on the flask; flask irregularities have a similar effect, both resulting in a loss of solvent, a loss aggravated by the fact that the condenser is necessarily tightly coiled and its effective surface area small. In addition, a condenser cut from block tin is expensive and its cheaper equivalent, dipped or plated brass, has been known, in one author's experience, to contain pin-holes in the plating, polymerization of acetone being traced to this cause. The all-glass construction with standard ground-glass joints obviates these defects and which of the two forms shown is used depends more upon personal taste and availability than anything else. Both forms are equally suitable for extraction with solvents more volatile than acetone which exaggerate the disadvantages of the 'Underwriters' apparatus although it should be noted that even ground-glass joints (unlubricated as they must be) can pass vapour from boiling ether or carbon disulphide unless sleeves of polytetrafluoroethylene are used to cut down solvent losses.

Sets of apparatus are conveniently erected as permanent fixtures with an electrically heated water-bath incorporating a constant level device and a condenser, fitted to the bath to minimize steam rising and condensing on the outside of the condensers of the extraction apparatus. The bath should be installed in a metal tray of sufficient size to contain the water of the bath since eventually corrosion will cause a leak which, through sheer contrariness, will inevitably develop catastrophically at night. Adequate inspection and maintenance should minimize such catastrophes, but human nature is fallible. BS 903-1950 introduced the use of a heat exchanger between the top of the extraction cup and the condenser. This arose from the work of Milligan (1931) but has been shown in an extensive study by Wake (1969) to be unnecessary.

The examples illustrated in Fig. 3.3 (a)–(c) usually comprise extraction cups with a capacity of 20–30 cm^3, the round-bottomed flask (Fig. 3.3(a)) being 150 cm^3, and the conical ones (Fig. 3.3(b) and (c)) 250 cm^3.

Although not a design recommended in existing Standards, Fig. 3.3(d) illustrates the routine extraction apparatus used in the authors' laboratory. A 150 cm^3 round-bottomed flask is used for convenient heating on a bank of heating mantles, and to enable the solvent to be removed under reduced pressure, at low temperature. The Soxhlet cup, although having a capacity of only 8 cm^3, will hold the weight of sample usually extracted (3 g) and has considerable advantages in the length of time required for extraction (see later).

Raw rubbers, unvulcanized compounded rubbers and some thermoplastic materials often become tacky during extraction and tend to coalesce, thus invalidating the extraction data. This can be overcome with sheeted samples by placing them between filter paper, lens tissue or linen prior to extraction, whilst for cut-up samples BS 1673: Part II-1954 suggests the use of silver sand to dilute the polymer. In all cases the materials used should have been pre-extracted.

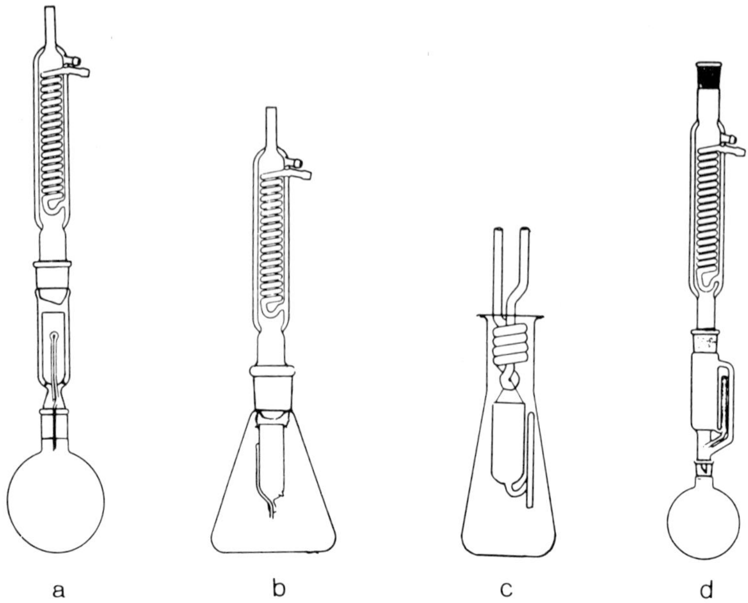

FIG. 3.3. Four basic types of extraction apparatus suitable for the extraction of rubbers and rubber-like polymers.

A useful technique with thermoplasts and compounded rubbers is to prepare a thin film using a hot laboratory press. Temperatures of up to 180 °C are required for many thermoplasts but it is a simple matter to press for a few seconds and obtain a sheet 0·2–0·5 mm thick. Similarly a compound can be lightly cured by holding at *ca*. 150 °C for 5 min and the sheet will then have a much reduced tendency to flow during the extraction although the extract will then contain cure residues, together with the original curatives. If the edges of the sheet are discarded there is no evidence for degradation of the polymer in this time scale.

CHOICE OF SOLVENT

For the extraction of natural rubber, whether vulcanized or raw, acetone is usually specified as it fulfils most of the criteria for a good extractant. These are that the polymer should be swollen slightly by the solvent but should not be soluble in it; it is convenient that it should boil well below any extracted material so that it can be removed easily from the extract without loss of any extracted liquid or heat damage to any solids; in addition, the solvent should be inert to any possible ingredient of the extract and not objectionable by virtue of excessive toxicity (all solvents are toxic to some extent), inflammability or odour. It is advantageous if the solvent used is cheap, as the re-use of recovered solvent carries with it a certain element of risk and should be avoided.

TABLE 3.2

SOLVENTS FOR THE EXTRACTION OF RUBBERS AND RUBBER-LIKE POLYMERS

Elastomer	*Raw/compounded*	*Vulcanized*
NR	Acetone	Acetone
	Methanol	Methanol
	2-Propanol	Butanone
Synthetic polyolefins (BR, SBR, IR, etc.)	Acetone	Acetone
	2-Propanol	Butanone
IIR	Acetone	Acetone
	Butanone	Butanone
CR	Light petroleum (BP range 60–80 °C)	Light petroleum (BP range 60–80 °C)
	2-Propanol	2-Propanol
	Methanol	Methanol
NBR	Methanol	Methanol
	2-Propanol	2-Propanol
	Light petroleum (BP range 60–80 °C)	Light petroleum (BP range 60–80 °C)
PVC	Diethyl ether	—
	Methanol	—
Thermoplastic block copolymers, SIS, etc.	Methanol	—
NR/PP/PE type EP/PP/PE type	Methanol	—

Table 3.2 indicates solvents which have been found generally acceptable in the extraction of common elastomers, but attention is also drawn to Table 3.3 which gives a much broader picture of the resistance of a range of rubbers to various solvent types, and may help in the selection of an unusual solvent for a specific application.

Acetone used for extraction should be free from its polymers and from water and this means that acetone, other than from a newly opened container which has been stored in the dark, should be redistilled before

TABLE 3.3

RESISTANCE OF RUBBERS TO VARIOUS LIQUIDS

Rubber	*Ali-phatic*	*Aro-matic*	*Halo-genated*	*Ketones*	*Alcohols*	*Animal and vegetable oils*	*Water*
	hydrocarbons						
Natural	P	P	P	G	G	P–G	E
cis-Polyisoprene	P	P	P	G	G	P–G	E
SBR	P	P	P	G	F	P–G	G–E
Butyl	P	P	P	G–E	E	E	G–E
cis-Polybutadiene	P	P	P	G	F–G	P–G	E
Ethylene/propylene	P	P–F	P	G	P	P	E
Polychloroprene	G	F	P	P	F	G	F–G
Nitrile	E	F	P	P	E	E	F–G
Polyurethane	E	F	F	P–F	G	G	G
Polysulphide	E	E	G	G	G	E	F
Silicone	P–G	P–F	F	F–G	G	F	F
Fluororubbers	E	E	G	P	E	E	E
Chlorosulphonated polyethylene	F	F	P	P	G	G	G
Acrylic	E	F	P	P	P	E	E

Ratings: E = excellent, G = good, F = fair, P = poor. After Lindley (1966).

use for all accurate work, although it would be pedantic to insist on this for ordinary factory routine. Methanol is also a useful solvent for the qualitative extraction of natural rubber, raw or vulcanized, for subsequent chromatographic examination of the extract. Whilst additives are generally extracted quantitatively, the extract appears cleaner than when obtained with acetone.

Extraction of polar synthetic rubbers may be with ether or with light petroleum of specified boiling range. If the latter is preferred it should be noted that the Soxhlet-type extraction apparatus sometimes gives trouble due to the lower boiling components of the solvent creating vapour locks in the syphon tube. The remedy is to use a Soxhlet with an external syphon (see Fig. 3.3(d)). Ether, being of constant boiling point, is free from this trouble provided that excessive heat is not applied to the flask, preventing the condensate from running back.

Methanol has been found to be an acceptable solvent for selectively extracting plasticizers from block copolymers such as SBS or SIS, and those based on a polyolefin/polypropylene blend, whilst it also affords a good separation of the polymeric plasticizers used with acrylonitrile rubbers and chloroprenes (Williamson, 1957). Robertson and Rowley (1960) recommend the carbon tetrachloride–methanol azeotrope for the removal of polymeric plasticizers from PVC but this is unlikely to find favour today because of the toxicity of carbon tetrachloride.

Because of the vast range of solvents suggested by different authors it is imperative that any quantitative results, or specifications, define fully the solvent system used and the extraction procedure.

TIME OF EXTRACTION

Rubber extractions with acetone or the other solvents listed are usually carried out overnight, with BS 1673/2-1967, ASTM D297-1981 and ISO 1407-1976 tending to agree on some 300 cycles through the extraction cup, although ISO allows as few as 160. If the sample has been properly comminuted, the longer times required by these standards are probably unnecessary although when new apparatus is used the extraction rate should always be checked as also should its behaviour with any unusual solvents before leaving extractions overnight.

A simple rubber compound containing only a minimum of added materials is usually quantitatively extracted in periods less than 8 h but this gives the necessary margin of safety to allow for the presence of unusual substances.

RAPID EXTRACTION

The apparatus already illustrated in Fig. 3.3(d), using the low-volume Soxhlet extractor, enables relatively rapid extractions to be carried out. The cycle time is of the order of one minute (with 3 g rubber) and thus after three hours the minimum number of 'ISO passes' has been exceeded and, in the authors' experience, essentially quantitative extraction of general-purpose vulcanizates/compounds etc. has occurred. The particular advantage of reducing the extraction time from eight to three hours is that an extraction with subsequent examination of the extract and residue can be carried out within one working day.

An even more rapid method has been proposed by Kress (1956), is covered by ASTM D297-1981 with slight differences, and is about to be incorporated into a revised edition of ISO 1407. The procedure is as follows:

> Sheet out 1 g of the vulcanizate with the mill set as tight as possible. Make several passes with a single unfolded sheet until the sample is as coherent as possible. Where this is not possible place the milled crumbs into a filter paper envelope and for uncured stock sandwich between filter paper. Cut, with scissors, a test portion of between 80 and 110 mg and drop the weighed test piece into the boiling solvent and continue rapid boiling for 30 min. Kress uses 20 cm^3 of solvent in a 250 cm^3 conical flask with a condenser but states that a beaker of solvent on a hot-plate and covered with a watch glass answers perfectly well. After extraction, press the test piece between 'folds of absorbent paper towels' to remove excess solvent and dry at 105–110 °C for 10 min.

Kress advocates the use of a mixture of methyl ethyl ketone and ethanol, 75:25 by volume, and weighing the test portion before and after in order to determine the extract quantitatively. This mixed solvent is chosen empirically to give results in line with the standard ASTM acetone extraction, MEK itself giving too high a figure. However, we are not here concerned with Kress's suggestion that his method should replace the existing quantitative procedure but only that it provides a convenient means of obtaining quickly an extracted sample for qualitative examination and for this purpose the use of MEK without ethanol is probably preferable.

A completely different philosophy was adopted by Higgins (1978) who used a high-speed macerator which generated ultrasonic pulses to extract quantitatively 2·5 g samples of raw rubbers in a matter of minutes. Samples of SMR 5, 20, 50 and 5-LV rubbers, together with an experimental set of oil-extended NRs (OENR) containing up to 25% oil were examined by the rapid method and ASTM D297. The results are illustrated graphically in Fig. 3.4 and clearly show that the method allows a rapid determination of the extract level with extraction times of less than 5 min.

Subsequent unpublished work has shown that this technique is equally valid for the analysis of compounded or vulcanized rubber and thus has a particular significance in that it allows the cold extraction of unvulcanized samples permitting the identification of the added curatives themselves, rather than their decomposition products, particularly important for the identification of thiazole and thiuram accelerators. It is, of course, quite possible to use a cooled Soxhlet extractor, but this is extremely time consuming and cumbersome.

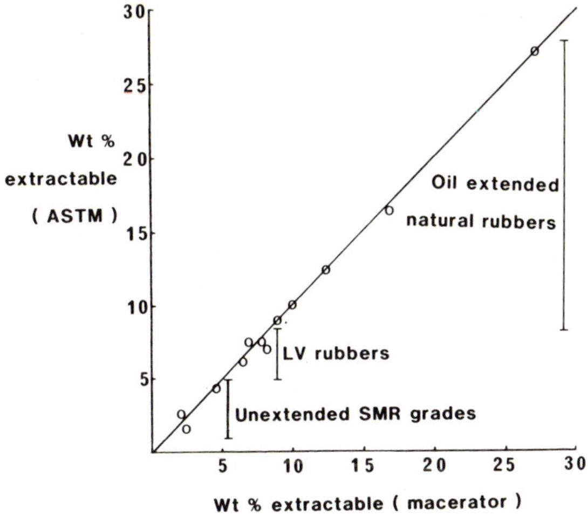

FIG. 3.4. Percentage extracted by the ultrasonic macerator v. percentage extracted using the ASTM method. (Courtesy MRPRA.)

MICROSCALE EXTRACTION PROCEDURE

The first published account describing applications and a procedure for carrying out extractions of rubber on the microscale is that of Wyatt (1941). An illustration of his apparatus was included in the second edition of this volume (Wake, 1969) but it appears obsolete today, being commercially unavailable, whilst the practising analyst would tend to use standard 'micro glassware' not available in 1941.

The size of the test portion is some 20 mg and by strict definition, therefore, the method should be considered a semi-micro one as the sample is over 10 mg in weight. Nevertheless the extract will normally be between 1 and 5 mg and thus the use of the word 'micro' is acceptable.

Wyatt's results indicate the rapidity of extraction using this apparatus, 2 h being sufficient for an extraction with acetone as compared with 7 h for the conventional modern macro method. The shorter time equates well with the time required using the small extraction apparatus illustrated in Fig. 3.3(d) when 3 h is generally adequate for a 3 g sample. Wyatt also gives data for chloroform and alcoholic 'potash' extractions, and again illustrates that the more rapidly obtained results have accuracies comparable with those of the slower macro method.

The test portion size of 20 mg is also a point which requires consideration. If a piece of material of adequate weight is available (ISO 1407-1976 requires 3–5 g) then there is little point in carrying out a micro extraction unless there is a specific reason as discussed in Chapter 2. In this case the problem of obtaining a representative sample does not arise as one is specifically looking for differences on the microscale. Should a micro extraction be required of a 5 g sample such that it is representative of the whole, then it should be appreciated that the 20 mg constitutes only some 1/250 part of this, and it is advisable to carry out a two-part homogenization by taking a 0·1–0·2 g sample from the initially homogenized 5 g and carrying out a further homogenization before taking the 20 mg samples.

The statistical rationale behind this argument and the validity of micro sampling techniques are given in Chapter 14.

THERMAL EXTRACTION

Recently, McSweeney (1970) has demonstrated a thermal extraction method coupled with subsequent TLC separation for the identification of compounding ingredients in rubbers. The apparatus, which is particularly suitable for the examination of small amounts of rubbers (see Fig. 3.5), consists of a small furnace surrounding a tube through which nitrogen is

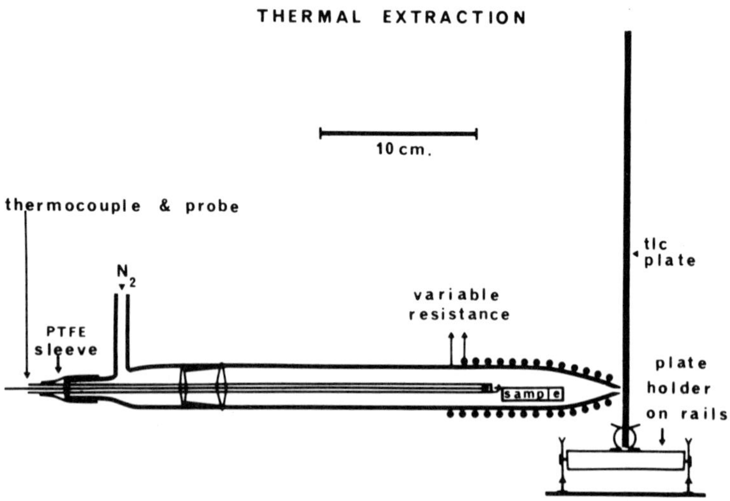

FIG. 3.5. Thermal extraction. (Courtesy Plastics and Rubber Institute.)

passed. As the temperature of the oven is slowly raised to 250 °C, the nitrogen sweeps volatilized material directly on to the startline of a TLC plate. By progressively moving the TLC plate while the oven temperature is raised, thermally fractionated material is deposited along the start line so that after development of the plate information concerning the relative volatilities of the compounds can be obtained.

Samples of a few milligrammes of a vulcanizate are quite adequate for the qualitative identification of curatives, cure residues, protective agents and extender oils by this procedure, although the technique makes no pretence to quantitative accuracy.

MULTIPLE EXTRACTIONS

In rubber technology frequent use is made of factice and mineral rubber with both natural and general-purpose synthetic rubbers whilst, as has already been made clear for PVC compositions, polyesters are increasingly used as plasticizers. This is also true of oil-resisting synthetic rubbers. Factice, also known as rubber substitute, exists in two broad classes distinguished by colour and known as 'light sub.' and 'dark sub.' in the jargon of the mill room. Actually, the colour difference signifies a chemical difference of some importance to the analyst. Brown or dark sub. is formed by reacting together a mixture of vegetable oils and sulphur and is a polymeric material of moderate molecular weight (about 7000, see Stamberger and Knight, 1928), whereas white or light sub. is formed from the same oils by reaction with sulphur chloride (S_2Cl_2) and the molecule, probably of similar molar mass, contains chlorine as well as sulphur. Factice is not extractable from vulcanized rubber by acetone although the presence of factice usually leads to a slight increase in the acetone extract due to the presence of small quantities of free oils; neither does a simple change of solvent, to chloroform for example, enable this to be done, so that recourse is had to a degradative extraction with alcoholic potash after the rubber has been extracted with acetone.

Mineral rubber is the trade name given to asphaltic hydrocarbons used as 'extenders' or cheap filling material or processing aids for rubbers where high-grade mechanical properties are not required. Asphalt, whether derived from native asphalt or obtained as a distillery residue from some petroleum sources, consists of a mixture of oils, resins, and asphaltenes, the latter two being of medium high molar mass, and it is part of these which resists extraction with acetone. Chloroform extraction until colourless liquid is obtained from the extraction cup is used, and to establish the nature of the material it is usually reckoned sufficient to precede the chloroform extraction by one with acetone, no other materials in commercial use in rubber compounds being insoluble in acetone but soluble in chloroform. However, it does not suffice to quote the chloroform extract so obtained as the mineral rubber content, since part of the mineral rubber will have been extracted by acetone and this part will depend on the source of the mineral rubber and probably also on the temperature of vulcanization of the compound. In matching a specification for a compound to the analytical figures this must be allowed for.

Rather similar to the factice problem in natural rubber is that of polymeric plasticizers in the oil-resistant synthetic rubbers. Typical materials are polypropylene adipate and polypropylene sebacate. The molar mass is not very high and, as with condensation polymers generally, the molar mass distribution is rather broad. This causes difficulty in finding a solvent which will extract ordinary monomeric plasticizers without extracting any of the polymeric plasticizer and, in fact, this cannot be done. Ether, which is normally used, definitely takes out some of the polymeric plasticizer. Difficulty is also experienced at the other end of the molar mass distribution in finding a solvent which will completely remove the high molar mass material without removing the base polymer. Acetone will successfully remove all the polymeric plasticizer but will also remove some of the rubber, particularly that of the butadiene-acrylonitrile type.

As already mentioned, Williamson (1957) has shown that methyl alcohol can be used successfully in certain cases for the analysis of polymeric plasticizers.

LATEX

ISO documentation provides three standards concerned with the conversion of the 'liquid' latex to the 'solid' rubber. For our purposes two of them—ISO 124-1974 and ISO 498-1974—can be considered the same in that they both afford total solids in the form of a thin sheet of dry rubber. The difference is that whereas ISO 124 is solely concerned with obtaining a total solids value, and dries a layer of latex at 70 °C (16 h) or 100 °C (2 h), ISO 498 is designed to prepare a smooth thin film of rubber and thus dries at 35 °C to constant weight, a procedure taking several days. The third document (ISO 126-1974) describes the determination of the dry rubber content by coagulation with acetic acid, separation from the latex liquids, and subsequent drying at 70 °C. It is obvious that this sample will be quite different from the other two in that most of the non-rubbers will have been separated from the rubber itself, and it should not therefore be used for further general analysis if this could be significant.

The dried latex may now be extracted by any appropriate solvent in the same way as a raw rubber or vulcanizate, but its physical state can be used to advantage in preparing a suitable sample for extraction. Chin *et al.* (1975) used a rotary evaporator to spread a thin film of NR latex (LA-SPP) over the sides of a 150 cm^3 round-bottomed flask which was then simultaneously coagulated and extracted by the addition of a methanol–acetic acid mixture. After a few minutes of rotating the solution in contact with the latex film, effectively quantitative extraction of pentachlorophenol was achieved in the form of a solution amenable to direct analysis.

A similar procedure was adopted by Edwards (1981) who prepared a latex (LATZ) film as described above and used pure methanol to extract tetramethyl thiuram disulphide (TMTD) by rotating the flask containing the solvent and latex film for 1 h before analysis of the extract by liquid chromatography. It should be commented that no TMTD was found in a number of samples of latex examined (<1 ppm) although quantitative recovery was obtained from freshly 'doped' latices.

SPECIFIC EXTRACTIONS

Both of the above-mentioned techniques represent examples of specific extractions in that the overall quantitative success of the extraction is irrelevant provided that the species being analysed is recovered either quantitatively or reproduceably. Other examples are the aqueous acid extraction of raw NR for the estimation of hydroxylamine and the 'saline solution' extraction of rubber products to isolate nitrosamines. The possibility of using a specific or selective solvent should always be considered if a request for a particular estimation is received. Apart from any other consideration, the cleaner the extract, the less likely is interference in the chosen analytical technique.

REFERENCES

Alexander, A. E. and Johnson, P. (1949) *Colloid Science*, Clarendon Press, Oxford.

Barrer, R. M. (1941) *Diffusion In and Through Solids*, Cambridge University Press, London.

Caspari, W. A. (1914) *India-Rubber Laboratory Practice*, Macmillan, London.
Chin, H. C., Singh, M. M. and Higgins, G. M. C. (1975) *Internat. Rubber Conf.*, Kuala Lumpur.
Cooper, W. and Smith, R. K. (1962) *J. Appl. polym. Sci.* **6**, 64.
Edwards, A. D. (1981) Unpublished work at MRPRA.
Henriques, R. (1892) *J. Soc. Chem. Ind.* **11**, 477.
Higgins, G. M. C. (1975) *Internat. Rubber Conf.*, Kuala Lumpur.
Higgins, G. M. C. (1978) *NR Technol.* **9**, 68.
Kress, K. E. (1956) *Rubb. World* **134**, 709.
Lindley, P. B. (1966) *Rubber. 'Kempe's Engineers Year Book'* **1**, 1341.
McSweeney, G. P. (1970) *J. Inst. Rubb. Ind.* **4**, 245.
Milligan, A. G. (1931) *J. Soc. Chem. Ind.* **50**, 144T.
Robertson, M. W. and Rowley, R. M. (1960) *British Plastics* **33**, 26.
Stamberger, P. and Knight, B. C. J. G. (1928) *J. Chem. Soc.* 2791.
Wake, W. C. (1969) *The Analysis of Rubber and Rubber-like Polymers*, 2nd ed., Maclaren, London.
Weber, C. O. (1894) *J. Soc. Chem. Ind.* **13**, 987.
Weber, C. O. (1902) *The Chemistry of India-Rubber*, Griffen, London.
Williamson, A. G. (1957) RABRM Laboratories, unpublished work.
Wyatt, G. H. (1941) *Analyst* **66**, 362.

Chapter 4

Analysis of Extracts

Extracts of compounded rubbers may contain materials added as curatives, protective agents, processing aids, property modifiers (e.g. resins) and plasticizers, separately or in any combination. In addition to these classes will be chemicals added during manufacture of synthetic rubbers (e.g. catalyst residues, antioxidants and surfactants) or naturally occurring chemicals retained during processing of natural rubber. Vulcanized rubber will also contain degradation products arising from the curatives, and may also contain materials originating during the actual service life of the product such as butterfat in milking machine tubes or detergents in washing machine gaskets.

In view of the diversity of possible chemicals present in a rubber extract, and the multiplicity of possibilities within each class of compound, it is likely that a separation stage will need either to precede any determination or to be an integral part of the procedure. There are occasions when chemical spot tests can be applied without carrying out any separation, but the possibility of interferences from other materials is always present and should be noted.

Any elemental sulphur present in the compound or vulcanizate will also be found in the extract, but consideration of the methods available for quantifying this will be deferred until Chapter 6 when the whole range of sulphur analyses will be considered.

In this chapter, we initially consider various selective analyses and spot tests for curatives, protective agents and other additives, and then discuss the much more specific identifications possible when a chromatographic separation forms part of the analytical procedure.

IDENTIFICATIONS WITH NO SEPARATION

The use of more or less specific chemical reagents to treat unchromatographed extracts is sometimes able to provide adequate information, particularly when known systems are being checked, possibly for quality control purposes. Thus it is frequently possible to check whether a *para*-phenylene diamine antioxidant is present, although it is not usually possible to distinguish between particular compounds within this general class. A few of the more useful chemical tests are presented below, but more extensive compilations are available in the literature (see Hummel and Scholl, 1981; Haslam *et al.*, 1972; Crompton, 1971).

Plasticizers

Dependent upon the type of polymer, two completely distinct types of plasticizer may be used. For hydrocarbon polymers (NR, IR, SBR, BR, etc.), various grades of mineral (hydrocarbon) oil are used. Polyvinylchloride, polyurethane rubbers, and to a lesser extent nitrile rubber, on the other hand, are usually plasticized with oxygenated compounds, such as esters. Although dibutyl and dioctyl phthalates are two of the most commonly used plasticizers, other phthalate esters (see Table 4.2), flame-retardant phosphate esters (see Table 4.3) and other aliphatic esters, useful for good flexibility at low temperatures (see Table 4.4), are also used not infrequently.

Except where they are added primarily as processing aids, as is the case for low levels of hydrocarbon oils in black-filled hydrocarbon rubbers, plasticizers are usually present in quantities far in excess of other additives. Under these circumstances the general type of plasticizer may be identified by evaporation of the solvent from the total solvent extract, and comparison of the infrared spectrum of the residue with those of commercial plasticizers. For extensive examples of published spectra, see Hummel and Scholl (1981) and Haslam *et al.* (1972). However, it should be noted that the main classes of plasticizer are characterized by strong bands in their infrared spectrum as given in Table 4.1.

The use of plasticizer mixtures is a continuing aspect of polyvinylchloride technology but is uncommon in nitrile or polychloroprene rubbers. It is therefore a useful preliminary step to investigate whether the extracted material is a single substance or a mixture. Haslam and Willis (1965) recommend a preliminary examination by dissolving the extract in carbon tetrachloride and collecting fractions from a silica gel/Celite column as it is eluted with carbon tetrachloride mixed with 1·5, 2·0, 3·0 and

TABLE 4.1

CHARACTERISTIC IR BANDS OF PLASTICIZERS

Plasticizer type	*Characteristic approximate major band positions* (cm^{-1})							
Phthalate ester		1725	1580,1600	1280	1120	1070	740	700
Phosphate ester		—		1270		1000		
Polyglycol ether		—			1100			
Hydrocarbon	2950	—	1470	—	—	—	—	
Aromatic hydrocarbon		—				830–710		
Aromatic chlorohydrocarbon		—				830–710		
Adipate ester		1725		1240	1170	1140		
Sebacate ester		1725		1240	1170	1130		
Succinate ester		1725		1210	1160			

4·0% isopropyl ether. After removal of the eluting solvent, density, refractive index, UV fluorescence, boiling point under reduced pressure, or other simple means can confirm if the fractions are different. Alternatively vacuum distillation could be used to determine complexity of the extract at the same time as making the separation. Given a material thought to be a single substance, a systematic approach involves elemental analysis, principally looking for phosphorus or chlorine (see Chapter 6), and then spot tests for phthalates or phosphates as seems appropriate. The following details, taken from Haslam *et al.* (1951), give convenient tests.

Tests for Presence of Phthalate

(i) Add 3 drops of plasticizer and 1 drop of concentrated sulphuric acid to 0·05 g of resorcinol in a test-tube. Heat for 3 min at 160 °C. Cool, add 2 cm^3 of water and 2 cm^3 of 10% (w/v) sodium hydroxide solution and agitate. A green fluorescence indicates the presence of phthalate.

(ii) Add 3 drops of plasticizer and 1 drop of concentrated sulphuric acid to about 0·05 g of phenol in a test-tube. Heat for 3 min at 160 °C and, after cooling, pour the contents into 5% aqueous sodium hydroxide where the familiar phenolphthalein colour will indicate the presence of phthalate.

Test for Phenolic and Cresylic Plasticizers

Dissolve 10 mg of the plasticizer in 5 cm^3 of 0·5 M alcoholic potassium hydroxide in a beaker. Immerse the beaker in a boiling water-bath to remove most of the alcohol, add 2 cm^3 of water and 2·5 cm^3 of 1 M hydrochloric acid to neutralize. Transfer 1 cm^3 of the solution to a clean test tube and add 2 cm^3 of buffer solution and 5 drops of indicator solution. A blue indophenol colour is immediately apparent if phenolic or cresylic groups are present. The buffer solution is prepared from 23·4 g of $Na_2B_4O_7.10H_2O$ dissolved in 900 cm^3 of warm water to which is added 3·27 g NaOH as a 20% solution. The buffer solution is made up to 1 litre when cool. The indicator solution consists of 0·1 g of 2,6-dibromoquinone-chlorimide freshly dissolved in 25 cm^3 of alcohol.

Test for Phosphate Ester Plasticizers

Carry out the test for the presence of phosphorus as described in Chapter 6. If positive, it is probable that phosphate esters are present.

At this stage in the analysis it may be found useful to isolate small quantities of plasticizer in as pure a state as possible. Simple distillation on a semimicroscale, preferably under reduced pressure, may well be adequate. The possibility that separation of closely related plasticizers may not have been achieved must not be overlooked, though Collins (1955) has stated that mixtures of phthalate esters can definitely be separated by this means. A few drops of distillate will suffice to establish refractive index or solubility in water or other solvents, and to enable comparisons to be made with published physical properties of plasticizers.

Tables 4.2, 4.3 and 4.4, taken from the above references, supplemented by Doolittle (1954), give physical properties for some of the commoner plasticizers, although a more extensive compilation may be found in a review by Kline (1967).

Unequivocal identification of a plasticizer or of a mixture of plasticizers can only be achieved by the use of a separative technique, usually chromatography, in conjunction with identification procedures. These will be considered in more detail below.

Factice

Although factice is, strictly speaking, not a plasticizer, being added to mixes for a whole range of reasons including the improvement of dimensional stability and rigidity of unsupported articles in hot air cures, the improvement of extrusion and calendering, and the improvement of surface finish and electrical insulation, it does nevertheless act as a softener, improving filler dispersion and decreasing the energy requirements of mixing. It may therefore be appropriately discussed at this point.

TABLE 4.2

PROPERTIES OF PHTHALATE PLASTICIZERS

Phthalate	*Density*/°C	*Refractive index*/°C	*Boiling point* (°C) *or range*
Dimethyl	1·189/25 1·195/15·5	1·514/25 1·517/20	282–285/760 mm
Diethyl	1·120/25 1·120/15·5	1·500/25	290–300/760 mm
Di-*n*-butyl	1·045/25	1·491/25	340/760 mm 320/160 mm
Diamyl	1·024/15·5	1·487[1]	336–340/760 mm
Dihexyl	1·0085/20	1·487/20	340–350/760 mm 220/5 mm
Dioctyl:			
2-ethylhexyl	0·986/20	1·486/20	230/5 mm
2-methylheptyl	0·986/20	1·486/20	228–237/4 mm
capryl	0·966/25	1·480/25	229/4·5 mm

[1]Temperature not quoted.

TABLE 4.3

PROPERTIES OF PHOSPHATE PLASTICIZERS

Phosphate	*Density*/°C	*Refractive index*/°C	*Boiling point* (°C) *or range*
Octyl diphenyl	1·090/25	1·508/25	375
Tricresyl	1·162/25 1·180/15·5	1·553/25 1·560/20	260–275/10 mm
Trioctyl (2-ethylhexyl)	0·926/20	1·443/20	216/5 mm
Triphenyl	1·25/15 (solid)	—	MP 45 °C

Dark factice is prepared by heating vegetable oil with 10–25% of its weight of sulphur at 130–150 °C. White factice, on the other hand, is produced by reacting vegetable oil at 30–50 °C with sulphur monochloride. If, therefore, analysis of an unknown rubber product indicates a higher level of organically bound sulphur than might reasonably be expected for its crosslink density, then the presence of factice will be suspected.

TABLE 4.4

PROPERTIES OF ADIPATE AND SEBACATE PLASTICIZERS

	Density/°C		*Refractive index*/°C		*Boiling point* (°C) *or range*	
Adipate or sebacate	*Adipate*	*Sebacate*	*Adipate*	*Sebacate*	*Adipate*	*Sebacate*
Di-isobutyl	0·957/20		1·428/25		145–163/4 mm	
Di-*n*-hexyl	0·929–0·936/25		1·439/25		143–183/3 mm	
Dioctyls:						
(i) dicapryl	0·915/20	0·907/20	1·440/25	1·444/25	211–217/4 mm	230–240/4 mm
(ii) 2-methylheptyl	0·928/20	0·917/20	1·448/25	1·447/25	213–223/4 mm	248–255/4 mm
(iii) 2-ethylhexyl	0·927/20	0·911/25	1·446/25	1·451/25	208–218/4 mm	256/5 mm
		0·913/20		1·450/20		264/6 mm
Dinonyl	0·914/25		1·445/25		230/5 mm	
n-Octyl, *n*-decyl (mixture)	0·918/20		1·447/25		220–254/4 mm	
Di(butyl carbitol)	1·02/15·5		—		—	

The detection and measurement of factice content depends upon the fact that the origin of the factice is a vegetable oil, which consists of glycerides (glyceryl esters of fatty acids). After the rubber has been thoroughly extracted, it is treated with alcoholic potassium hydroxide to hydrolyse the glycerides. The liberated fatty acids are then estimated gravimetrically.

Determination of Factice

2 g of finely ground or chopped vulcanizate is weighed accurately and extracted for 8 h in a Soxhlet extractor with *n*-hexane. The rubber is then refluxed for 4 h in 50 cm^3 of a 1:1 mixture of toluene and 1 M ethanolic potassium hydroxide solution. The ethanolic–toluene mixture is removed by filtration. The residual rubber is washed twice with 25 cm^3 hot 96% ethanol and then three times with 25 cm^3 hot water. The combined ethanol–toluene extract and washings are evaporated to small bulk, but not to complete dryness. The residue is dissolved in water, transferred quantitatively to a separating funnel, and acidified to about pH 3 with 20% hydrochloric acid. The resultant suspension is extracted five times with 25 cm^3 ether. The combined ether extracts are washed three times with 25 cm^3 water, evaporated on a water-bath, and then dried at 105 °C for 3 h. The fatty acid is weighed.

The wt % of factice is then given by $1{\cdot}18 \times$ wt % fatty acid.

This procedure is recorded in the German Standard DIN 53 588. A variant of it, in which the rubber is extracted first with acetone and then with chloroform instead of *n*-hexane, is given in ASTM D 297-1981 Method No. 21.

Tests for Some Specific Additives

In addition to the above tests on unseparated (or unchromatographed) extracts, it is also possible to check the presence of, and determine quantitatively, a variety of other additives which are put into raw rubber or raw latex. A hydroxylamine salt (sulphate or chloride) is added to some grades of raw natural rubber and raw natural rubber latex in order to confer constant viscosity properties on the rubber by preventing the occurrence of storage hardening reactions. Although their usage is currently diminishing rapidly, boric acid and sodium pentachlorphenate (SPP) are sometimes added to NR latex to prevent bacterial degradation of the latex. Methods for the direct quantitative determination of all three in rubber extracts are available and are given below.

Determination of Hydroxylamine (after Berg and Becker, 1940)

5 g of thinly sheeted raw rubber is extracted by refluxing for 16 h in 45 cm^3 of 0·8 M sulphuric acid. The acid extract is quantitatively transferred to a 50 cm^3 volumetric flask, and made up to the mark with 0·8 M sulphuric acid. A mixture of 1 cm^3 extract, 1 cm^3 of *fresh* 8-hydroxyquinoline solution (1% w/v in ethanol) and 3 cm^3 of 1 M sodium carbonate solution is placed in a test tube of 30 cm^3 capacity, and oxygen or air bubbled through for 5 min. The tube is stoppered and warmed to 40 °C for 20 min in a water-bath. The solution is diluted to 10 cm^3 with water in a volumetric flask, and the green colour measured in a colorimeter or spectrophotometer set to 700 nm.

Zinc ions interfere with the above method which cannot therefore be used without modification for compounds or vulcanizates. Interference by zinc ions can be prevented by addition of 3 cm^3 of a 10% solution of DCTA (1,2-diaminocyclohexane-*N,N,N',N'*-tetraacetic acid) in 1 M sodium carbonate solution, as complexing agent. This modification of the above method is due to A. M. Petric of the MRPRA Laboratories.

Determination of Boric Acid

Boric acid, in aqueous extracts of rubber or rubber products, is simply determined by neutralization of the extract to phenolphthalein, addition of mannitol, and titration with 0·02 M sodium hydroxide solution.

Total boron in rubbers can be determined using a modification of the method due to Hayes and Metcalf (1962).

Finely chopped rubber sample (0·1 g) is covered with sodium carbonate (0·1–0·2 g) in a closed platinum crucible and ignited until fusion is complete (*ca.* 3 min). The cooled residue is dissolved in 1 cm^3 50% acetic acid in water, transferred to a 10 cm^3 graduated flask, and made up to the mark with glacial acetic acid. Aliquots (1 cm^3) of this solution are placed in polythene bottles and 3 cm^3 of curcumin–acetic acid reagent (0·12% w/w in glacial acetic acid) and 3 cm^3 of 1:1 acetic acid/sulphuric acid added. After 20 min at 20 °C, 50 cm^3 absolute ethanol is added, and the volume adjusted to 100 cm^3 in a 100 cm^3 volumetric flask, using absolute ethanol. The optical density is measured at 555 nm.

Determination of Sodium Pentachlorophenate

Sodium pentachlorophenate can be determined rapidly and accurately in Hevea latex concentrate by reaction with 4-aminophenazone to give a blue complex which is measured at 600 nm. The method is described in detail by Chin *et al.* (1975). The presence of sodium salicylate added to freeze–thaw stabilized natural rubber latex concentrate does not interfere, but ammonia in high ammonia grades of latex must first be reduced in concentration to 0·2% (w/w) by the addition of formalin or boric acid solutions.

Ultraviolet Spectroscopy

When a molecule contains double bonds and elements with electron pairs in p (or higher) orbitals then the energy range of electronic transitions falls in the ultraviolet or even the visible region of the spectrum. Since these electronic absorptions are usually very intense, they provide a highly sensitive means of measuring concentrations of suitable molecules. 'Suitable molecules' include most antioxidants, many vulcanization accelerators and even sulphur (see Fikhlengol'ts *et al.*, 1966).

Ultraviolet absorption bands are, however, usually very broad, and often relatively featureless. The technique is therefore of limited diagnostic value, particularly in situations where two or more materials may be present simultaneously, as is frequently the case for unseparated extracts from compounded rubbers.

There are nevertheless occasions when much useful information may be obtained by combining ultraviolet spectroscopy with treatment of the extract with a selective chemical reagent.

If, for example, the ultraviolet spectrum of an alcoholic extract is recorded both before and after treatment with a selective chemical reagent, the disappearance of one band or the appearance of a band at a new position is relatively easy to detect, even in the presence of severely overlapping bands. Indeed by measuring the ultraviolet spectrum of the treated extract *relative* to that of the untreated extract, by placing one solution in the sample beam and the other in the reference beam, ultraviolet absorption bands which are unaffected by the chemical treatment cancel out.

Some of the more useful treatments for commonly observed compounding ingredients are shown in Table 4.5. The information is compiled from papers by Lloyd (1962) and Wexler (1963).

The relatively low specificity of the ultraviolet method can be put to use as a method for obtaining a crude estimate of the total antioxidant present. Indeed Blois (1958) demonstrated that antioxidants of a wide variety of types discharge the intense violet colour from the stable free radical α,α-diphenylpicrylhydrazyl (DPPH), replacing it by a yellow colour. Measurement of the optical density at 517 μm, at which wavelength only

TABLE 4.5

Compounding ingredient	*UV* λ_{max}	*Chemical reagent*	*Reaction product*	*UV* λ_{max}
Mercaptobenzothiazyl sulphenamide	282	$SnCl_2$/HCl	MBT + amine	327 —
Sulphur	261	Sodium borohydride	H_2S	—
Phenolic AOs	280–290	Potassium hydroxide	Phenolate anion	300–310
Dithiocarbamates	273	Copper sulphate	Copper complex	435
Mercaptobenzothiazole (MBT)	327	Copper sulphate	Copper complex	— (insoluble)
Tetramethylthiuram-disulphide	282	Copper sulphate	Copper complex	434

the purple free radical absorbs, gives an estimate of the total antioxidant present. Glavind (1963) used this method to determine total antioxidant in biological systems (mammalian blood and liver). Although the method appears to be relevant to polymers, it does not yet seem to have been used for this purpose. The method would have the advantage that naturally occurring antioxidants in natural rubber (such as the sitosterols) would be determined, although thiol-containing proteins also present might interfere.

Analysis for Antioxidants

Although antioxidants have been determined directly on the rubber itself, for example by Luongo (1965) and Miller and Willis (1959), the method is applicable only to uncured mixes not containing carbon black, and then only to materials of known composition, and will not be considered further. Many techniques are available for the detection of antioxidants after separation from the polymer by extraction (e.g. Wheeler, 1968). Some have been mentioned above in the section on ultraviolet absorption spectroscopy. Others, highly specific in nature, will be discussed below in the sections on gas, thin layer and liquid chromatography.

It is, however, frequently possible to determine at least the class of antioxidant used (if any) by the use of inexpensive colorimetric spot tests which can indicate whether the antioxidant is staining, non-staining, anti-ozonant, or protects against metal catalysed degradation. Often this will be sufficient to determine whether the mix or vulcanizate is protected for a given application. A few of the more useful tests are given below, although for additional information the reader is referred to the more specialist literature (e.g. Hilton, 1958; Wheeler, 1968; Crompton, 1971; Haslam *et al.*, 1972; Hummel and Scholl, 1981).

Paraphenylene Diamines

(a) To the extract is added a fresh 1% solution of 4-nitrobenzene diazonium fluoroborate (nitrazole) in methanol containing a few drops of concentrated hydrochloric acid. Aromatic amines give red, violet, or blue colours.

(b) To the extract is added a 4% solution of benzoyl peroxide in benzene. Aryl substituted *para*phenylene diamines give yellow to orange yellow colours which may turn red–violet to blue on the addition of stannous chloride.

Phenols

(a) To the extract is added a few drops of dilute sodium hydroxide solution. Addition of a few drops of a 1% solution of 4-nitrobenzene-diazonium fluoroborate (nitrazole) in methanol gives a highly coloured azo-dye. Phenols substituted in the *ortho* or *para* positions do not react. In this case the following test (b) may be applied.

(b) Addition of an equal volume of Millons reagent* to the extract dissolved in methanol gives a yellow to orange colour with phenols.

Analysis for Accelerators

Accelerators are active vulcanization ingredients, many of which are chemically altered during the vulcanization process. It is therefore important to recognize that whilst most accelerators can be recovered unchanged from raw mixes, only the 'vulcanization residues' can be recovered from cured products. Sulphenamide accelerators give amine and mercaptobenzothiazole (MBT) or ZMBT, mercaptobenzothiazyl disulphide (MBTS) gives MBT, and thiurams give the corresponding dithiocarbamates. Guanidines give, in part, the aromatic amine, but can also be detected unchanged.

It is important to note that chemical changes may also take place during extraction of raw mixes. Hot extraction can allow vulcanization reactions

*Available from laboratory chemical suppliers.

to proceed and so cold extraction, whilst possibly not being quantitative, is to be preferred for the identification of accelerators in a compounded but unvulcanized mix. Acetone extraction is known to cause the decomposition of thiuram disulphides to give dithiocarbamates (Wake, 1969). This can be avoided by using solvents such as 2-propanol or methanol and it is thus essential to use alcohols if a distinction between these is required. An alternative procedure is to use a cold extraction technique. If the minute traces of thiuram remaining after cure are being sought, it is advisable to use an alcoholic extraction regardless of temperature.

Once again, only a very few chemical tests are given in outline here, following procedures given by Hofmann and Ostromov (1968, 1969). The interested reader is referred to the specialist literature for more detailed descriptions, or for additional methods, as for example work by Brock and Louth (1955) and Crompton (1971).

Mercaptobenzothiazole (MBT)

Addition of a chloroform solution of copper oleate gives a yellow colour. Subsequent addition of a few drops of a freshly prepared chloroform solution of tetramethyl thiuram disulphide turns the chloroform layer brown.

Dithiocarbamates

Addition of a 1% solution of copper oleate to a chloroform solution of the extract gives a dark brown or black precipitate or colour due to the formation of copper dithiocarbamate.

Thiurams

A chloroform solution of copper oleate gives a yellow green colour, which is changed to red–brown to dark brown by the addition of mercaptobenzothiazole.

Thiurams in the Presence of Dithiocarbamates

This procedure follows that of Zijp (1956). The rubber mix or vulcanizate is extracted with 2-propanol, and the solvent removed by evaporation under reduced pressure. The residue is dissolved in toluene and the solution divided into two portions, A and B, one of which is used to demonstrate the presence of dithiocarbamates, and the other to demonstrate the presence of small amounts of any thiuram accelerators that may have been used.

To fraction A is added dropwise a slight excess of a 10% solution of cobalt oleate in toluene. This reagent forms coloured complexes with dithiocarbamates but not with thiurams. The solution is placed on an alumina chromatographic column and developed with toluene. Any green bands are eluted by continual development with toluene and the eluted material is recovered by evaporation of the solvent. An infrared spectrum is obtained on the residue, as a mull.

The alternative is chromatography of the copper complex on a silica gel–Celite column.

10 cm^3 of the toluene solution is shaken with a 1% solution of copper sulphate in 15% aqueous sodium sulphate. The sodium sulphate is present to assist breaking of the emulsion formed on shaking. The toluene solution is then separated, dried with sodium sulphate, and then evaporated to dryness before being taken up again in a small quantity of carbon tetrachloride, 1 cm^3 of which is applied to the column. These copper complexes are light-sensitive and should not be left exposed between operations.

Fraction B is used to differentiate thiurams from dithiocarbamates. If no dithiocarbamate is present in fraction A there is no need to use fraction B since thiurams also will be absent. If a dithiocarbamate is present then it may be important to check whether this has been derived from a thiuram, the nature of the dithiocarbamate indicating the need for a test.

Fraction B is first evaporated and then dissolved in acetone and divided into two lots of 5 cm^3 each in test-tubes. Both are treated with a few drops of concentrated phosphoric acid and boiled to eliminate carbon disulphide. (A spot test with copper sulphate solution will confirm this elimination.) The solutions are then neutralized with 20% sodium hydroxide solution and reduced with sodium hydrogen sulphite. The acetone is evaporated and a few drops of a concentrated ammonia solution, ammonium citrate solution and chloroform are added to both and a few drops of 2% copper sulphate solution to one only. The two tubes are now compared and if the chloroform layer shows the presence of the dark brown copper dithiocarbamate, the presence of a thiuram may be inferred. Some coloration is usually present, hence the need for comparison with the blank.

Guanidines

Shake vigorously a chloroform solution of a rubber extract with a few ml of water to which have been added a few drops of 6% sodium hypochlorite solution. In the presence of diaryl guanidines, the chloroform layer becomes brown coloured, whilst the aqueous layer remains colourless. In the presence of *o*-tolylbiguanide, the aqueous layer becomes yellow, whilst the chloroform layer remains colourless.

Analysis for Other Additives

Detailed descriptions of the chemical analysis for other additives such as processing aids, blowing agents or peptisers is beyond the scope of this book. Such analyses are reviewed by Hummel and Scholl (1981).

IDENTIFICATIONS AFTER SEPARATION

The complexity of modern mixes, coupled with the large number of accelerators, antioxidants and other additives which are commercially available and the ever-increasing need to carry out analyses in ever shorter times, has made it inevitable that separative techniques have become of prime importance in the analysis of rubber extracts.

Chromatography is the general term covering a wide range of separative techniques all of which have the same fundamental principles of operation, and all of which can be used, under differing circumstances, for the separation of chemical constituents of rubber extracts.

All chromatographic techniques involve a stationary phase and a mobile phase which passes through, around, or over it, depending upon its nature. If a mixture of components is placed at one end of the system, each individual component will be carried with the mobile phase, at a rate dependent upon its relative affinity for the two phases. Different chemicals will thereby become separated from each other.

The different chromatographic techniques which will be considered here are column and paper chromatography, thin layer chromatography, gas chromatography and liquid chromatography. Gel permeation chromatography is considered in detail in Chapter 9 and is based on the sieving action of a porous stationary phase to separate materials which, although chemically similar, have different molecular shapes or molar masses. It will not be considered further here.

The earliest comprehensive separative analytical scheme was that of Bellamy *et al.* (1947). This depended on chromatography on an alumina column of the acetone extract dissolved in benzene, and elution of components from the column. The various eluted fractions were then rechromatographed after the addition of cobalt oleate. Later workers favoured treating the acetone extract to effect a preliminary separation by chemical means into acidic and alkaline components prior to separation of the components by chromatography either on a column or on paper. An example of the latter is that of Zijp (1956) whose procedure was given in some detail in the second edition of this book (Wake, 1969). The procedure will not be repeated here, although the chemical differentiation between thiurams and dithiocarbamates which it contained has been given above.

Unfortunately the rate of movement of the solvent through the paper tends to be very slow by the ascending method, and overnight runs used to be the norm. Downward development is faster, but the resolving power of the method becomes inferior. Other means of speeding up the process, such as circular paper chromatography, are available, but for the chromatography of organic materials, thin layer chromatography has now displaced paper chromatography almost entirely.

Thin Layer Chromatography

The ease of use of this technique, its speed of operation and low cost make it particularly suitable for the rubber-laboratory, both for identification of unknown cure systems and for quality control purposes.

The technique is closely related to paper chromatography, having a stationary phase, which may be cellulose, silica or alumina for example, applied as a thin layer on a rigid support which is commonly glass or plastic. It does not, however, suffer from the disadvantages of paper chromatography in that it is quick, with a run taking perhaps half an hour, and it has a much higher resolving power than has paper chromatography. Indeed high performance grades of surface coatings having particle sizes of only a few micrometres diameter give plates which will separate materials adequately within the space of only 5 cm or so, still further speeding up the process.

Pre-prepared plates can be purchased readily although, if a large number of thin layer chromatographic analyses are to be carried out, they will prove to be more costly than the alternative of coating one's own plates. For this purpose commercial spreaders are available, although plates can be coated using spacers and a glass rod.

The operation of the chromatographic technique is spelled out, for example, by Stahl (1965), Zweig and Sherman (1972), and by many others; its application to rubber antidegradant analysis is given in some detail in ASTM D 3156 and in ISO/DIS 4645.2. In essence, the procedure is to place a small spot of a dichloromethane solution of the rubber extract on to the coating on the plate, about 2 cm from the bottom, and then to stand the plate in a glass tank, with a lid, containing the desired solvent system in the bottom at a depth of about 1 cm. The solvent travels up the layer by capillary attraction. When the solvent front has reached the required height, the plate is removed from the tank and air dried prior to visualization of the separated spots with a reactive chemical spray, or repeated development of the plate either in the same or different solvent. Components of similar structure chromatographed under the same conditions will have similar mobilities. The mobility is measured in terms of R_F which is given by the equation

$$R_F = \frac{\text{distance travelled by the compound}}{\text{distance travelled by the solvent front}}$$

Thin layer chromatography is usually regarded as being semi-quantitative; the logarithm of the spot area is proportional to the quantity of material in the spot. In practice, however, visual comparison of spot size with those of a series of spots containing standard quantities of the authentic material will suffice to give an estimate of quantity which is usually adequate for the examination of mixes and vulcanizates.

More accurate quantitative results can in theory be obtained by the use of a scanning densitometer. In practice, however, extracts from rubbers are so complex that it is rarely possible to measure components without serious interference from other components.

During the development of the chromatogram it is necessary to preserve a uniformly homogeneous environment for the TLC plate by using a glass tank lined with absorbent paper, dipping into the solvent at the bottom of the tank. If this is not carried out, then under certain conditions of temperature and humidity, plates will take an abnormally long time to develop, and apparent compound mobilities will be much higher than expected.

Some favour the use of a short column chromatographic pretreatment of the extract to remove any oil present prior to TLC examination. However, this procedure is not to be recommended as polar materials might be retained on the column, while elemental sulphur may well be eluted rapidly with the oil. The alternative method of removing oil is to develop the plate initially in light petroleum, prior to elution with normal chromatographic solvents. The light petroleum carries the oil near to the solvent front and out of the way of other components. The plate is then developed in the required solvent. This procedure prevents the oil from interfering with the separation of the other components of the extract, whilst ensuring that all other materials are on the plate. The initial run in petroleum has the additional advantage of separating free sulphur at an R_F of about 0·6. If the

TLC plate has a fluorescent layer, then inspection of the dried plate under ultraviolet light renders any free sulphur present visible as a dark spot on a fluorescent background, as described by McSweeney (1971) and Davies and Thuraisingham (1968). The presence of free sulphur may be confirmed by spraying with aqueous silver nitrate, which gives a brown-coloured spot. Alternatively, after examination in ultraviolet light, the plate may be developed in the normal chromatographic solvent. As will be shown in Chapter 6, the presence of free sulphur in a cured product may well indicate that the product is undercured.

Since Yuasa and Kamiya (1964) published their paper on the use of TLC for the identification of additives in acetone extracts of rubber, a veritable flood of papers on the subject has appeared, and paper chromatographic techniques have been almost entirely displaced.

Comprehensive reviews of the TLC of rubber chemicals have been given by Kreiner and Warner (1969), Crompton (1971), Wheeler (1968), Kreiner (1971), Hummel and Scholl (1981) and Hofmann and Ostromov (1968, 1969). Hofmann and Ostromov (1968) have also used thin layer chromatography extensively in their compilation of methods for the investigation of rubbers in relation to German health regulations.

A useful review for the rubber-technologist is that of McSweeney (1970) who gives the practical details needed regarding technique and spray reagents useful for the analysis of rubber extracts. In a later paper, Higgins and McSweeney (1974) have given a method whereby amine residues, produced by decomposition of sulphenamide accelerators during cure, can be identified and hence the nature of the original sulphenamide deduced.

Rubber extracts tend to contain rather a large number of chemicals. In the authors' laboratory, it has been found profitable to standardize on a fluorescent grade of silica gel G as stationary phase and on a single solvent system following the light petroleum pre-run (light petroleum/diethyl ether, 60:40 v/v), so that the positions of all the various spots gradually become familiar to the analyst. In cases where particular problems arise, such as two components not being adequately resolved, or polar materials being present which do not move off the startline (e.g. DPG), then an alternative solvent system can be chosen and the separation repeated. Thus the guanidines separate if, after running the plate in the above solvent, the plate is re-run in a mixture of acetone and ammonia (98:2 v/v). They can be visualized by spraying with sodium hypochlorite solution.

The most generally useful spray reagents are 2,6-dibromo-*para*-benzoquinone-4-chlorimine and the corresponding dichloro-compound. Both are popularly known as Gibbs' reagent and are widely reactive, giving colours for different chemical types which range from a greyish white for certain phenolic antioxidants, through brown for dithiocarbamates, orange for MBT, to greens and purples for amine antioxidants and a wide range of other nitrogen-containing materials. Dithiazone spray reagent is particularly useful for visualizing zinc compounds, with which it gives a rich pink colour. An aqueous solution of copper sulphate is particularly sensitive towards dithiocarbamates, whilst anisaldehyde and sulphuric acid (5% each) in ethanol is a useful general spray which nevertheless shows some specificity.

Davies (1967) described how natural rubber may be distinguished from synthetic polyisoprene by the detection of β-sitosterol, a naturally occurring compound which survives most normal rubber treatments including vulcanization. McSweeney (1970) improved the procedure by using the more selective spray reagent, cupric acetate (3% in 8% phosphoric acid). On heating with sterols, this gives a red colour which, on further heating, turns first to blue and then to dark grey. The background does not darken, spot colours are stable, and very few other compounds react with the reagent. The copper acetate spray can be applied even after the plates have been sprayed with Gibbs' reagent.

Higgins and McSweeney (1974) showed how the amines produced from sulphenamides during vulcanization can be reacted with an alkaline solution of 7-nitrobenzo-2,1,3-oxadiazol (NBD) in methyl ethyl ketone solution. The NBD derivatives thus formed are extracted by toluene, transferred to chloroform and chromatographed on a fluorescent grade of silica gel in toluene/ethyl acetate (4:1, v/v). Comparison of the TLC spot position with the positions of spots due to the NBD derivatives of authentic amines enables the amine portion of the sulphenamide accelerator to be identified.

Thin layer chromatographic procedures have been developed for the identification of a wide range of rubber additives, and the interested reader is referred to the specialist literature, as detailed in the reviews listed above. Only two others will be mentioned briefly here, the first being the determination of castor oil in crumb grades of natural rubber. This method, originally developed by Davies and Tunnicliffe (1967), was improved by McSweeney (1972) and the method has been adopted by the International Organization for Standardization (see ISO/DIS 6225/1.2). The main improvement introduced into the original method was the use of the anisaldehyde spray reagent. This reagent is much more selective than the phosphomolybdic acid spray used in the original method, and avoids interference by overlapping compounds which are usually present. The anisaldehyde spray reagent gives mauve spots which turn green on heating. If, after heating, the plate is resprayed with anisaldehyde there is an immediate increase in the green coloration.

The second is the use of thin layer chromatography for the determination of the type of hydrocarbon extending oil. Extending oils used in the rubber industry contain oils of paraffinic, naphthenic and aromatic types. Those oils, known loosely as paraffinic, naphthenic and aromatic oils, do in fact consist of mixtures of different proportions of the three molecular types as shown in Table 4.6.

TABLE 4.6

COMPOSITION OF HYDROCARBON OILS

	Composition, %		
	Paraffinic oils	*Naphthenic oils*	*Aromatic oils*
Paraffinic carbon	50–68	40–60	34–44
Naphthenic carbon	26–40	20–60	10–34
Aromatic carbon	2–7	21–29	30–50
Specific gravity	0·86–0·90	0·90–0·92	0·96 1·03

Courtesy John Wiley and Son.

Thin layer chromatographic analysis of oil type has been investigated by Killer and Amos (1966), Gilchrist *et al.* (1972), Peurifoy *et al.* (1970) and Hellmann (1974). An alternative column chromatographic procedure (ASTM D2007) has been published by the American Society for Testing and Materials.

McSweeney (1970), in the authors' laboratory, has used two procedures for the analysis of oil type. The first, and simpler of the two, relies upon straightforward comparison of the thin layer chromatogram of a rubber extract with those of authentic samples of the three types of oil. A fluorescent grade of silica gel is used as the stationary phase, whilst the mobile phase is light petroleum of boiling range 40–60 °C. Two methods of visualization of the partly separated oils are possible. The plate is allowed to dry, and then examined in ultraviolet light. In light of wavelength 254 nm, light naphthenic and aromatic oils appear as ultraviolet absorbent spots (i.e. dark spots on a fluorescent background), with the naphthenic oils having higher mobility. In light of wavelength 366 nm, aromatic components show a blue fluorescence. In addition McSweeney (1972) found that the plate may be sprayed with a freshly prepared solution of anisaldehyde and concentrated sulphuric acid (5% each) in ethanol. On gently heating with a hot air blower, naphthenic oils give a predominantly pink colour, whilst the aromatic oils give a blue colour. Aliphatic oils are not observed. Additional resolution of the various types of oil can be obtained by developing the chromatogram twice.

A refinement of the above method, also due to McSweeney (1972), uses two-dimensional development of the thin layer chromatogram. The unknown sample is spotted on to a square TLC plate about 4 cm diagonally from one corner, and the chromatogram run in one direction in light petroleum as above. The plate is then allowed to dry, turned through 90°, and then redeveloped in a mixture of light petroleum/diethyl ether (60/40). The resultant pattern of colours obtained after spraying as above is used to determine the type of oil.

In both the above methods, paraffinic components (which do not react to the sprays) can be seen as fleeting spots of high mobility as the solvents are drying, and as a colourless area defined by the crescent shaped naphthenic oil spot after spraying. It is also frequently possible to detect the paraffinic oil as a non-wetted area, if the plate is sprayed with a fine mist of water.

As mentioned earlier, thin layer chromatograms can be extremely complex, particularly when one remembers that many commercial antioxidants (and sometimes also accelerators) consist of a number of different components. In addition there are numerous occasions when two materials have identical or nearly identical mobilities and each masks the characteristic colour reaction of the other. On some occasions the problem can be overcome by the use of specific sprays which react to only one of the overlapping components. On other occasions more sophisticated 'tricks' must be used. Examples of these are as follows (McSweeney, 1981).

Agerite White

The antioxidant Agerite White has the same mobility and similar spray reaction to Gibbs' reagent as has mercaptobenzothiazole (MBT). The materials can readily be separated by making use of the acidic nature of the MBT. After the initial development in light petroleum, the plate is exposed to ammonia vapour in a covered tank. The plate is then immediately placed in the second development tank, containing the light petroleum/diethyl ether solvent. The MBT, being present now as the highly polar ammonium salt, remains on or near the startline of the TLC plate, whereas the Agerite White travels with its normal mobility.

Nonox ZA

The converse procedure may be carried out in the case of the overlapping Nonox ZA (*N*-2-propyl-*N'*-phenyl-*para*-phenylenediamine) and MBT. *Para*phenylenediamines, being strongly basic, react readily to form hydrochloride salts when the TLC plate is held in hydrogen chloride vapour. The amine will then have very low mobility, and the MBT will travel at its normal speed.

Cyclohexyl Benzothiazyl Sulphenamide (CBS)

When an unvulcanized compound has been cold extracted, CBS will frequently be present in the extract. This has a high mobility, but is very difficult to detect on spraying with Gibbs' reagent. Brief exposure of the TLC plate to hydrogen sulphide gas reduces the compound to MBT which then gives the characteristic orange colour of MBT.

Antioxidant 445

Antioxidant 445 has a very weak response to spray reagents including Gibbs' reagent. If the plate which has already been sprayed with Gibbs' reagent and which shows the pale blue colour due to AO 445 is oversprayed with chromic acid, then the spot colour turns to a deep reddish brown.

Hexamine

When sprayed with Gibbs' reagent the activator hexamine gives a mauve spot which is reminiscent of phenylene diamine antioxidants with which it may be confused. Spraying a fresh chromatogram with the anisaldehyde reagent mentioned above gives a bright orange colour.

Thiourea

Thiourea sometimes occurs in mixes and vulcanizates, but it is difficult to detect since it has only a weak reaction to spray reagents. If the developed plate is first held in ammonia vapour and then immediately sprayed with Gibbs' reagent, thiourea gives a very strong purple spot.

Infrared Spectroscopy

Infrared spectroscopy is considered in detail in Chapter 8, but it must be recorded here that it can prove invaluable for the identification of small amounts of material separated from rubber extracts by thin layer chromatography. Various methods have been used for transferring the separated material to be identified from the TLC plate to the infrared spectrometer so that its infrared spectrum can be recorded. A common procedure is to scrape the spot (and adsorbent) from the TLC plate, or to suck it off with a mini glass vacuum cleaner, and then wash the material from the adsorbent using a polar solvent, for example in a small glass tube. Examples of a variety of designs of apparatus for this purpose are given by Crompton (1971). The method does, however, suffer from the disadvantage that it

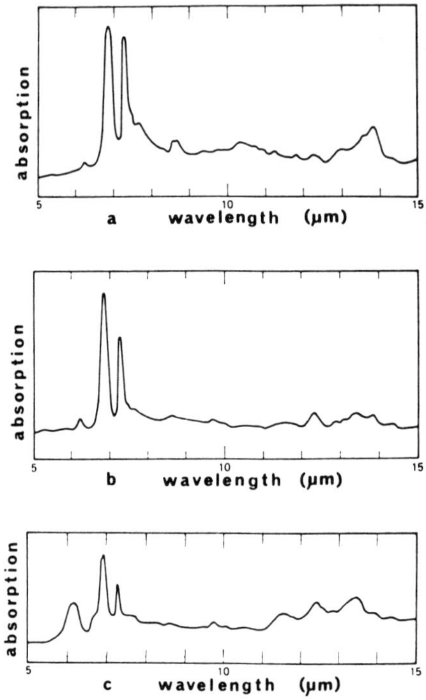

FIG. 4.1. Infrared spectra of extender oils: (a) paraffinic, (b) naphthenic, (c) aromatic. (Courtesy John Wiley & Sons.)

is difficult to avoid contamination of the sample by minute traces of adsorbent. These can interfere with the infrared examination.

This problem can, however, be avoided by 'chromatography' of the separated substance on a (commercially available) porous wedge of potassium bromide powder (see for example Rice, 1967; Garner and Packer, 1968). Material scraped from the TLC plate is placed into a glass tube with a flat base. The wedge of potassium bromide is placed on the sample and is prevented from touching the sides of the glass tube by means of a glass or metal ring. The tube is closed with a plastic cap in which a 3 mm diameter hole has been drilled centrally. About 0·5 cm^3 of a suitable solvent, e.g. methanol, carbon tetrachloride or toluene, is added to the tube. The solvent rises up the wedge, carrying the sample with it. Solvent evaporates from the top of the wedge so that the sample becomes concentrated at this point. This process may take 1–5 h depending on the volatility of the solvent and on the room temperature. After drying, the tip of the potassium bromide wedge is removed, crushed, and compressed into a thin disc which is examined in the infrared spectrometer. The spectrum obtained can then be compared with those of known compounds, an enormous compilation of which is given by Hummel and Scholl (1981).

Infrared examination of complete extracts can also be useful, particularly if considerable quantities of plasticizer or extender oil are present. Once again, Hummel and Scholl (1981) have provided a large range of such spectra.

The determination of the type of hydrocarbon oil used as an extender can also be carried out by infrared spectroscopy. Thus Cooper *et al.* (1971) have demonstrated (see Fig. 4.1) how the relative quantities of paraffinic, naphthenic and aromatic hydrocarbons can be determined by comparison of the relative intensities of bands at 13·9, 12·3 and 6·25 μm. White (1967) showed how this determination may be made more precise, using the comparison of bands at these positions with bands at the same positions in the spectra of *n*-heptane, phenanthrene and toluene respectively.

Gas Chromatography

This technique uses a stationary phase which is packed into a glass or metal column and a mobile phase which is simply an inert gas. Compounds travel through the column at a rate which depends upon the relative volatility of the compound and its solubility in an involatile chemical adsorbed on the stationary phase. Sensitive detectors are available which make the technique almost universally applicable, and also highly accurate as a quantitative tool, in contrast to thin layer chromatography which is only semi-quantitative.

Gas chromatography is particularly useful for the characterization and quantification of the more volatile substances such as amine and phenolic antioxidants, accelerators, plasticizers, fatty acids, stabilizers and other additives. Wise and Sullivan (1962), Gaeta *et al.* (1969) and Ostromov and Hofmann (1978) have used gas chromatography to determine anti-degradants in extracts from mixes and vulcanized rubbers, Patel *et al.* (1973) have used it to identify accelerators in vulcanizates, and both

Hummel and Scholl (1981) and Crompton (1971) have reviewed the use of gas chromatography for these purposes.

The identification of unknown peaks eluted from a gas chromatograph is often not easy, although more information on the approximate size of the compound is available than from examination by infrared spectroscopy or thin layer chromatography. However, just as the collection of separated components from a thin layer chromatogram has proved invaluable, so too has it for components eluted from a gas chromatograph. Components may be condensed in a cold trap and subsequently examined by any one of several techniques, including elemental analysis, infrared spectroscopy, or even nuclear magnetic resonance spectroscopy (see below). Commercial equipment is available for trapping individual peaks from a gas chromatograph and examining the infrared spectrum of the component directly in the gas phase, and very sensitive spectrometers which can obtain mass or infrared spectra 'on the fly' are also now available. Although very powerful, the required instrumentation is very costly, and these techniques will not be considered further. Access to the literature on this area may be obtained through one of the review articles mentioned above.

Nuclear Magnetic Resonance

Nuclear magnetic resonance spectroscopy (see Chapters 8 and 9) is another powerful, if costly, technique which can be used to advantage for the examination of chemicals which have been separated from extracts by such techniques as column, gas, thin layer or liquid chromatography.

High Performance Liquid Chromatography

The advent of microparticulate stationary phases and high-pressure pumps coupled with microprocessor control, has brought about the rebirth of the classical technique of column chromatography, which is at present the most rapidly expanding of the chromatographic techniques. The high separating power even of short columns (20–50 cm long), the availability of efficient high-pressure pumps and the development of highly sensitive detectors, operating by measurement of refractive index, ultraviolet absorption or electrical conductivity to name but three, make the technique very powerful indeed. It is almost universally applicable, provided the material can be brought into solution; it is extremely rapid (analyses taking perhaps 2–30 min), it is highly quantitative, and it is very reproducible. Although the technique is not cheap, it must be the method of choice in a quality control situation when a large number of virtually identical analyses have to be performed in a given period of time. Using temperature programming and/or elution with mixed solvents in which the relative concentrations change with time (gradient elution), mixtures ranging from very low to very high polarity can be eluted in a single run. Gross and Strauss have used high-performance liquid chromatography for the study of plasticizers (1977), stabilizers (1976) and antidegradants and accelerators (1979). Reviews of its application to the rubber industry have been given by Sidwell (1980) and Sullivan *et al.* (1976).

Ion Chromatography

A particular type of liquid chromatography, called ion chromatography, should also be mentioned. The technique is discussed in detail in Chapter 6, but here it is worth recording its existence and its ability to separate, identify and quantify both cations and anions present in aqueous solutions such as latex sera, or aqueous extracts of rubber products.

REFERENCES

Bellamy, L. J., Lawrie, J. H. and Press, E. W. S. (1947) *Trans. IRI.* **22**, 308; **23**, 15.
Berg, R. and Becker, E. (1940) *Ber.* **73**, 172.
Blois, M. S. (1958) *Nature* **181**, 1199.
Brock, M. J. and Louth, G. D. (1955) *Analyt. Chem.* **27**, 1575.
Chin, H. C., Singh, M. M. and Higgins, G. M. C. (1975) *Proc. Int. Rubber Conf.* Kuala Lumpur, 299.
Collins, J. H. (1955) *Testing and Analysis of Plastics, Part 1, The Identification of Plastics*, 2nd Ed., Plastics and Rubber Institute, London.
Cooper, W., Poulton, F. C. J. and Sewell, P. R. (1971) in *Encyclopaedia of Industrial Chemical Analysis*, Snell, F. D. and Ettre, L. S., Eds. **12**, 81, Interscience, London.
Crompton, T. R. (1971) *Chemical Analysis of Additives in Plastics*, Pergamon Press, Oxford.
Davies, J. R. (1967) *J. Chromatog.* **28**, 451.
Davies, J. R. and Thuraisingham, S. T. (1968) *J. Chromatog.* **35**, 513.
Davies, J. R. and Tunnicliffe, M. E. (1967) *J. Chromatog.* **30**, 125.
Doolittle, A. K. (1954) *The Technology of Solvents and Plasticizers*, Chapman Hall, London.
Fikhlengol'ts, V. S., Zolotareva, R. V. and L'vov, Yu. A. (1966) *Ultra Violet Spectra of Elastomers and Rubber Chemicals*, Transl. by Stubbs, A. E., Plenum Press, Data Divn., New York.
Gaeta, J. L., Shulter, E. W. and Attma, A. G. (1969) *Rubber Age (N.Y.)* **101**, 47.
Garner, H. R. and Packer, H. (1968) *Appl. Spectrosc.* **22**, 122.
Gilchrist, C. A., Lynes, A., Steel, G. and Whitham, B. T. (1972) *Analyst* **97**, 880.
Glavind, J. (1963) *Acta Chem. Scand.* **17**, 1635.

Gross, D. and Strauss, K. (1976) *Kaut. u Gummi Kunstst.* **29**, 741.
Gross, D. and Strauss, K. (1977) *Kunstst.* **67**, 426.
Gross, D. and Strauss, K. (1979) *Kaut. u Gummi Kunstst.* **32**, 18.
Haslam, J., Soppet, W. W. and Willis, H. A. (1951) *J. Appl. Chem.* **1**, 112.
Haslam, J. and Willis, H. A. (1965) *Identification and Analysis of Plastics*, Iliffe, London.
Haslam, J., Willis, H. A. and Squirrell, D. C. M. (1972) *Identification and Analysis of Plastics*, 2nd ed., Iliffe, London.
Hayes, M. R. and Metcalf, J. (1962) *Analyst* **87**, 956.
Hellmann, H. (1974) *Z. Anal. Chem.* **272**, 30.
Higgins, G. M. C. and McSweeney, G. P. (1974) *Rubber Chem. Technol.* **47**, 1206.
Hilton, C. L. (1958) *Rubber Age* **84**, 263.
Hofmann, W. and Ostromov, H. (1968) *Kaut. u Gummi Kunstst.* **21**, 244, 318, 322, 368, 432, 481, 560, 620, 693.
Hofmann, W. and Ostromov, H. (1969) *Kaut. u Gummi Kunstst.* **22**, 14.
Hummel, D. O. and Scholl, F. K. (1981) *Atlas of Polymer and Plastics Analysis*, 2nd ed., Vol. 3, Carl Hanser Verlag, Munich.
Killer, F. C. A. and Amos, R. (1966) *J. Inst. Petroleum* **52**, 315.
Kline, G. M. (1967) *Mod. Plast.* **44**, 129.
Kreiner, J. G. (1971) *Rubber Chem. Technol.* **44**, 381.
Kreiner, J. G. and Warner, W. C. (1969) *J. Chromatog.* **44**, 315.
Lloyd, D. G. (1962) *Photoelectric Spectroscopy Group Bulletin* No. 14, 395.
Luongo, J. P. (1965) *Appl. Spectrosc.* **19**, 117.
McSweeney, G. P. (1970) *J. IRI* **4**, 243.
McSweeney, G. P. (1971) Unpublished work at MRPRA.
McSweeney, G. P. (1972) Unpublished work at MRPRA.
McSweeney, G. P. (1981) Unpublished work at MRPRA.
Miller, R. G. T. and Willis, H. A. (1959) *Spectrochimica Acta* **14**, 119.
Ostromov, H. and Hofmann, W. (1978) *M.V.P. Berichte der Max von Rettenkoferinstitut des Bundesgesundheitsamtes* **2**, 1.
Patel, S. M., Hively, R. A. and Cole, J. O. (1973) 103rd Mtg. of A.C.S. Rubber Div., Michigan. See *Rubber Chem. Technol.* (1973) **46**, 1325.
Petric, A. M., Unpublished work at MRPRA.
Peurifoy, P. V., O'Neal, M. J. and Woods, L. A. (1970) *J. Chromatog.* **51**, 227.
Rice, D. D. (1967) *Analyt. Chem.* **39**, 1906.
Sidwell, J. A. (1980) *High Performance Liquid Chromatography—Analytical Applications in the Rubber and Plastics Industries*, RAPRA Members Report No. 49.
Stahl, E., Ed. (1965) *Thin Layer Chromatography*, Academic Press, London.
Sullivan, A. B., Kuhls, G. H. and Campbell, R. H. (1976) *Rubber Age* **3**, 41.
Wake, W. C. (1969) *The Analysis of Rubber and Rubber-like Polymers*, 2nd Ed., Maclaren, London.
Wexler, A. S. (1963) *Analyt. Chem.* **35**, 1936.
Wheeler, D. A. (1968) *Talanta* **15**, 1315.
White, D. W. (1967) *Rubber J.* **149**, 42.
Wise, R. W. and Sullivan, A. B. (1962) *Rubber Chem. Technol.* **35**, 684.
Yuasa, T. and Kamiya, K. (1964) *Bunseki Kagaku* 966 (see also *Chem. Abstr.* (1965) **62**, 2895).

Zijp, J. W. H. (1956) *Rec. Trav. Chim. Pays-Bas* **75**, 1129.
Zweig, G. and Sherman, J. (1972) *Handbook of Chromatography*, CRC Press, Cleveland, Ohio.

Chapter 5

Chemical Analysis for Polymer Type

The material to be used for chemical analysis is the polymer after extraction and will, of course, still contain any inorganic powders and carbon black originally present. Characterization of the polymer may depend on some property or properties it possesses as a whole or on the identification, by some chemical reaction, of one or more of the monomer units or of functional groups or atoms forming parts of such units.

The most common methods of identification are those of pyrolysis of the test piece followed by infrared spectroscopy or gas chromatography of the pyrolysate. These are the methods of choice and are dealt with in detail in Chapter 8. Here we are concerned with the older methods of non-instrumental chemistry which still have value for the non-specialist laboratory whilst the general background of this classical approach to chemical analysis of polymers helps the understanding of problems in quantitative analysis.

Convenience and prior knowledge will determine the procedure to be adopted, for it is never necessary to work right through an analytical scheme since the analyst is never completely without information concerning his sample. It is necessary to consider a total analytical scheme, however, because the precise stage at which it is entered or left depends upon what information is available or is required, and hence the scheme must allow for as many combinations of knowledge and ignorance as possible.

The first essential is a list of the polymers included together with their salient chemical and obvious physical properties. This list is given as Appendix II.

PRELIMINARY EXAMINATION

No one today, not even the newly qualified analyst in an industrial laboratory, can possibly handle the materials we are discussing without some prior knowledge of them. Translucent polythene kitchenware is recognized by the housewife as 'plastic' and therefore different from 'rubber'. The mere handling of the sample must therefore give the analyst some information. Other information can be obtained by attempting to ignite the sample and noting whether it burns and also the odour of the smoke. This should be done cautiously and assistants should be shown how to do this without inhaling. Some polymers can give off fluorine in quantity, or cyanogen and hydrogen cyanide, whilst acrylonitrile and styrene monomers are toxic if inhaled. Heating the polymer in a small ignition tube results in pyrolysis, the distillate condensing around the mouth of the tube whence litmus or other indicator paper will give a ready indication of any acid present. Table 5.1 summarizes the differentiation of polymers that can be made by such simple means.

Many analysts will be tempted to carry out these preliminary tests on a piece of the sample as received, i.e. without waiting for solvent extraction to be done. This procedure is fairly safe in experienced hands but should not be attempted by the inexperienced, since high percentages of plasticizer can markedly affect the burning behaviour and mask the smell. Even when extracted, however, care should be taken with burning tests on loaded polymers or rubbers as the filler also can influence the test. For example, antimony sulphide when used in large quantities can make a natural rubber compound self-extinguishing and carbon black will, in general, impart a smokiness to a flame which would otherwise be clear. These simple tests will usually eliminate a number of possibilities and frequently suggest the identification. In the latter case an independent confirmatory test should be used followed by quantitative estimation of the polymer identified. The latter may be required as part of a full analysis but even if identification only is the aim of the analyst, quantitative estimation is useful with a completely unknown sample as it removes the possibility of the unsuspected presence of a second polymer which has been missed.

QUALITATIVE EXAMINATION FOR ELEMENTS PRESENT

The number of elements which can occur in an unknown sample of a rubber-like material is quite large because of the very varied substances

TABLE 5.1

GUIDE TO PRELIMINARY EXAMINATION OF EXTRACTED POLYMERS

	Polyethylene	*Chlorosulphonated polyethylene*	*Butyl rubber polyisobutene Ethylene propylene rubbers*	*Natural rubber Butadiene/styrene Polybutadiene*	*Polychloroprene*
Appearance	Can be white, translucent or have delicate colouring	Usually loaded with pigments and not black	Colour range of the rubber limited. PIB when not admixed with other polymers is very pliable and flows fairly easily	Opaque with limited colour range. Natural rubber can be obtained transparent but only with straw to dark amber colours	Opaque with limited colour range
Burning test	Burns with clear flame unless loaded with carbon black which causes a smoky flame	Burns with clear yellow flame but then extinguishes	As polyethylene	Smoky flame and distinctive odour. Can be loaded with flame retarders which make it self-extinguishing	Extinguishes on removal of flame. Odour of HCl
Pyrolysis (pyrolysate tested with bromocresol green)	Clear neutral liquid distils which solidifies to wax on cooling	Acid liquid distils	Clear neutral liquid distils	Liquid distils, usually of dark colour	Acid liquid distils, usually of dark colour, often no liquid distillate
Boiled with 80% sulphuric acid	Inert	Inert	Inert	Decompose	Decompose

that are used as fillers, plasticizers, vulcanization agents and so forth. A full analysis would need to include within its scope the possibility that, for example, part of the sulphur in a vulcanized rubber is replaced by selenium. Yet the possibility is so improbable that it would be pedantic to allow for it in ordinary analytical practice. For the purpose of identifying the polymer present, only a limited number of the elements occur as part of the polymer and of these fewer still have diagnostic significance. In order of atomic number these are: nitrogen, fluorine, sulphur, chlorine and bromine. It might be expected that silicon should be included in this list since it occurs in silicone rubbers but, unfortunately, silicon-containing fillers (silica and silicates) are in very common use in all rubbers and the presence of silicon is, therefore, without diagnostic significance.

TABLE 5.1—*contd.*

Polyvinylchloride	*Acrylonitrile rubbers*	*Thioplasts*	*Silicone rubbers*	*Polyurethane rubbers*	*Fluoro polymers*
Can be glass transparent, and coloured or opaque with wide range of bright colours	Opaque with limited colour range	Rubber smells of sulphur compounds. Very limited colour range, and, in fact, usually black	Can be smoky-brown transparent rubber but is usually filled with TiO_2 or Fe_2O_3 giving the corresponding colours. Carbon black cannot be used in silicones	Can be transparent when it is usually an amber colour or can take a wide range of colour. Foamed urethane rubbers have closed (non-interconnecting) cells and are usually brightly coloured	Can be transparent water white but is usually loaded with white fillers
Usually extinguishes on removal of flame but very flexible PVC, containing a large amount of plasticizer, may be inflammable	Smoky flame and sickly odour	Blue sulphur flame without smoke. Sulphurous odour	Burns with flame only lightly smoky. Swells markedly on burning giving a white ash. Glass surface held in flame is coated white with silica	Burning behaviour varies. Some melt in a flame w/o burning; others burn gently or burn after melting	Burns brightly in a flame but extinguishes with immediate suddenness when removed. The sudden extinction is very noticeable
Acid liquid distils, usually clear or only slightly coloured	Neutral or alkaline liquid distils, usually of dark colour	Acid liquid distils, usually of dark colour			Acid brown liquid distils
Inert	Decompose	Decompose	Decompose	Decompose	Inert

The behaviour of thermoplastic rubbers will depend upon their constituent parts as described in the introduction. If the sample is supplied as a finished product, which one would expect to be cured, a hot pressing at 180 °C will rapidly distinguish a thermoplast from a vulcanizate.

Many analysts will prefer to make Lassaigne's (sodium fusion) test the basis of their examination for elements present and this should normally be carried out on a previously extracted sample.

Lassaigne's Test

The solution obtained from the fusion can be divided into four parts and used in the following ways.

For nitrogen: Boil in a test-tube with a small quantity of freshly prepared ferrous sulphate solution. Acidify with dilute hydrochloric acid and add a few drops of ferric chloride solution. A substantial blue precipitate indicates the presence of nitrogen in the polymer. If the precipitate is scanty or if filtration and the presence of the blue colour on the filter paper is necessary to be sure that the test has produced a positive result, then nitrogen, as a constituent of the polymer, can be ruled out.

If fluorine is present and only a trace of nitrogen then the former will mask the latter unless special precautions are taken. The addition of ferrous sulphate gives FeF_2 which is not ionized and which will have removed the bulk of the iron and prevent it from reacting with the cyanide formed by the nitrogen. If nitrogen is recorded as absent and later fluorine is found, it is worth while repeating the nitrogen test with a larger excess of ferrous sulphate. In addition, it is advantageous to acidify with sulphuric instead of hydrochloric acid.

For chlorine and/or bromine: Acidify with dilute nitric acid and add silver nitrate solution. A white precipitate soluble in ammonia is given by chlorine; a yellow precipitate insoluble in ammonia is given by bromine. Iodine will not be met and silver fluoride is soluble.

For fluorine: Prepare a test paper by immersing filter paper in an aqueous solution of alizarin red S 0·1% and zirconium nitrate 0·1% and subsequently drying. When required, moisten this with 50% acetic acid. Neutralize the test solution with dilute hydrochloric acid and place one drop on the moistened test paper. The red spot will turn yellow in the presence of fluoride.

For sulphur: Add a few drops of sodium nitroprusside solution when a violet coloration will indicate the presence of sulphur.

An alternative procedure which has much to commend it is the use of the Beilstein test to detect halogens followed by the soda-lime test for nitrogen, the Lassaigne test being used only if, after the preceding tests, it is still necessary to test for sulphur. This use of the Beilstein test is only possible if the absence of fluorides can be assumed as their high thermal stability usually precludes their reacting with the copper wire (the chlorine in chlorofluorides gives a positive result).

Beilstein Test for Halides (Excluding Fluoride)

A piece of copper wire of $\frac{1}{2}$–1 mm diameter, i.e. 18–24 SWG, is cleaned and heated to redness in a bunsen flame until insertion into the flame fails to produce any coloration. The red-hot wire is then pushed into or on to the sample for an instant and then returned to the flame when a bright green flash will indicate the presence of halogen.

Soda-lime Test for Nitrogen

A small amount of the sample at the bottom of a fusion or test-tube is covered with soda lime to a depth of, say, 1 in. Heating the base of the tube results in the release of ammonia if nitrogen is present in the sample, and moistened litmus paper is used to detect it at the mouth of the tube. This test is independent of the presence of fluorine.

DIFFERENTIATION OF POLYMERS AFTER ELEMENT TESTS

Completion of the preliminary examination and of suitable tests for the elements present brings the analysis to the state in which the possible polymers can be named and a course of action decided on to distinguish between the possibilities indicated.

Halogens, Sulphur and Nitrogen Absent

The absence of all these characterizing elements can occur only with hydrocarbon or acrylate-based polymers and silicone rubbers. The hydrocarbon polymers can be raw rubbers, i.e. unvulcanized or, exceptionally, rubber vulcanized with peroxide. Materials now in production which come in this class are: Butadiene–styrene copolymers, butyl rubber, polyisobutene, ethylene–propylene and certain acrylate rubbers, polyethylene, silicone rubbers, and certain thermoplasts such as the block copolymers. Natural rubber is excluded since it usually contains, as protein, sufficient nitrogen to give slight coloration in Lassaigne's test.

Reference to Table 5.1 shows that polyethylene, polyisobutene and butyl rubber burn with a clear flame unless loaded with carbon black, whilst silicone rubber swells on burning and gives a white ash. The appearance and feel of polyethylene, even when black loaded, will distinguish it from the rubbers. Specific tests (see later in this chapter) can be applied for styrene copolymers and isobutene, while acrylate rubbers can be saponified by alcoholic potassium hydroxide. Silicone rubbers are soluble (by hydrolysis) in 80% sulphuric acid. The behaviour of thermoplastic rubbers depends upon their constituent components.

Nitrogen Present in Traces Only

The presence of traces only of nitrogen, the other characterizing elements being absent, can occur only with unvulcanized natural rubber, natural rubber vulcanized with peroxides, or with ester or urethane rubbers cross-linked by isocyanates. The latter vary in their burning behaviour whereas, unless specially compounded, natural rubber burns freely with a smoky flame. A specific test can be applied for natural rubber—the Weber colour reaction (see below).

Nitrogen Present

There are two classes of elastomer which contain appreciable amounts of nitrogen. The more important is that containing the butadiene–acrylonitrile copolymers which have nitrogen contents ranging from the order of 5%, to materials containing almost 14% corresponding to equimolar quantities of the two monomers. The other class contains terpolymers of butadiene, styrene and vinyl pyridine, but with only a small proportion of the last. The presence or absence of sulphur will be solely a question of vulcanization.

Chlorine Present

In the absence of other characterizing elements, chlorine occurs in polyvinyl chloride or its copolymers with vinyl acetate or vinylidene chloride, rubber hydrochloride and chlorinated natural rubber. Unvulcanized polychloroprene contains traces of sulphur and vulcanization frequently adds more sulphur through the use of accelerators although, theoretically, it can be crosslinked by metal oxides alone. Methods of distinguishing between the various chlorine-containing polymers are described later in this chapter.

Chlorine and Sulphur Present

The presence of both these elements excludes chlorinated thermoplastics and the choice thus remains between polychloroprene, chlorosulphonated polyethylene, and natural rubber vulcanized by sulphur chloride in the cold curing process. This last process is used only for certain surgical goods or textile proofings, and chlorine and sulphur occur only in rather small quantities, say a few per cent. A vulcanized acrylate rubber will also contain small, but not trace, quantities of both chlorine and sulphur as it contains chlorovinyl ether in the polymer. Like polychloroprene but unlike chlorosulphonated polyethylene, acrylate rubber can be decomposed by nitric acid. Low levels of either chlorine or bromine, together with sulphur, could also indicate the presence of halobutyl rubber or surface halogenated hydrocarbon polymers, a situation which will be considered below when the halogenated polymers are discussed in more detail.

Sulphur Present
Sulphur is, of course, the standard vulcanizing agent for most rubbers and, during vulcanization, is incorporated into the crosslinked structure; it is then inextractable. Its occurrence with chlorine has been mentioned. Occurrence with traces of nitrogen is more or less specific for vulcanized natural rubber and this can easily be confirmed by the Weber reaction, although the possibility of a mixed urethane/sulphur cure as described by Baker (1977) must be considered. Occurrence in quantity, in the complete absence of either nitrogen or chlorine, limits choice to the thioplast materials, a choice which is adequately confirmed by smell and by burning.

Fluorine Present
The presence of fluorine indicates the most difficult group of rubbers and rubber-like materials with which we are concerned and, even using more sophisticated instrumental techniques such as infrared spectroscopy, it is not always easy to distinguish between them. The major fluorocarbon monomers are:

tetrafluoroethylene	TFE
chlorotrifluoroethylene	CTFE
vinylidene fluoride	VF_2
hexafluoropropylene	HFP

any of which may be met copolymerized with each other or with 1-hydropentafluoropropylene (HPTFP). Representative examples of currently available materials are shown in Appendix II.

ACID RESISTANCE

The acid-resistance test devised by Parker (1944) is simple to carry out and requires no equipment other than test-tubes, a vessel for heated water, a thermometer, a watch with a seconds hand, and some mixed acid.

The mixed acid consists of equal volumes of fresh concentrated nitric acid (*d.* = 1·42) and sulphuric acid (*d.* = 1·84) and is poured into a clean, dry test-tube to a depth of about ½ in. This is placed in water at 70 °C (158 °F) and a small piece or sliver of the material freshly cut to avoid wax bloom (about 1 mm^3, e.g. ‘a slender thread or thin flake’ is quoted in the original paper) is dropped into the acid. The sample at first floats and the reaction may not be apparent immediately, but this stage is followed by one in which bubbles of gas appear on the sample surface, growing in number until the material begins to disintegrate and disperse in the liquid. The first appearance of gas bubbles and the complete disintegration of the samples are difficult to locate accurately in time, but the period that elapses before disintegration starts (i.e. the period from insertion of sample to the first trace of dispersion) can generally be determined with reasonable accuracy, and this is accordingly taken as the 'period of reaction'. Reactions should be carried out in duplicate in separate tubes. The following periods apply to polymer materials and to the vulcanized rubbers:

Polyvinylchloride	more than 30 s
Rubber hydrochloride	less than 5 s
Natural rubber	0–5 s
Polychloroprene (Neoprene GN)	5–8 s
Styrene–butadiene copolymer	10–25 s
Butyl rubber	more than 60 s

SPECIFIC OR NEAR-SPECIFIC TESTS

Provisional identification of the polymer having been made as a result of the preliminary examination, the examination for elements present might be eliminated in favour of a specific test for the rubber believed to be present. A number of such tests have been described in the literature and, whilst their significance in these days of instrumental analysis could be questioned, it is a rare analyst who has never had to resort to one or more of them at some time, in some perplexing or ambiguous situation.

The tests are described in tabular form, in Table 5.2, and then in a degree of detail.

Azo Dye Reaction for Styrene

This test is dependent on the presence of phenyl groups (i.e from the styrene part of the polymer only).

Boil 1–2 g of dried acetone-extracted material under reflux with 20 cm^3 of nitric acid ($d.$ = 1·42) for 1 h. Dilute by pouring into 100 cm^3 of distilled water, extract with ether (50, 25 and 25 cm^3). Combine the ether extracts and wash twice with 15 cm^3 of distilled water and reject water washes. Extract the ether with 3 × 15 cm^3 of 1 M NaOH and finally with 20 cm^3 of distilled water. Reject the ether layer. Combine the alkaline extracts and washing, make just acid with concentrated hydrochloric

TABLE 5.2

SPECIFIC OR NEAR-SPECIFIC TESTS

Functional group	*Applicable polymers*	*Test name or description*	*Reference*
Styrene	SBR, PS, SBS SIS	Azo dye reaction (qualitative)	Min. of Supply (1944)
		Nitration (quantitative)	Hilton *et al.* (1959)
			Smith (1976) ISO 4655-1977
Polyisoprene	NR, IR, Guttapercha, Guayule	Weber reaction	Kirchhof (1928) Stern (1944)
		Chromic acid oxidation (Kuhn–Roth) (quantitative)	ISO 5945-1982
Isobutene	Polyisobutylene, butyl, halobutyl rubbers	Mercuration (qualitative)	Min. of Supply (1944) Burchfield (1945)
Unsaturation	All unsaturated polymers	Iodine monochloride (quantitative)	Wake (1969)

ISO 4655-1977 also allows the use of carbon and hydrogen analysis for the determination of the styrene content of SBR latex. This technique is equally applicable to any polymer system, the components of which are known, and its significance will be considered in Chapter 8 under 'Elemental Analysis'.

acid, and then add a 20 cm^3 excess. Heat on the steam-bath and reduce by adding granulated zinc (5 g), cool and add 2 cm^3 of 0·5 M sodium nitrite. Pour the diazotized solution into an excess of alkaline β-naphthol solution when formation of a red dye indicates the presence of styrene.

The presence of other rubbers has not been found to interfere with this test but colours other than a strong scarlet should be ignored. It is probably best carried out on a pyrolysate.

Styrene Content of Copolymers

Chemical methods for the direct determination of styrene have had a chequered history. A method introduced during the 1940s depended on nitration of the polymer and separation of a crystalline product determined as *para*nitrobenzoic acid. It was later shown that separate estimation of nitro and acid groups gave discordant results and the product, far from being a single pure material, was of variable composition as far as substituent positions and amounts were concerned. The difficulty was avoided by Hilton *et al.* (1959) who separated the nitrated products and used UV absorption to determine the total phenyl groups independent of the exact constitution of the absorbing molecule, which may then be a mixture of isomers and homologues containing nitro and carboxylic acid groups without interfering with a valid result. Spectra in fact confirm that the product differs from *para*nitro benzoic acid. Hilton *et al.* recommend measuring the absorbance at 265, 273·75 and 285 nm. Making three measurements in this way is something of an 'insurance policy' when the method is applied to a properly extracted sample of a known copolymer, and most analysts will, as a routine, be content with measurement at 273·75 nm, the absorption peak. The instructions for the method are as follows.

Weigh accurately 0·25 g of the extracted SBR rubber (or a proportionately smaller quantity if a high styrene-resin is present) and reflux for 16 h with 20 cm^3 of concentrated nitric acid (*d.* = 1·42) in a 100 cm^3 flask fitted with a reflux condenser. At the end of this period cool and wash down the condenser with water, transfer the contents to a beaker and partially neutralize the acid by adding 50 cm^3 of 5 M sodium hydroxide solution. After this addition cool the solution thoroughly (below room temperature), transfer to a separating funnel and extract with three separate 50 cm^3 portions of ether. Combine the ether extracts and wash with water to remove nitric acid. Transfer the ether extract to a beaker containing a few grams of anhydrous sodium sulphate. After a few minutes transfer the ethereal solution to a separating funnel and extract it with at least three portions of 0·1 M sodium hydroxide. Dilute this solution in the ratio 1:10 by withdrawing 25 cm^3 with a pipette into a second 250 cm^3 standard flask and make up to the mark again with 0·1 M sodium hydroxide.

Measure the absorbance of this solution at the three wavelengths (or only at 273·75 nm if it is safe to do so), using 1 cm cells with 0·1 M sodium hydroxide solution occupying the blank position. The percentage composition is given by:

At 265 nm	Styrene	= 1·52*A*–0·56
273·75 nm	Styrene	= 1·44*A*–0·45
285 nm	Styrene	= 1·60*A*–0·42

where A = measured absorbance of the solution corrected proportionately for deviation of the weight taken from 0·25 g.

There is some slight interference when mixtures with natural rubber are analysed but this can be ignored unless the natural rubber is present in larger quantities than the SBR.

Weber Reaction for Polyisoprene

This reaction is based on the early observation by Weber (1902) that brominated rubber reacts with phenols to give coloured compounds. It is a useful test and is easy to carry out but, although often quoted as being specific for natural rubber, it is not.

Place approximately 0·05 g of the dried acetone extracted rubber in a clean, dry test-tube and add 5 cm^3 of bromine solution (10% v/v in carbon tetrachloride). Place in a water-bath and raise the temperature slowly to the boiling point, continuing the heating until not more than a trace of bromine remains; 5–6 cm^3 of a 10% solution of phenol in carbon tetrachloride is then added and the heating continued for a further 10–15 min. A violet colour is given by natural rubber, usually within a few minutes. The colour takes somewhat longer to develop when only a small amount of natural rubber is present in admixture with other rubbers.

Kirchhof (1928) expanded the method somewhat by suggesting that a few drops of the mixture may be poured into various organic solvents, e.g. chloroform, ether or acetic anhydride, and the new colour observed.

From this procedure Stern (1944) produced Table 5.3.

More recently, Gupta and Ghosh (1966) described how the Weber colour reaction could be made quantitative for the determination of the natural rubber content of blends, or grafts, of this with polymethylmethacrylate.

Bahrani *et al.* (1974) adopted the different approach of pyrolysing the elastomer and brominating the pyrolysate, rather than the rubber itself. This enables the whole reaction to be completed in a quarter of an hour and, provided the temperature of pyrolysis is maintained in the 400–470 °C region, there is no interference from polymers containing butadiene as a constituent. No mention is made of interference from the other elastomers listed in Table 5.3, but the authors do point out that examination of the pyrolysate removes the possibility of coloured fillers masking the violet colour of the test.

TABLE 5.3

ORIGINAL WEBER REACTION WITH VARIOUS SUBSTANCES

Substance	*Colour of phenol fusion*	*Colours with the following solvents*			
		Chloroform	*Acetic anhydride*	*Ether*	*Alcohol*
Smoked sheet	Violet	Light violet	Light violet	Light violet	Light violet
Plantation crèpe	Violet	Light violet	Light violet	Pale brown	Light violet
Balata belting	Red–brown	Yellow–brown	Brown	Blackish brown	Black–brown
Latex rubber (glove)	Brown–violet	Red–orange	Dirty yellow	Yellow–brown	Orange–yellow
Balata	Deep red	Light violet	Red–violet	Greyish-brown	Blackish-brown
ABR (Hycar)	Orange–brown	Yellow	Yellowish-orange	Colour-less	Lemon yellow
Polysulphide	Pale straw	Very pale yellow	Pale yellow	Light yellow	Very pale straw
ABR (Perbunan)	Yellowish-brown	Dark yellow	Yellow with white ppt.	Yellow–white ppt.	Yellow with brown ppt.
Neoprene	Red–brown	Reddish-violet	White with brown ppt.	Brown with black ppt.	White with brown ppt.
SBR	Greenish-grey	Almost colourless with slight turbidity			

Polyisoprene Content by Methyl Group Determination (Kuhn–Roth Method)

The method is based on the oxidation of isoprene to acetic acid which is steam distilled from the digestion solution and estimated by titration.

Take an accurately weighed test portion of about 0·25 g, extract for 16 h with acetone and, if the presence of bituminous material or factice is suspected, carry out a further extraction with chloroform for 4 h. Place the extracted test portion in the digestion flask (100 cm^3), add the digestion mixture (25 cm^3 of a solution of 20 g chromium trioxide, 50 cm^3 water and 15 cm^3 conc. sulphuric acid) and digest at 100 °C until complete digestion has occurred. Steam distil from the digestion solution 150 cm^3 of distillate, cool and aerate the distillate at 30 cm^3s^{-1} for 30 min after which titrate the acetic acid with standard sodium hydroxide solution (0·1 M) with phenolphthalein as indicator. Calculation of the polyisoprene content requires a correction for the fact that only 75% of the theoretical yield of acetic acid is generated.

The following interferences are quoted in ISO 5945-1982 and ASTM D297, whilst they are referred to in ASTM D1278 and BS 903 Pt. B12. The data appear originally to have been published by Burger *et al.* in 1943. Natural rubber, synthetic polyisoprene, balata, gutta percha, chicle and guayule are indistinguishable by this reaction.

TABLE 5.4

INTERFERENCES IN THE KUHN–ROTH METHOD FOR ISOPRENE DETERMINATION

Constituent	*Effect*
Reclaimed NR/IR, etc.	Generally a little low; multiply by 1·05
Ebonite of NR/IR, etc.	Gives low value; multiply by $(1 + 0{\cdot}015S)$ where S is sulphur content as % of NR/IR
Polysulphide rubbers (thiokol RD)	Gives high value, increased by 18% of polysulphide rubber loading
Polyolefins BR, SBR, NBR, etc.	Marginally increases the calculated value by *ca.* 3% of their loading
IIR	No contribution but can prevent complete oxidation of isoprene rubber
PVA and copolymers	Liberates acetic acid which invalidates the results
CR	Increases the value by some 3% of its loading if procedure modified to remove effect of chlorine[a]
Protein	Increases measured value by one quarter of its loading
Cellulose	Negligible
Carbon black	Negligible
Factice	Negligible (if extracted properly)

[a]If a chlorinated rubber is known to be present, add an excess of potassium iodide solution (100 g l^{-1}) to the distillate after aeration and remove the liberated iodine (yellow colour) with sodium thiosulphate solution (100 g l^{-1}) before commencing the titration.

Detection of Isobutene by Mercuration

Isobutene is formed when polyisobutene (polyisobutylene) or butyl rubber is pyrolysed and it is the pyrolysis product that is used for the test. If there is any possibility that a mixture of polymers is present it is better to carry out a separation and use only the suspected polyisobutene or butyl for the pyrolysis. Polyisobutene itself can usually be extracted and if in doubt sufficient can always be separated for pyrolysis. Extraction is preferably by light petroleum (boiling range 40–60 °C). Butyl rubber, if vulcanized, is insoluble but boiling with nitric acid (*d.* = 1·42) will destroy any other polymers likely to be mixed with it. The acid-treated residue can then be heated to dryness and then further heated for pyrolysis. Assuming the sample is suitable for carrying out the pyrolysis the following instructions outline the method.

A test paper, which is used without drying, is first prepared by immersing filter paper in a solution made by dissolving 1 g of yellow mercuric oxide in 20 cm^3 of boiling 5 M sulphuric acid and then cooling before use. Pyrolyse a small quantity of the polymer to be examined and allow the pyrolysis vapours to impinge against the test paper. Vapour of isobutene causes this to turn yellow whereas diene vapour causes a brown coloration.

If there is any uncertainty then the mercurated butene compound should be prepared in the following way:

Place about 1 g of the polymeric material in a test-tube provided with a cork and a bent delivery tube. The end of the delivery tube passes through a cork almost to the bottom of a second test-tube which has a side arm. Cool the second test-tube by surrounding with ice. Connect a delivery tube to the side arm of the second test-tube and lead into a solution of 0·5 g mercuric acetate in 10–15 cm^3 methyl alcohol contained in a small test-tube. Heat the rubber strongly so that it is virtually all decomposed and distilled off. Reject the liquid which collects in the bottom of the second test-tube. Evaporate off the methyl alcohol on the water-bath but do not heat any longer than is necessary. Boil the residue with 25 cm^3 light petroleum (BP 40–60 °C) and filter from insoluble matter. Evaporate the filtrate to low bulk, chill in ice, and scratch to induce crystallization. Dry the mercury derivative at 30–40 °C and determine the melting point (about 55 °C). For final confirmation, carry out a mixed melting-point determination with the mercurial made from authentic isobutene. The derivative is methoxy-isobutyl-mercuri acetate ($CH_3O.C.(CH_3)_2.CH_2Hg.COOCH_2$) which is a vesicant and should be handled with care. Ethylene–propylene-based rubbers can colour the test paper, but only faintly.

UNSATURATION

This is an extremely complex area which cannot be summarized without possibly being misleading. Because of its relative infrequency of use in these days of instrumental techniques it is felt that the space can be better used in this edition, and that the interested reader will find a full description of the available techniques and problems in the second edition (Wake, 1969).

MISCELLANEOUS IDENTIFICATION PROBLEMS

Three areas of polymer analysis are considered worthy of expanding somewhat because of their regular occurrence and importance; these are (1) the unambiguous identification of the various chloropolymers; (2) the fragmentation and subsequent fragment identification of polyurethanes; and (3) certain aspects of natural rubber analysis.

Chloropolymers

Two series of colour reactions, one due to Wechsler (1953) and the other to Winterscheidt (1954), provide a great deal of information which, together with a few other tests, enable the various chloropolymers to be identified.

Wechsler Tests

Polyvinylchloride: A few mg of the sample are dissolved in pyridine (*ca.* 1 cm^3) which is then boiled for a minute, cooled and 1 cm^3 of dilute (0·5 M) alcoholic potassium hydroxide solution added. In the presence of polyvinylchloride there will be the rapid development of a brown–black colour. Subsequent addition of 1 cm^3 of 0·1% solution of β-naphthylamine in 20% aqueous sulphuric acid followed by 5 cm^3 of amyl alcohol and vigorous shaking will lead in a few hours to a pink coloration of the organic layer. Separation of the organic layer and basification (10 cm^3 1 M NaOH) changes the colour to yellow, and acidification changes it back to pink.

Postchlorinated polyvinylchloride: Dissolve the sample as above; a bright yellow colour forms on boiling. The addition of alcoholic potassium hydroxide gives a yellow–red precipitate which subsequently turns brown.

Polyvinylidenechloride (and copolymers): These react very readily *in the cold* with pyridine and alcoholic potassium hydroxide with the sample rapidly turning dark brown–black. This test can also be used as a spot test on the surface of the sample.

Chlorinated rubber: Allow a few mg of the sample to stand in pyridine at room temperature for a few minutes, then add methanolic KOH: the result is a yellowish precipitate that gradually darkens to a yellow–brown colour.

Rubber hydrochloride: As with chlorinated rubber allow the sample to stand in pyridine at room temperature, and add methanolic KOH. The reaction is much weaker, giving a yellow–red precipitate.

TABLE 5.5

WINTERSCHEIDT COLOUR REACTIONS OF POLYVINYL COMPOUNDS WITH CHLOROACETIC ACIDS

Polymer	*Colour reaction with monochloro-acetic acid*	*Colour reaction with dichloro-acetic acid*
Polyvinylchloride	Blue	Purple
Post-chlorinated polyvinylchloride	No colour reaction	No colour reaction
Vinyl chloride–vinyl acetate copolymer	Bluish purple	Reddish purple
Vinyl chloride–vinyl isopropyl ether copolymer (80:20)	Yellow–red to purple	Blue to purple
Vinyl chloride–vinyl acetate–butyl acrylate terpolymer (1:1:1)	Light red, then blue	No colour reaction

Neither polychloroprene, chlorosulphonated polyethylene, chlorinated butyl rubber nor surface chlorinated polymers react significantly to the colour tests described and need further examination.

Polychloroprene: This is the only one containing more than a few percent chlorine and may be distinguished from rubber hydrochloride by its rapid evolution of hydrogen chloride at 200 °C.

Immerse the sample (in a test tube) in an oil bath at 210 °C, with the mouth of the tube covered with a Congo-red paper moistened with dilute hydrogen peroxide solution; the paper will turn blue if polychloroprene is present.

Chlorosulphonated polyethylene: Esposito (1960) describes a test, specific for chlorosulphonated polyethylene.

About 10 mg of the sample is heated with pyridine (3 h) and 25 mg of 2-aminofluorine* added. A reddish brown colour develops within one hour which Esposito claims to be due to reaction with the SO_2-Cl group.

Chlorobutyl; surface chlorinated hydrocarbons: Chlorobutyl and surface chlorinated hydrocarbon polymers undergo the same reaction as their 'parent' polymers whilst in the absence of instrumental techniques, the simplest way to test for surface chlorination as opposed to bulk chlorination is to carry out either the Lassaigne or Beilstein test on samples from the surface and bulk of the sample.

Polyvinylidenechloride: One further test is worthy of note; Verdcourt (1947) describes how the heating of a small piece of polymer with morpholine gives a dark red–brown colour within two minutes and rapidly turns black if polyvinylidenechloride is present. This rapidly distinguishes it from polyvinylchloride.

Polyurethanes

The class of polymers known as urethanes is unique amongst the polymers which we are considering, in that not only does it include two types of copolymer, the polyester and polyether urethanes (as described in the Introduction), but also within each type there are many different polyurethanes depending upon the specific diisocyanate, polyester, glycol and polyether. Also present are amine or hydroxy chain extenders.

Generally, therefore, it is necessary to hydrolyse the polyurethane to identify the components from which it was made, although Corish (1959) describes in great detail how infrared spectra of microtomed sections of polyurethane rubbers may be used to identify the components directly.

If the polyurethane is based on a polyester a useful colour test has been described by Baumann and Steingiser (1959). It depends on the local hydrolysis of the polyester by caustic soda and the formation of a hydroxamic acid which reacts with ferric chloride to give an intense violet- or purple-coloured chelate compound.

*Carcinogen.

The polymer is comminuted and placed in 2 M methanolic potassium hydroxide solution containing phenolphthalein, or this solution may be spotted on to the sample if it is a foamed material. Add a few drops of a saturated methanolic solution of hydroxylamine hydrochloride but not sufficient to discolour the phenolphthalein. Heat for less than one minute to not more than 50 °C. After one minute acidify with 1 M aqueous hydrochloric acid and add one drop of a 3% aqueous solution of ferric chloride. The colour should appear at once. This is a general reaction for esters and prior removal of any plasticizer is important.

Swann and Esposito (1958) used *para*dimethylaminobenzaldehyde in a colour test for polyurethanes based on the yellow colour developed with the isocyanate group but this may not be applicable to the fully cured rubbers. It was intended to identify urethane resins which are still soluble in solvent.

According to Schröder (1962, 1963), pyrolysis yields some isocyanate which, if the pyrolysate is taken up in acetone, gives a yellow to deep red but usually brown colour with a 1% solution of sodium nitrite. This pyrolysate is always weakly acid and contains nitrogen whereas all other nitrogen-containing pyrolysates of high polymers are alkaline.

If a fuller qualitative analysis is required then the steps which must be taken (Schröder, 1963) comprise the following:

(a) Saponification with water at high temperature and pressure.
(b) Isolation of the dicarboxylic acid and identification by paper chromatography of the dimethylesters.
(c) Detection of polyalcohols by paper chromatography.
(d) The isocyanate used is identified from the corresponding aryl amine, which is again identified by paper chromatography. Suitable development of the chromatogram yields characteristic colours, as well as R_f values illustrated in Table 5.6.

More recently Dawson *et al.* (1970) have described a systematic scheme (Table 5.7) for the hydrolysis of both polyether and polyester urethanes, with subsequent identification of a range of both aliphatic and aromatic diamines by thin layer chromatography using a variety of eluants and sprays for visualization.

The examination of diols and triols was similarly covered but not the diacids. There appears, however, to be no reason not to include the paper chromatographic method of Schröder (1963) which requires conversion of

TABLE 5.6

R_f VALUES AND COLOURS OF ARYL AMINES

	para *Phenylene-diamine*	*1,5-Naphthylene diamine*	*4,4',4'-Triamino-triphenyl methane*	*4,4'-Diamino diphenyl methane*	*Tolylene diamine (2,4 and 2,6 mixture)*
R_f	0·34	0·65	0·53	0·63	0·50
Coloration	Red	Grey	Pale yellow	Bright yellow	Pale yellow

the diacids to the dimethyl esters followed by development with *n*-propanol:conc. ammonia:water (6:3:1) and visualization using bromothymol blue.

TABLE 5.7

THE 'DAWSON' SCHEME FOR THE SEPARATION OF POLYURETHANE FRAGMENTS

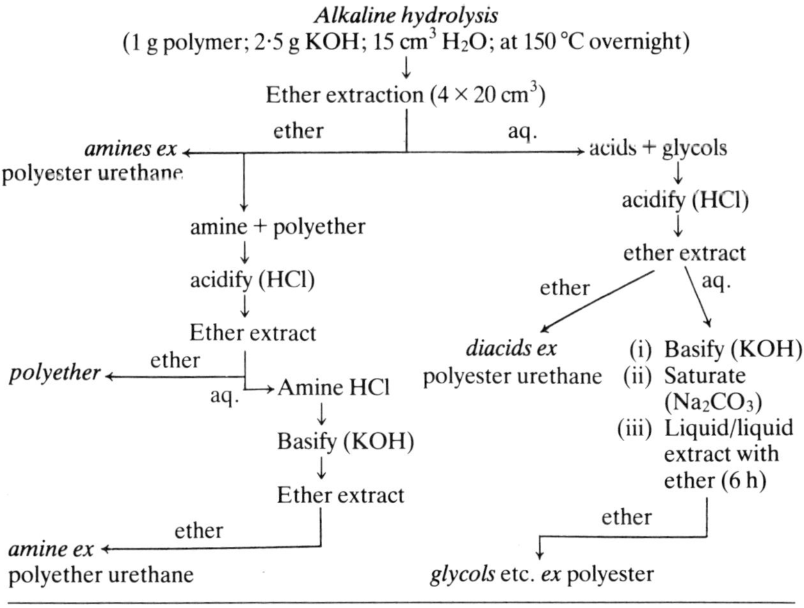

The interested reader is also referred to the German Standard DIN 53 623 Pt. I (Testing of Rubber; Identification of Polymers; Identification of Polyurethanes).

Natural Rubber

Rubber obtained from *Hevea brasiliensis* exudes from the tapping cut as latex containing about 40% total solids. The serum forming the continuous phase is a complex solution and the interface between rubber and serum is stabilized by a mixture of surface active materials in which proteins predominate. For use as latex, this 40% latex is concentrated by one of several possible processes and stabilizers are added; for use in the form of sheet or crumb rubber the latex is coagulated and, after suitable treatment, dried. Rubber goods prepared from latex are thus generally associated with a larger proportion of the solids initially in the serum or at the interface than are sheet or bale rubbers. However, normal commercial rubber still contains detectable quantities of the various impurities which betray its natural origin. The most convenient of these from an analytical point of view are protein and β-sitosterol, although it should be mentioned that one form of purified natural rubber is enzymatically deproteinized and only contains some 10% of the normal protein content whilst skim rubber contains very substantially more. It is not intended here to distinguish between the various grades of natural rubber nor to consider the various additions to latex, but to concentrate on that regular question: 'Is the polyisoprene natural or synthetic?' and to extend this to cover both deproteinized and skim rubbers. It is once again mentioned that these methods are non-instrumental, and that many other methods will be covered under the various instrumental techniques described in subsequent chapters.

β-Sitosterol

One major difference between protein and β-sitosterol is that the former is not extracted by a solvent such as acetone or methanol, whilst the latter is. Its separation by thin layer chromatography and subsequent visualization by a specific spray reagent has already been considered in Chapter 4 and the practical aspects will not be repeated, but two points merit comment. The first is the question of reliability of observation; the experiences of a colleague at MRPRA (McSweeney, 1970) over many years have been that the analyst has never failed to find β-sitosterol in samples, raw or vulcanized, which contain more than *ca.* 10% by weight of natural rubber, provided that the sample has not been subjected to extraction with an organic solvent before being received for analysis—if any other component is found to be present by TLC examination, β-sitosterol will be observed if originally present. These analyses include samples cured at up to 200 °C and an eighty-year-old sewer pipe seal which must have suffered everything other than organic solvent extraction!

A few samples of reclaim have been examined and here too β-sitosterol has been found although, if it is compounded with new natural rubber, this may be of limited value. The second point is the quantification of the β-sitosterol content, to enable the natural rubber level to be determined. There is intrinsically no difficulty in quantifying the analysis, but unpublished studies at MRPRA, using a silylation/gas chromatographic technique, showed that there was a substantial variation in sitosterol contents for different samples of rubber and thus the analysis is not particularly useful. It is probable that a visual estimation of the natural rubber content, based on the size of the TLC spot, *and experience*, gives an equally valid figure.

Protein

A method was developed by Loadman and McSweeney (1970) enabling the protein in extracted samples containing natural rubber to be hydrolysed, and the resulting amino acids separated by paper chromatography.

> A sample of the extracted sample (1·0 g), lightly milled or cut into slivers, is placed in a thick-walled glass tube (10 cm × 1 cm i.d.), 6 M hydrochloric acid (3 cm^3) added, the tube sealed and heated to 100 °C for 16 h after which it is cooled, opened and the solution examined directly by paper chromatography using *n*-butanol–acetone–water (60:20:20) as eluant, and ninhydrin as the visualizing spray. A characteristic series of spots clearly indicates the amino acids from the natural rubber protein.

Deproteinized Natural Rubber

Both low-protein and enzymatically deproteinized natural rubber still contain β-sitosterol and thus in the unextracted state cause no difficulty in analysis. If the sample has been previously extracted with an organic solvent, then the 'cleanliness' of the sample, together with a nitrogen content of typically below 0·05%, precludes a non-instrumental technique but several instrumental methods remain valid.

Skim Rubber

Skim rubber has a much higher protein (hence nitrogen) content than the normally available grades of rubber. The analysis of many hundreds of samples of various grades of Standard Malaysian Rubber has shown nitrogen contents to be independent of grade, in the range 0·2–0·5%, whilst the International Standard ISO 2000-1975 specifies a maximum of 0·6%. There is a much greater variability in the nitrogen content of skim rubbers but the lower limit appears to be about 1·5%, with an upper limit of 3·5%. There should not therefore be a problem in identifying skim rubber in the raw state, and as it is generally used in the form of sheets the problem of dilution with 'normal' rubber does not often occur.

The Rubber Research Institute of Malaysia (RRIM) (1954) has published a method to distinguish between the more usual grades of natural rubber and skim rubber or, indeed, to detect adulteration of the former with the latter. The method relies on the difference in specific gravities between the two, and the fact that normal grades of raw natural rubber have specific gravities below 0·93 (Davey, 1973).

A solution of ethanol (AR) 40·8% w/w in distilled water at 25 °C has exactly this density and is used for the flotation test. In this solution, skim, or natural rubber adulterated with more than about 5% skim, will sink whilst normal grades will float. The sample is pre-wetted by dipping it in a solution of detergent : water : ethanol (5 : 45 : 55 v/v) and rinsed with some of the test solution prior to immersion in a fresh portion of the test solution which is used for the actual test.

Particular care should be taken to ensure that the samples contain no entrapped air bubbles and it is advised that at least half a dozen small samples of the order of 1–2 mm^3 be examined in order to minimize this possibility.

REFERENCES

Bahrani, M. L., Chakravarty, N. K. and Saxena, A. K. (1974) *Analyt. Chem.* **46**, 446.

Baker, C. S. L. (1977) *Ind. Gomma* **21**, 40.

Baumann, G. F. and Steingiser, S. (1959) *J. Appl. Polym. Sci.* **1**, 251; (1945) *Canad. Paint Varn.* **33**, 50.

Burchfield, H. P. (1945) *Ind. Engng. Chem. Anal.* **16**, 424; (1945) *ibid.* **17**, 106.

Burger, V. L., Donaldson, W. E. and Baty, J. A. (1943) *Rubber Chem. Technol.* **16**, 660.

Corish, P. J. (1959) *Analyt. Chem.* **31**, 1298.

Davey, J. E. (1973) Unpublished work at MRPRA.

Dawson, B., Hopkins, S. and Sewell, P. R. (1970) *J. Appl. Polym. Sci.* **14**, 35.

Esposito, G. C. (1960) *Off. Dig. Fed. Paint Varn. Prod. Cl.* **32**, 67.

Gupta, P. K. S. and Ghosh, P. (1966) *Analyt. Chem.* **38**, 505.

Hilton, C. L., Newall, J. E. and Tolsma, J. (1959) *Analyt. Chem.* **31**, 915.
Kirchhof, F. (1928) *Kautschuk*, **4**, 190; (1960) *Kaut. u Gummi* **12**, WT402.
Loadman, M. J. R. and McSweeney, G. P. (1970) Unpublished work at MRPRA.
McSweeney, G. P. (1970) *J. IRI* **4**, 243.
Ministry of Supply (1944) *Identification and Estimation of Natural and Synthetic Rubbers.*
Parker, L. F. C. (1944) *J. Soc. Chem. Ind.* **63**, 378.
Rubber Research Institute, Malaysia (1954) *Planters' Bulletin*, New Series **14**, 97.
Schröder, E. (1962) *Plast. u. Kaut.* **9**, 121 and 186; (1963) **10**, 25.
Smith, K. J., Jr. (1976) *Polym. Eng. Sci.* **16**, 168.
Stern, H. J. (1944) *India-Rubber J.* **106**, 431, 449, and 491.
Swann, M. H. and Esposito, G. C. (1958) *Analyt. Chem.* **30**, 107.
Verdcourt, B. (1947) *PATRA Journal* **10**, 135.
Wake, W. C. (1969) *The Analysis of Rubber and Rubber-like Polymers*, 2nd ed., Maclaren, London.
Weber, C. O. (1902) *The Chemistry of India-Rubber*, Griffen, London.
Wechsler, H. (1953) *J. Polym. Sci.* **11**, 233.
Winterscheidt, H. (1954) *Seifen-Oele-Fette-Wachse* **80**, 239.

Chapter 6

Quantitative Elemental Analysis

Quantitative elemental analysis of an extracted polymer, compound or vulcanizate can give extensive information on many topics of interest to the rubber technologist, ranging from the quantitative analysis of blends, the homogeneity of mixes and the nature of various degradative processes, to the state of cure of rubber products. Furnace tube combustion for the determination of sulphur and halogens is still used, but is being increasingly superseded by the oxygen flask combustion technique, largely because of the much greater speed of the latter.

A still newer technique is X-ray fluorescence analysis. This can give an analysis, typically for sulphur, zinc or halogen, within the space of 1–2 minutes, ideal for repetitive analyses such as quality control. The elements which are of major interest to the rubber analyst, and which will be considered below, are carbon, hydrogen, nitrogen, oxygen, chlorine, bromine, fluorine, silicon, phosphorus and sulphur.

Although the level of a given element in a polymer repeat unit may be substantial, calculation of the amount of polymer present from the element estimated always involves the use of a factor greater than unity, and the errors are, naturally, similarly increased.

CARBON AND HYDROGEN (WITH NITROGEN)

Although it is routine to check for the presence of 'hetero atoms' in an elastomer and to quantify the levels of those such as nitrogen and halogen, if found to be present, relatively little attention is paid to the analysis of carbon and hydrogen. This is surprising in view of the relevance of these elements to the determination of a bulk formulation and their usefulness in checking polymer blend analyses. However, it should be noted that there is one standard (ISO 4655) which allows the use of carbon and hydrogen determinations to estimate the amount of styrene in a raw styrene–butadiene copolymer.

It is usual to determine carbon and hydrogen together, by oxidation to carbon dioxide and water which are then collected and measured. The quantitative nature of the reaction was studied and confirmed well over 150 years ago by Berzelius, thereby laying the foundations for organic analytical chemistry. In 1910 Pregl (see Roth, 1937) showed how the combustion could be carried out on the microscale, using less than 5 mg of sample, and the method has remained basically unchanged since.

Numerous commercial instruments are now available for carrying out carbon, hydrogen (and nitrogen) analyses. In general these rely on heating the sample in oxygen and driving the volatile products through a tube packed with a solid oxidant which ensures 100% conversion to carbon dioxide and water. Combustion products from interfering elements such as sulphur and halogen are removed with appropriate reagents packed sequentially in the tube whilst oxygen is removed and the oxides of nitrogen converted to nitrogen itself in a heated reduction tube filled with copper powder. The evolved gas is then analysed gravimetrically or instrumentally for carbon dioxide, water and nitrogen so that the original carbon, hydrogen and nitrogen loadings may be calculated. It is also possible to automate most equipment so that the throughput can be doubled by operating the instrument through the night.

For the analyst who does not wish to purchase the commercial equipment, complete details for the construction of a manual gravimetric unit are illustrated by Ma and Rittner (1979) whilst Childs and Henner (1970) have compared some instrumental and classical methods of analysis. Much has been published on the problems of ensuring complete oxidation, and the choice of suitable tube-packing chemicals, but these generally relate to difficulties encountered with specific organic chemicals, and reflect an attempt to produce a 'universal tube' suitable for all chemicals. The structural environments of carbon, hydrogen and nitrogen in polymers do not normally provide problems but of particular relevance could be papers dealing with the problem of perfluoro samples by Lebedeva *et al.* (1961), MacDonald and Turton (1968) and Olson (1968). Padowetz (1969) has designed an elegant piece of apparatus for carrying out comparative studies on reagents used in the removal of sulphur and halogens in carbon, hydrogen and nitrogen analysis.

The applications of carbon, hydrogen and nitrogen analyses to the rubber analyst reach far beyond the routine confirmation of molecular formulae. One simple application is in the analysis of nitrile–butadiene rubbers, whether raw, compounded, cured or black filled. Provided that the sample has been extracted, an automatic analysis of carbon, hydrogen and nitrogen, together with a weighing of the incombustible ash and a sulphur analysis, should account for nearly 100% of the sample. (There is always some oxygen content.) The nitrogen level can then be used to calculate the acrylonitrile loading together with the carbon and hydrogen contained therein. The 'excess' hydrogen must be derived from butadiene and hence its loading can be calculated, together with its carbon contribution. Any residual carbon is from carbon black. A very realistic analysis of the polymer in terms of its black and filler loadings, together with the nitrile content of the rubber, can thus be obtained from a sample as small as 0·5 mg.

Whilst there are dangers in analysing very small samples, and assuming them to be representative of the bulk, Loadman (1974) used this to advantage in the examination of degraded vulcanizates to distinguish between *abrasive damage*, when there was no change in the polymer: black:ash ratio, *oxidative degradation* when there was some loss of polymer relative to black, but also an increase in oxygen content, and *pyrolytic degradation* which resulted in loss of polymer relative to black but with no increase in oxygen level.

Ieki and Daikatsu (1972) have devised a scheme whereby the molecules rendered volatile during a differential thermal analysis experiment are analysed directly for carbon and hydrogen, thus enabling data on the decomposition routes to be obtained.

NITROGEN

There are, in essence, only two methods which need be considered for the analysis of nitrogen in rubber. The first is the Dumas method, in which the rubber is destroyed by oxidation and the oxides of nitrogen reduced to nitrogen itself using copper powder, whilst the second, due to Kjeldahl (1883), relies on the action of concentrated sulphuric acid to form ammonium hydrogen sulphate which is then treated with alkali to liberate ammonia. A third method, proposed by ter Meulen (1924), appears never to have become popular. The Dumas method is that used by automatic C, H, N analysers and was considered in more detail with the analysis of carbon and hydrogen. A comparison of the two major methods has been made by Dunke (1967) who used them to measure the nitrogen content of nitrile–butadiene rubbers and preferred the Kjeldahl method due to the smaller standard deviation of its results.

This method, in its current microform, was described by Ma and Zuazaga in 1942 and has changed little since. The procedure is fully described in ISO 1656-1974 and in the Rubber Research Institute of Malaysia (RRIM) Test Methods, SMR Bulletin No 7 (1973). The amount of sample taken should be adjusted if possible to give 1–2 mg of ammonia and broadly speaking the samples will fall into three categories: natural rubber (for protein content), nitrile–butadiene rubbers (for acrylonitrile content), and others, including polyurethanes and styrene–butadiene–vinyl pyridine copolymers.

This final group highlights one problem of the method. For most nitrogen-containing compounds the catalyst system which is added to the digestion solution consists of selenium, copper sulphate and potassium sulphate, in the weight ratio 1:4:30. If, however, urethane rubbers, or those containing heterocyclic aromatic compounds with nitrogen in the ring (cf. vinyl pyridine), are similarly treated, a low result will be obtained unless the catalyst is changed to mercuric oxide: potassium sulphate (30:6). In this case sodium thiosulphate should be added at the distillation stage to decompose the mercury: ammonia complex.

The nitrogen content of natural rubber is related to the protein level and, although the nitrogen content of proteins varies, a conversion factor for nitrogen to protein of 6·25 has long been used and is still considered by the RRIM (1973) to be the most generally acceptable value. The protein content of natural rubber varies depending upon its source: a representative sample of smoked sheet, Standard Malaysian Rubber, Standard Indonesian Rubber, etc. will have a nitrogen content in the range 0·3–0·6% but rubbers prepared by the concentration of latex will usually have one of two nitrogen levels, the normal grades, being lower than usual, with nitrogen contents of about 0·2%, and 'skim' rubber being higher with much more protein and nitrogen levels of 1·5–2·5% or occasionally higher. Recently another type of natural rubber has come on to the market from Malaysia, one in which the protein has been removed by enzymes prior to coagulation of the latex. This can show nitrogen levels as low as 0·04% but a more usual value would be about 0·1%. This natural rubber is designated DPNR (deproteinized natural rubber). It will be appreciated that to generate 1–2 mg of ammonia, a typical sample requirement for a normal grade of natural rubber would be 200 mg.

The acrylonitrile level of a nitrile–butadiene rubber varies between some 27% (Krynac 802) and 50% (Krynac 806) with the level reflecting the degree of oil resistance possessed by the copolymer. Also available commercially are isoprene–acrylonitrile copolymers such as Krynac 833 (34% acrylonitrile) and polyvinylchloride/nitrile–butadiene blends, with the latter containing 34% acrylonitrile (typically Krynac 850, 870). Nitrogen levels *in the raw polymer* are thus in the 7–13% range and a 20 mg sample will generate the appropriate level of ammonia.

The use to which the measured nitrogen value is put will depend upon the other data available and illustrates the interrelationships between techniques which should always be used, if possible, to check data. The polymer type (say NBR) will be known from pyrolysis–infrared or pyrolysis–gas chromatography, the temperature of the glass transition (T_g) will indicate the acrylonitrile content of the nitrile–butadiene rubber, and thus the polymer loading may be calculated. On the other hand, thermogravimetric analysis could have given the polymer loading and the nitrogen content used to obtain the acrylonitrile level. Indeed it only requires polymer identification and ash content, together with carbon, hydrogen or nitrogen analysis to give a complete analytical breakdown of the system.

Polyurethane rubbers contain small amounts of nitrogen but these will be easily distinguished by qualitative analysis; the nitrogen content can be about 5% of the polymer loading. Copolymers of vinyl pyridine with butadiene and styrene are also in use but mainly for the very special purpose of bonding rubber to textiles.

It is emphasized again that these analyses should always be carried out on extracted samples in order to remove added nitrogenous materials such as amine antioxidants and cure residues. A thiazole cure inevitably results in some accelerator fragments becoming chemically attached to the polymer chain, but these contribute negligible additional nitrogen. In the case of a reversion-resistant urethane cure, however (Baker *et al.*, 1970), this will add between 0·2% and 0·4% of the polymer weight in the sample to the measured nitrogen content.

OXYGEN

In spite of the importance of oxygen to the organic chemist, it was probably the last element for which an accurate quantitative procedure was developed. Historically it has been measured by difference, a risky business as not only does it entail a summation of all the errors of the measured element concentrations, but also there is the possibility of an unexpected element being missed and therefore not included in the analysis.

The first satisfactory procedure was developed by Schültze (1939) and modified to the microscale by Zimmermann (1939). In 1947 Aluise and co-workers used a method developed by Unterzaucher (1940) to analyse a range of synthetic rubbers and in 1948 Chambers published a detailed appraisal of the same method in the analysis of natural rubber.

The principle is simple and consists of pyrolysing the sample in a stream of pure nitrogen and leading the products over carbon heated to 1100 °C or more. All the oxygen is converted to carbon monoxide which is then oxidized to carbon dioxide by iodine pentoxide. The equivalent amount of iodine liberated is oxidized to iodate and determined titrimetrically with sodium thiosulphate. Chambers claimed that interference from halogens and sulphur could be removed by the incorporation of a soda-asbestos-filled scavenging tube prior to the oxidation step. A system similar in principle is recommended by the Association of Official Analytical Chemists (AOAC) (1975) but this uses a tube filled with copper powder to remove interfering materials, copper oxide to convert the carbon monoxide to dioxide, and a gravimetric finish.

In view of the fact that automatic carbon, hydrogen and nitrogen analysers determine carbon content by measuring carbon dioxide, it is not surprising that several have been modified to determine the carbon dioxide obtained from the oxygen in a sample. The alterations entail changing the reaction tubes and temperatures, but using the same detector system as previously. It seems probable that at levels of oxygen greater than some 2% there is little to choose between the three basic techniques, iodometric, gravimetric and instrumental, but at lower levels the last becomes less accurate and the choice lies, depending upon individual preference, between the other two. Smith *et al.* (1972), however, claimed to have used a commercial elemental analyser in the oxygen mode to measure the oxygen contents of petroleum products at levels below 1%.

A problem arises if the analysis is carried out on a fluoropolymer, as hydrogen fluoride reacts with silica to give water and silicon tetrafluoride.

A number of modifications to the basic technique have been developed to circumvent this: Ehrenberger *et al.* (1963) used a nickel combustion tube and Cruikshank and Rush (1962) a platinum one, whilst Olson and Kulver (1970) preferred to use the completely different technique of isotope dilution analysis, first described by Grosse *et al.* as early as 1946.

An excellent review, which has dated little, for quantitatively measuring the oxygen content of organic materials was presented by Davies (1969).

More recently Ma and Rittner (1979) provided many useful data and techniques but without specific reference to polymers.

Oxygen analyses in the polymer field, be they on raw rubbers or vulcanizates, tend to fit into one of two categories—those involving relatively high levels of oxygen, when the oxygen is an element within the polymer repeat unit, and those when the oxygen content is low, being due to oxidation. One example of the former is the determination of the level of methylmethacrylate in methacrylate-grafted natural rubber, supplied as MG49 (49% w/w polymethylmethacrylate) in a blend with natural rubber, whilst examples of the latter are self evident. If studies of oxidative degradation are being made it is essential that a suitable control sample be analysed concurrently. This is not always easy to find by inspection and it is advisable to take a number of samples progressively deeper into the bulk of the sample so that the 'base' level may be determined with a degree of reliability.

Chambers (1948) illustrated the changes in the oxygen content of natural rubber brought about by drying exhaustively, solvent extracting and milling for various periods of time. He reported that no special precautions were observed in selecting the samples and would thus support the general contention that rubber is normally homogeneous at the 1–5 mg level unless there are specific reasons for its not being so.

CHLORINE AND BROMINE

The estimation of chlorine is of importance in the determination of polychloroprene rubber loadings and for the determination of the polymer content of PVC compounds. It is also important for establishing the nature of a PVC compound, since there is an increasing tendency to use copolymers of vinyl chloride with another component, and determination of the chlorine content of the separated and purified polymer may be an essential step for certain identification. This is, of course, particularly necessary when vinylidene chloride is the second component. Polyvinylidenechloride (PVDC) itself or a copolymer which contains mainly vinylidene chloride is not a rubbery material and falls rather outside the limits we have set to this book.

Bromine is normally only found in bromobutyl rubber or a surface brominated elastomer. In either case its level will be low, up to 2% of the polymer loading, and the methods described for estimating chlorine are equally valid, although perhaps with minor modifications, for bromine.

Conversion of 'Organic' Chlorine to Chloride

The simplest method of estimating chlorine in polymers is that described by Phillips (1948). The apparatus for this is very easily put together in the laboratory since all that is required is a borosilicate glass tube, some means of support and rolls of wire gauze to cover the tube at the points of heating. The method is semi-micro in scale and was taken from the account in Belcher and Godbert (1945) of a method due to Bobranski and Sucharda (1936).

The actual apparatus is shown in Fig. 6.1 and consists of a hard glass tube of 10 mm internal diameter, covered with rolls of wire gauze at A, B and C. Oxygen is passed into the tube at the end and the sample is contained in the platinum boat A placed at a short distance from Z-sectional platinum contacts B, which are heated with a long burner over the length of the wire gauze. The long porcelain boat C contains halogen-free barium carbonate

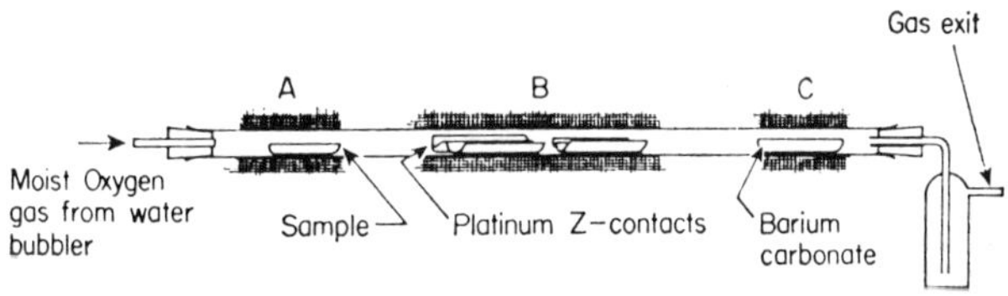

FIG. 6.1. Semi-micro combustion method for chlorine.

and is heated by a burner. Rubber bungs are used for the connections at the ends and a guard bubbler containing acidified silver nitrate solution is provided. Should the solution in this become turbid during combustion, the combustion must be abandoned, since it shows that the hydrogen chloride gas formed has not all been absorbed by the barium carbonate.

To carry out the estimation, support the combustion tube at both ends by retort stands, insert the platinum contacts after re-etching in warm dilute aqua regia, fill the long boat C level with halogen-free barium carbonate and insert through the end. Pass oxygen at 11–12 cm^3 per min, and heat the tube (surrounded by wire gauze) to about 600 °C at positions B and C. While the tube is heating, weigh out about 20 mg of sample, then disconnect momentarily and insert the sample in the boat. Burn slowly so that complete combustion takes about 1 h. After completion of combustion cool the tube in a stream of oxygen. When cool, withdraw the boat of barium carbonate and suspend its contents in water in a porcelain dish. Titrate the barium chloride formed with 0·025 M silver nitrate in the presence of a few cm^3 of starch solution and dichlorofluorescein indicator.

It should be realized that dry gas will give a low result. Thus, a combustion of laboratory reagent carbon tetrachloride (Phillips, 1949) in a moist oxygen stream showed 90·2% chlorine compared with a theoretical figure of 89·8%, but when oxygen was dried with anhydrone, values as low as 35% were obtained. The method should not really be used with compounded chloroprene rubbers as zinc oxide is almost invariably present but, fortunately, it is not used in compounding PVC. The presence of zinc oxide leads, of course, to the formation of zinc chloride which, at 600 °C, remains substantially in the combustion boat. However, the zinc oxide in neoprene rubbers is not usually more than 5% of the polymer loading and, unless the temperature rises too high, the error due to zinc chloride is small and for many purposes can be ignored. Zinc oxide reacts with chlorine at 700 °C, but, of course, hydrogen chloride gas will attack it at a much lower temperature. Much of this gas is formed at the catalyst and not at the sample, and this is one reason for the smallness of the error. With large amounts of zinc oxide present, considerable interference occurs as was shown by Stern and Hinson (1953).

The bomb/peroxide fusion method (see also Chapter 10) has been recommended by Haslam and Soppet (1948), who advise that 'chloride free' sodium peroxide must be used here so a reagent blank is advisable.

Probably the most common method in use today for converting the halogen in halopolymers to halide is the closed-flask, or *oxygen-flask*, technique. The apparatus is illustrated in Fig. 6.2, and shows a conical flask fitted with a stopper through which platinum electrodes pass, one of which terminates in a loop into which a platinum gauze cup fits.

Flasks of various dimensions and shapes have been proposed (see Hempel, 1892; Mikl and Pech, 1953; Schöniger, 1955), but a typical capacity, based on a polymer containing about 2 mg of chlorine, would be 500 cm^3 to 1 litre whilst the absorbing solution would consist of 1–5 cm^3 of 0·05 M potassium hydroxide solution, 1 cm^3 of 30% hydrogen peroxide and 10 cm^3 of distilled water.

The flask is thoroughly flushed out with oxygen, the sample wrapped in tissue paper or lens tissue and placed in the platinum cup, then the absorbing solution added prior to combustion by the application of a high voltage across the electrodes.

After the combustion the flask is shaken vigorously and allowed to stand, with intermittent shaking, for an hour before titration. The solution should be inspected to confirm that a 'clean burn' has occurred and if there are any black particles to be seen the analysis must be aborted. Opinions differ as to the reliability of this method for halogen analysis, Haslam and Willis (1965) regarding it as *semi-quantitative* (although interestingly Haslam *et al.*, 1972, claims it to be a very accurate method for determining the polyvinyl chloride content of a blend of this with polytetrafluoroethylene) and the French having a National Standard on the technique. Childs *et al.* (1963) point out that some bromine is produced during the

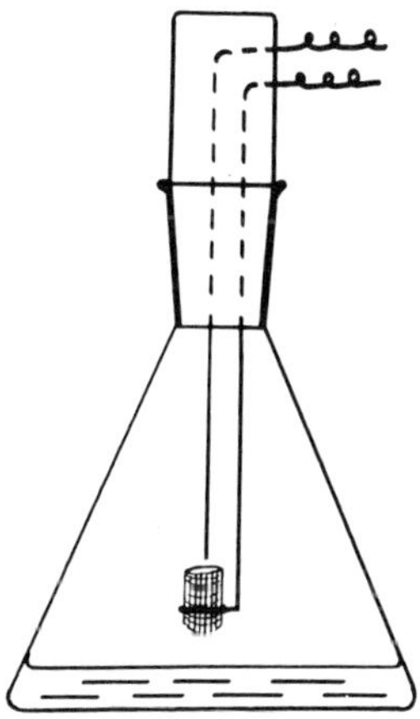

FIG. 6.2. Oxygen flask combustion apparatus.

oxygen flask combustion of bromocompounds and that the bromine must be reduced ($N_2H_4 . H_2SO_4$) to bromide prior to its estimation.

Experience in the authors' laboratory would indicate that, whilst 'bad burns' do occur with a significant regularity, these are readily observed and the time spent repeating the combustion is a negligible price to pay in view of the many advantages of the procedure.

Determination of Chloride and Bromide

The methods used for determining the halide level in the entrapping solution are gravimetric and titrimetric, the latter having either a 'visual' or electrometric finish.

The gravimetric procedure is simple and consists of acidifying the alkaline solution with nitric acid, adding an excess of silver nitrate solution, and obtaining the weight of the precipitated silver halide after filtration of the solution.

Stott (1952), working in the RAPRA laboratories, examined and compared the simple adsorption titration used by Phillips (1948) with an electrometric method, and with mercurous perchlorate in place of silver nitrate. The adsorption indicator used was *para*dichlorofluorescein and, used in conjunction with the combustion method, the titration is carried out in the presence of a large excess of barium carbonate crystals. For this reason some operators prefer to carry through the titration in a porcelain evaporating basin, stirring the liquid with a glass rod.

Theoretically, the pink colour developed at the end point and resulting from adsorption of the dye on the silver halide should appear on the precipitate. However, the presence of a relatively large amount of barium carbonate (used to absorb HCl in the combustion method) tends to obscure the observation. In any event, in the more dilute halide solutions where the error would be largest, the precipitate does not coagulate, but remains almost colloidally dispersed. In this case, the end point is very clearly and reproducibly defined by the appearance of a pink colour, apparently throughout the solution.

Steyermark *et al.* (1972) diluted the halide solution with aqueous ethanol, adjusted the pH with 0·5 M nitric acid to the yellow end point of bromophenol blue indicator (plus 3 drops) and then titrated with mercuric nitrate solution to a lilac end point using diphenylcarbazone as indicator. This represents the current favoured titrimetric method with a visual end point, although it *must not* be used in conjunction with the bromine reduction procedure as the lilac end point does not occur.

The most common electrometric titration procedure is the potentiometric one and of this group the one using silver nitrate as titrant with a mercurous sulphate electrode (after Huffman, 1962) in a solution of aqueous ethanol, acetone, or 2-propanol is favoured. This will allow one to distinguish between chloride and bromide, and estimate each individually should they both be present. A word of caution should be added in that, whilst good agreement with true values will be obtained for mixed samples of similar concentrations, the accuracy of the analysis will decrease substantially if the mole ratio falls outside the range 4:1–1:4.

Amperometric titrations appear to have received little attention in this field although techniques have been described by Olsen (1971) and Foulk and Bawden (1926). The application of colorimetric titration to the estimation of chloride and bromide has been discussed by Olsen and Krivis (1960) and Sundberg *et al.* (1958).

A micro method using a specially designed peroxide bomb and an automatic titrimeter has been described by Haslam and Hall (1958) in which extreme rapidity is combined with accuracy. Where large numbers of analyses have to be carried out the purchase of the special apparatus would pay for itself in a very short time but the bomb and the titrimeter both require special manufacture and reference should be made to the original paper for details and names of suppliers. The figures given by Haslam and Hall correspond to a standard deviation of 0·19% which compares with 0·24% given above for Phillips' results. The sample size in both methods is 20 mg but, whereas the analyst must carefully monitor the whole combustion in one method, thus taking at least 30 min, the newer method requires only a few minutes for the bomb combustion.

FLUORINE

The determination of fluorine in polymeric materials is of constant interest. There are two possible approaches, the burning of the compound in oxygen and its destruction by oxidative fusion. The combustion method is discussed by Freier and co-workers (1955) who worked mainly with highly fluorinated liquids but who also report a successful combustion with polytetrafluoroethylene. The combustion was carried out in a fused quartz combustion tube containing platinum contacts and quartz chippings. The latter participate in the reaction which may be represented as

$$(C_2F_4) + SiO_2 + O_2 \rightarrow SiF_4 + 2CO_2$$

The silicon tetrafluoride is absorbed in water which it hydrolyses with the production of hydrofluoric acid, which can be titrated directly using phenolphthalein as the indicator.

Haslam and co-authors (1972) distinguish between fluorinated polymers with less than 20% by weight fluorine and those with more. In the case of the former, combustion by the oxygen flask method following the works of Willard and Horton (1950) and Gel'man and Kiparenko (1965) has been shown to give excellent results.

The mixtures to be ignited were prepared from the samples (10–20 mg), together with polyethylene foam (25 mg), wrapped in filter paper (25 mg) impregnated with potassium nitrate. After combustion and absorption in 5 cm^3 of water the fluoride was titrated against thorium nitrate solution. Light and Mannion (1969), however, claimed that the oxygen flask gives good results with polytetrafluoroethylene (contrary to the experience of Haslam). They used a polycarbonate combustion flask and dodecyl alcohol as a combustion aid.

The fusion method is typified by the use of the Parr bomb (see Chapter 10). A sample of about 0·2 g is mixed with 1·0 g of potassium as an accelerator and placed in the bomb. The charge of sodium peroxide appropriate to the bomb size is then added followed by immediate sealing and firing. The fluorine is recovered as a soluble fluoride and can either be determined gravimetrically as calcium fluoride, or titrimetrically, using ceric nitrate. Ma and Gwirtsman (1957) describe a micro method using the Parr micro-bomb whilst Schröder and Waurick (1960) chose fusion with metallic sodium.

Haslam and Whettem (1952) preferred to use an electrically heated bomb, again charged with sodium peroxide but with starch as accelerator, to carry out the combustion, and a titration using alizarin red as indicator to determine the resultant fluoride loading. It is probable that this, and all titrimetric methods, can be improved by the use of a fluoride-specific ion selective electrode, as described by Light and Mannion (1969).

SILICON

The occurrence of this element in a rubber, or rubber-like material, could be due to its presence either in the elastomeric phase, as a silicone rubber, or in the filler as a silicate. The distinction between these is easily made by pyrolysis, followed by infrared spectroscopic examination of the pyrolysate (see Chapter 9). The analysis of inorganic ashes is covered in Chapter 10 so here we are concerned with the organosilicones.

The major problem confronting the analyst is that of destroying the organosilicone without losing volatile silicon-containing fragments, or producing the extremely stable silicon carbide. On both these counts dry ashing is to be avoided although the oxygen flask method has been used by the Schwarzkopfs (1969) who added concentrated sulphuric acid to the flask to dehydrate silicic acid and weighed the silica thus obtained.

Smith (1960) described a wet digestion procedure, using a mixture of fuming nitric and sulphuric acids, with a direct weighing of the residue as silica, but in view of the possibility of the rubber being filled with other insoluble non-silicaceous fillers, typically titanium dioxide, a subsequent treatment with hydrofluoric acid and measurement of the weight loss is the preferred procedure. The problem of removing non-silicaceous fillers was also considered by Shcherbacheva (1957) who used carbonate fusion, after a sulphuric acid digestion, to solubilize all but silicic acid.

Fusion in metal bombs has been advocated by Smith (1960) as the best way of destroying the organic part of the molecule; he preferred sodium peroxide as did Debal (1972) although Wetters and Smith (1969) advocated potassium hydroxide. The silicon in the fusion mixture may then be determined gravimetrically by complexing with molybdic acid and treating with 8-hydroxyquinoline (oxine) to give the three-component complex which is filtered and ignited at 500 °C to afford $SiO_2.12MoO_3$. Titrimetrically the silicon in the fusion mixture may be obtained by adding a known excess of oxine to the silicomolybdate complex and, after dilution to a known volume, and filtration, titration of an aliquot against standard bromide–bromate solution. Both techniques are described by McHard *et al.* (1948) who point out that an empirical standardization of the bromide–bromate solution should be made, using a pure organosilicone, to compensate for a slight deviation from the theoretical factor.

A further titrimetric method has been described by Bartusek (1973) whilst Debal (1972) reports a colorimetric method based on the silicomolybdate complex.

PHOSPHORUS

Phosphorus occurs in natural rubber latex as free *ortho*phosphate, sugar phosphates and phospholipids. It is also added to the latex, as diammonium phosphate, to precipitate magnesium phosphate from certain latices which have a high magnesium content, prior to centrifugation, and thus improve the stability of that latex. Unfortunately, if too much phosphate is added the stability will decrease again; thus in any situation where the stability of a natural rubber latex is suspect, the phosphorus content should be determined.

In order to be certain of analysing for the total phosphorus content a dried film should first be prepared from the latex, care being taken to adhere to the sampling procedures discussed in Chapter 2, and detailed in ISO 123.

The analysis of this dry rubber sample for phosphorus content can be divided into two parts; first, the removal of the polymer and the production of a solution of phosphate ions, and second the development of a colour (which can be measured spectrophotometrically) the intensity of which is proportional to the concentration of phosphorus.

The first stage may use any of the three procedures described above—Kjeldahl acid digestion, oxygen flask combustion, or fusion in a sealed bomb. In all cases it is necessary to ensure that the phosphorus is present in the final solution as ortho*phosphate* by boiling the derived aqueous solution for a few minutes prior to the development of the colour.

The fusion procedure (using sodium peroxide) for the determination of phosphorus in organic compounds has been described by Fennell *et al.* (1957) and Christopher *et al.* (1964) who found no problems with the technique although they did comment that the sodium peroxide had a significant phosphorus content. The oxygen flask method can result in a low value being obtained due to absorption on to the platinum basket. One way of avoiding this is to replace the basket with a quartz spiral to locate the sample, as described by Corner (1959).

The authors' experience, however, is that it is the acid digestion procedure which is most frequently used and this is described in detail below. Two colorimetric methods are available; one generating a yellow colour, due to a vanadiphosphomolybdate complex, is preferred for milligram quantities of phosphorus whilst the other, which produces a blue or blue/green colour (molybdenum blue), is more sensitive and used for determinations in the microgram range. There are marked differences between the two methods; the yellow colour is quite stable and relatively insensitive to slight variation in experimental procedure whilst the molybdenum blue colour is extremely sensitive to small variations in pH, concentration, temperature, light, etc., and indeed spectrophotometric examination of the colours produced under different conditions has shown maxima between 650 and 900 nm (see Ma and Rittner, 1979).

It is thus essential that a standard, reproducible procedure is used to develop the colour, and that calibration standards, prepared from stock solutions of (say) potassium dihydrogen phosphate are run concurrently under these conditions.

Unfortunately the rubber analyst is normally concerned with the measurement of relatively low levels of phosphorus in rubber and will tend to use the molybdenum blue method. For this reason the following procedure, used in the authors' laboratory, is given in detail.

Phosphorus Determination in Rubber (Courtesy of MRPRA)

A weight of sample (0·1–0·5 g) such that the phosphorus content is less than 100 μg is placed in a micro Kjeldahl flask (10 cm^3 capacity), 2 cm^3 of concentrated sulphuric acid added, and the solution warmed until charring just begins. Portions of concentrated nitric acid (0·2 cm^3) are added, the solution being heated after each addition until reaction has ceased and then cooled, until a total of 1·0 cm^3 has been used. A final addition of 0·25 cm^3 is made after which heating is continued until there is no further reaction. After cooling, distilled water (10 cm^3) is added carefully and the solution heated gently to boiling. It is boiled until acid fumes are observed at the mouth of the flask. This is repeated after the addition of a further 5 cm^3 of distilled water. The solution should now be colourless; if not 60% perchloric acid (0·5 cm^3) is added and the solution gently heated further, taking particular care and *using a safety screen.* The final clear solution is diluted to 50 cm^3 with distilled water in a volumetric flask.

A suitable aliquot (initially 5 cm^3) is neutralized to Congo Red paper with concentrated ammonia solution and transferred to a 50 cm^3 flask, distilled water being used to dilute the solution to about 25 cm^3. Ammonium molybdate solution (see end of method) (5 cm^3) is added, the solution shaken and the reducing solution (see end of method) (5 cm^3) added. The flask is then placed in a boiling water bath, for 30 min, ensuring that the solution is below the water line. It is then cooled to ambient temperature and the final dilution to 50 cm^3 made in the volumetric flask.

The absorbency of this solution is then measured at 700 nm against a 'blank' solution prepared by taking all the reagents (with the exception of the rubber) through the complete procedure. The phosphorus content is calculated by reading from a calibration graph, prepared by measurement of the absorbencies of a series of standard phosphate solutions which have had their colours developed concurrently.

Ammonium molybdate solution. A solution of 10 g ammonium molybdate in 100 cm^3 distilled water is poured slowly into a cooled solution of 300 cm^3 50% aqueous sulphuric acid. This solution is stored in the dark.

Reducing solution. Sodium metabisulphite (40 g), sodium sulphite (1 g) and Metol (0·2 g) are dissolved in 100 cm^3 distilled water. This solution has a shelf life of no more than one week and ideally should be prepared freshly for each analysis.

It is emphasized that many other procedures are published for developing the molybdenum blue colour and these are equally valid but could well give an absorbency maximum at a wavelength *different from* the 700 nm found for this one.

SULPHUR

Sulphur estimations may need to be carried out for the determination of the polymer present, as with thioplast rubbers or chlorosulphonated polyethylene, but the commonest use is as part of the analysis of a sulphur-vulcanized rubber.

Much work has been carried out over the years by various authors, including Bateman *et al.* (1963), Moore (1964), Craig (1957), Campbell and Wise (1964), and Scheele (1961), when studying the mechanism of vulcanization. As a result, not only is the analysis of a rubber for its total sulphur content required, but so too is the analysis for free (elemental) sulphur, and for sulphide sulphur. The level of free sulphur in a vulcanizate gives an indication of undercure, while the level of sulphide sulphur can give an indication of overcure.

Determination of Total Sulphur

The determination of total sulphur and combined sulphur is carried out by the same procedure, the only difference being that combined sulphur is the total sulphur remaining after solvent extraction.

Methods of determining the total sulphur content of an organic material have a long history of change and development, during which numerous methods have been proposed and adopted only to pass out of fashion after a few years. Johnson and Messenger (1933) give a very complete historical review starting at the original work of Henriques (1892) and Weber (1894). Methods for determining the total sulphur content of rubbers have been reviewed by Auler (1961a).

The method adopted must depend on what is needed from the analysis. Some products will contain sulphides or sulphates of zinc, barium or calcium, and it must be decided in advance whether the analytical method to be chosen is to include any of these.

Although the methods to be described below are fully covered in both International and British standards they will nevertheless be given here in some detail because of their unique importance to the rubber analyst. Methods available include combustion, both furnace tube and oxygen flask, oxidative fusion and X-ray fluorescence. Wet oxidation, popular for many years, is being phased out of the imminent revision of BS 903, and will not be published as an International Standard method.

Furnace Tube Combustion Method

Combustion of the test portion in a stream of oxygen or air followed by absorption of the products is an elegant method in the tradition of classical organic analysis. This method was used by Eaton and Day as early as 1917 but, even then, they emphasized the need to limit its application to certain classes of compound. The difficulty lay in the range of decomposition temperatures covered by the various substances which could be present. The temperature of combustion needs to be sufficiently high to decompose all the organic material present and, ideally, not high enough to cause decomposition or volatilization of any inorganic sulphur compounds.

The principal causes of trouble are the occurrence of lithopone, which is a mixture of barium sulphate and zinc sulphide, of barium sulphate as barytes, of calcium sulphate, and the possibility of the formation of sulphides and sulphates from the zinc oxide always present and calcium carbonate or magnesium carbonate which may also be present.

A thorough study and testing of a modified combustion method was carried out in various laboratories of the Dunlop Rubber Co. Ltd, and the method finally published from the Dunlop Research Centre has probably been more thoroughly tried out under routine conditions than other methods which have been in use for many years. The method is a semi-micro one derived from the Grote procedure. Since it was first described (BS 903-1950) it has been modified (Bauminger, 1955, 1956a, b), written up as an International Standard (see Appendix I) and as a British Standard due shortly to replace BS 903, Parts B6-10, and, as described below, can be used to determine total or unextracted sulphur in fully compounded natural and synthetic rubbers including those containing chlorine and/or nitrogen.

The method is relatively fast, with results being available within an hour, the operating procedure is simple, and estimations are well within the capabilities of a younger assistant.

Simple combustion in oxygen requires a temperature of 1350 °C if all the sulphur likely to be present in rubber compounds is to be included, but the

TABLE 6.1

RECOVERY OF SULPHUR FROM INORGANIC POWDERS

Powder	*Sulphur as % of theoretical for pure material*		
Pure calcium sulphate	99·6	100·0	99·8
Pure sodium sulphate	99·0	99·1	99·1
Barium sulphate (purified barytes)	99·3	99·5	99·2
Commercial barytes	88·8	89·0	89·0
Commercial blanc fixe	88·3	88·5	
Commercial lithopone	97·1	97·7	

After Bauminger (1956a).

use of a mixture of vanadium pentoxide and zinc oxide (Bauminger, 1955) enables quantitative recovery of all forms of sulphur by combustion at 1000 °C. The combustion tube must be of translucent or transparent silica and the furnace should be capable of being controlled to ±20 °C. If possible, because of its reliability, a resistance pyrometer should be permanently installed. Table 6.1 shows results (Bauminger, 1956a) obtained on organic powders which are used in rubber. The catalyst used was a mixture of 4 parts vanadium pentoxide with 1 part zinc oxide, 1 g of the mixture being used with between 10 and 15 mg of the polymer.

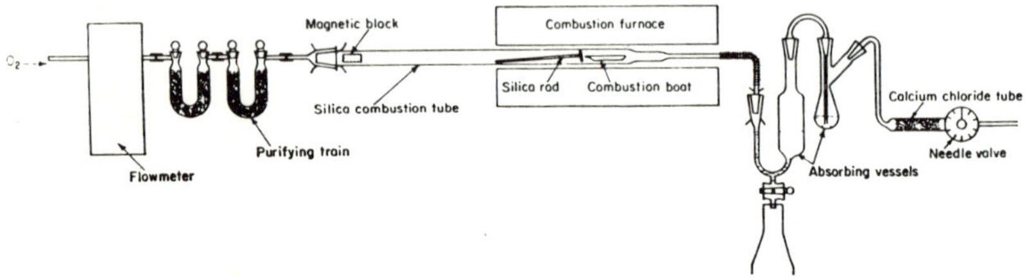

FIG. 6.3. Apparatus for the determination of total sulphur by the furnace tube method. (Courtesy BSI.)

The following procedure is set out in detail in ISO/DIS 6528/3, details of the apparatus required being shown in Fig. 6.3:

> The rubber sample (10–50 mg) is mixed with 1 g of catalyst consisting of 0·8 g dry vanadium pentoxide and 0·2 g zinc oxide, and burned in a stream of oxygen at a temperature of 1000 °C. Combustion products are absorbed by hydrogen peroxide solution, 3 cm^3 of 30% hydrogen peroxide being mixed with 30 cm^3 water, 15 cm^3 of this being added to the main absorption vessel and 5 cm^3 to the other. If nitrogen and halogen are known to be absent (the trivial nitrogen content of natural rubber can be ignored), then the sulphuric acid can be determined by titration with standard 0·02 M sodium hydroxide solution using a methyl red/methylene blue indicator. In all other cases a barium perchlorate titration is used. Small amounts of zinc chloride may distil over from rubbers containing chlorine, and would interfere with the barium perchlorate titration. This interference may be removed by passing the combined absorbing solutions slowly through a short ion exchange column. Sufficient 2-propanol is added to the absorbent solution to bring it to 70–90% alcohol by volume. A few drops of Thorin indicator solution is added, followed by sufficient methylene blue solution to

change the colour from orange to yellow. The solution is then titrated to a permanent pink colour using 0·01 M barium perchlorate in a mixture of 80/20 (v/v) 2-propanol and water.

Oxygen Flask Combustion Method

The oxygen flask combustion technique was described in detail above in relation to its use for determination of chlorine and bromine. The apparatus has the advantage of being less costly than that for the furnace tube combustion method. A single determination takes longer than does an analysis in the furnace tube but if multiple analyses are required then the oxygen flask method is much more rapid since a batchwise procedure can be used, whereas this is not available for the furnace tube method.

Another feature of the oxygen flask technique is that combustion takes place in an atmosphere of oxygen, but in the absence of catalyst. This means that any stable sulphur containing fillers such as lithopone barytes, calcium sulphate and sulphates from the zinc oxide always present will not be decomposed, and their sulphur contents will not be determined. This may of course be an advantage in that the rubber analyst is primarily interested in organically bound sulphur. Similar non-detection of sulphur in inorganic compounds by the combustion tube method may be achieved by simple omission of the catalyst during the combustion stage.

The procedure to be followed is spelled out in detail in ISO/DIS 6528/1. 40–80 mg of the finely milled test portion is wrapped in paper and placed in the combustion flask which is then filled with oxygen and sealed. The sample and paper are ignited, and any sulphuric acid produced during the combustion is absorbed in hydrogen peroxide solution contained in the bottom of the combustion flask.

The method used for the determination of the sulphuric acid formed is precisely the same as that described above for the furnace tube combustion method. Interfering zinc ions must in this case be removed, whether or not halogen is present, by passage of the absorbent solution through a short ion exchange column prior to carrying out the titration.

If the rubber sample contains whiting as a filler then results are often low because a part of the liberated sulphur trioxide reacts with the whiting to give the stable calcium sulphate. However, Davey (1979) has found that if one uses a mixture of 0·25 cm^3 concentrated hydrochloric acid, 2 cm^3 water and 1 cm^3 6% hydrogen peroxide as the absorbing solution, any calcium sulphate dissolves and, on passage down an ion exchange resin in the hydrogen ion form, becomes converted to sulphuric acid, which can be titrated in the usual way. The results shown in Table 6.2 were obtained.

TABLE 6.2

TOTAL SULPHUR DETERMINATIONS

Sample	*Absorbing solution*	*Theoretical % S*	*Found % S*
Masterbatch	H_2O_2	2·44	2·48, 2·49
Masterbatch + 10 pts. ZnO	H_2O_2	2·20	2·23, 2·21, 2·28
Masterbatch + 100 pts. $CaCO_3$	H_2O_2	1·22	0·54, 0·63, 0·07
Masterbatch + 100 pts. $CaCO_3$	H_2O_2/HCl	1·22	1·19, 1·18

Sodium Peroxide Fusion Method

In their review of methods for the determination of sulphur, Johnson and Messenger (1933) state that the first use with rubber of fusion with sodium peroxide was made by Alexander (1904) and Esch (1904), both dependent on earlier work on sulphur in coal by von Konek (1903) and, less directly, Asboth (1895), who used a mixture of sodium peroxide and sodium carbonate. Other fusion mixtures have been recommended, but sodium peroxide fusion is the only method still in use today and the violence of the reaction is contained by the use of a pressure vessel, such as the Parr bomb (see Chapter 10, Fig. 10.2). (Fusion in open crucibles can be carried out but is to be deprecated even if the reaction is confined by the immersion of the reaction crucible upside down in a second crucible, and the space between them packed with sodium carbonate.) The 'bomb' method is somewhat more dangerous than alternative ones, although it is apparently still in widespread use, notably in Germany. Because of this it is likely that it will be retained in the International Standard ISO 6528/2 which is currently in an advanced state of preparation, but it will not constitute part of the forthcoming revision of the corresponding British Standard (BS 903 Part B6-10) and the method will not be further described here.

X-ray Fluorescence Method

X-ray fluorescence analysis is a very costly, if very powerful, technique which is capable of measuring the concentration of a wide range of elements, in a given sample, simultaneously. In this technique (see Jenkins, 1974) the sample is bombarded with X-rays from an X-ray source, and this causes the sample to emit X-ray radiation of a different energy (i.e.

fluorescence). The energies and intensities of the emitted X-rays are measured, and used to provide a rapid, accurate identification and estimation of the elements in the sample.

Such instruments are not available to many rubber analysts. However, there have recently come on to the market small X-ray fluorescence analysers which have radioactive elements to excite the fluorescence of the elements in the sample. In place of the expensive dispersive analysers used for large installations, these analysers use a series of filters to isolate the X-ray fluorescence of the element of interest, and can provide a direct readout of the percentage sulphur content of a sample in under one minute.

Samples are best prepared as discs of 2–5 cm diameter by hot pressing against (for example) cellophane, which can be peeled off prior to the analysis. It must be remembered that this technique is relatively surface dependent so that surface contamination must be kept to an absolute minimum. Surface texture is also important. This does not cause problems when samples can be hot pressed; however, for vulcanizates this is not possible, and a standardized method of surface treatment is essential. Davey and Loadman (1977) found that a suitable method of sample preparation for cured products is to cut a plug of rubber with a cork borer. The plug end is then cut as smoothly as possible with a razor blade and smoothed with fine emery cloth. The ground surface is wiped clean with cotton wool dampened with methanol. Provided that the standards with which the unknown is compared are prepared in the same way very accurate results are obtained with a 50 s measurement time. Using this technique, a series of ten different analyses gave a sulphur content of 1·263% with a standard deviation of 0·013%.

Choice of a suitable radioactive element source, and use of the appropriate filters, enables the instrument to be applied to the analysis of other elements, such as zinc or halogen.

Although relatively expensive, the speed and ease of operation, without even needing to weigh the samples, makes this an ideal instrument for quality control monitoring, particularly of compounded rubber mixes.

Determination of Free (Elemental) Sulphur

Most rubber mixes contain elemental sulphur and a sulphur-containing accelerator. During vulcanization the sulphur gradually becomes combined with the rubber, as too does some of the accelerator (see Moore, 1964). Determination of the level of free sulphur in a raw mix is important for checking whether compounding has been carried out correctly.

Equally, if not more, important is the determination of free sulphur in a cured product, since any appreciable level found must indicate that the product is undercured. Such a finding would explain deficiencies in physical properties in the product. In addition, vulcanizates may contain higher levels of free sulphur than the rubber can dissolve, the result being an unsightly or even harmful bloom of sulphur.

Methods for the determination of free sulphur in rubber have been reviewed by Auler (1961b), and several methods have been published in BS 903 Part B7. The nitric acid and bromine methods of the latter are now little used, being replaced in the revised standard by the sulphite method (see Mackay and Avons, 1940). The copper spiral method, on the other hand (see Hardmann and Barbehenn, 1935), is retained, and remains the best of the older chemical methods, being subject to fewer interferences although, in the absence of sulphur donor accelerators, the copper spiral and sulphite methods are of comparable accuracy.

Copper Spiral Method

Full details of this procedure are given in ISO/DP 7269 and in the revision of BS 903 Part B7 which is shortly to be published.

A thinly sheeted or finely chopped sample of rubber (0·5 g) is acetone extracted in a Soxhlet extractor, with a coiled piece of clean copper gauze in the flask containing the boiling acetone. Sulphur extracted from the rubber reacts with the copper to form a black layer of copper sulphide. The acetone in the extraction flask is filtered off, and the copper spiral(s) washed with hot acetone. The extraction flask now containing the copper spiral(s), filter funnel and filter is assembled into the apparatus shown in Fig. 6.4, 50 cm^3 of dilute hydrochloric acid is added slowly to the extraction flask through the funnel, and the flask allowed to stand at room temperature for 5 min.

The solution is brought slowly to the boil, and boiled for 30–40 min. Any hydrogen sulphide formed is swept by a stream of nitrogen into the absorption flask which contains buffered cadmium acetate solution as also do the gas washing bottles. The hydrogen sulphide is trapped and forms quantitatively a precipitate of cadmium sulphide. Excess standard iodine solution (0·025 M) is added, and the wash bottle swirled gently until all the cadmium sulphide has been dissolved. Residual iodine is then back-titrated with standard (0·05 M) sodium thiosulphate solution.

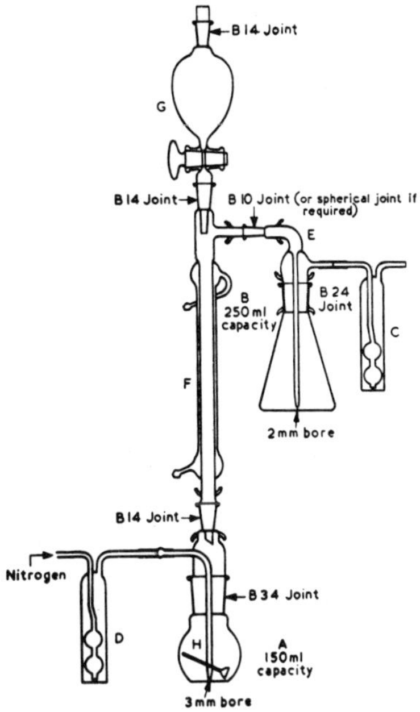

FIG. 6.4. Apparatus for determination of free sulphur by the copper spiral method. (Courtesy BSI.)

Interferences with this method have been shown by Davey (1981) to be relatively few (see Table 6.3 below) but since the method depends upon an extraction stage, low results will be obtained if raw mixes have been prepared using insoluble sulphur unless prolonged (up to 72 h) extraction periods are used. After cure, however, any sulphur remaining will be extractable. For uncured mixes, care should be taken to minimize heating, since this can cause curing to take place during the initial stages of the extraction. Later in the extraction, accelerators will have been removed and heating is unlikely to be harmful.

Sulphite Method

Approximately 2 g of finely divided or thinly sheeted sample is placed in a conical flask and 100 cm^3 of 0·05 M sodium sulphite solution, together with 3–5 cm^3 of liquid paraffin to prevent frothing, added. The mouth of the flask is covered with a watch glass, and the contents of the flask are boiled gently for 4 h. During this period, sulphur reacts with the sodium

TABLE 6.3

EFFECT OF VARIOUS COMMERCIAL ACCELERATORS ON THE COPPER SPIRAL AND SULPHITE METHODS OF ANALYSIS

	Abbreviations (see ISO/DIS 6472)	*Theoretical total S %*	*Total S% found*	*Apparent % free S Cu spiral method*	*Apparent % free S sulphite method*	*Apparent % free S sulphite without charcoal*
Dipentamethylene thiuram hexasulphide	DPTH	57·1	58·2	22·5	26·4	30·3
Dipentamethylene thiuram tetrasulphide	DPTD	50·0	45·8	20·0	20·0	33·1
4,4′-Dithiodimorpholine	DTDM	27·1	20·6	10·3	21·0	23·8
4-Morpholinyl-2-benzothiazolyl disulphide	MOR	33·8	34·9	6·1	20·0	23·7
Zinc isopropyl xanthate	ZIX	42·2	39·4	4·6	3·5	8·9
Tetramethyl thiuram disulphide	TMTD	53·3	45·8	1·4	38·6	
Thiocarbamyl sulphenamide		25·8	24·8	0·9	16·6	20·5
Tetramethyl thiuram monosulphide	TMTM	42·2	42·0	<·01	33·3	67·4
Zinc dimethyl dithiocarbamate	ZMDC(ZDMC)	41·9	40·2	<·01	21·4	26·5
Zinc 2-mercaptobenzothiazole	ZMBT	33·2	31·9	<·01	5·1	17·6
Ethylene thiourea	ETU	31·4	32·0	<·01	7·6	47·0
Dicyclohexylamine salt of di-*n*-butyl-dithiophosphoric acid		15·5	13·8	<·01	8·0	10·2
N-cyclo hexyl-2-benzothiazyl sulphenamide	CBS	24·2	24·8	<·01	0·7	5·2
2-Mercaptobenzothiazole	MBT	38·3	36·9	<·01	12·4	15·9
2,2′-Dibenzothiazyl disulphide	MBTS	38·5	36·8	<·01	11·2	13·1

sulphite to form sodium thiosulphate. After cooling, 5 g activated charcoal is added and the flask allowed to stand for 30 min during which time accelerator residues are adsorbed on the charcoal. Insoluble residues are removed by filtration, and to the filtrate is added 10 cm^3 of formaldehyde solution (400 g/l) to complex with the excess sodium sulphite. After standing for 5 min, 5 cm^3 glacial acetic acid is added, and the sodium thiosulphate formed is reacted with excess 0·025 M iodine solution. Excess iodine is back titrated with 0·05 M sodium thiosulphate solution using starch as indicator.

As can be seen from work due to Davey (1981) presented in Table 6.3, many accelerators give significant interference in this method, with substantial proportions of those accelerators acting as though they were free sulphur. Fortunately accelerators tend to be used at low percentage levels, but the level of interference is often so large that the analysis of unvulcanized mixes should not be undertaken by the bisulphite method. In addition to the interferences shown by accelerators, Davey (1981) has shown that other classes of compounds can also interfere with the sulphite method. Thus the antioxidant Santoflex 13 reacts as though 11% of it were free sulphur, and the peptizer Renacit VII reacts as though 32% of it were free sulphur. Neither compound interferes with the copper spiral method.

Other Methods
The above methods appear to be those most widely used at present, although others are available. One in particular due to Poulton and Tarrant (1951) measures the sulphur in an extract by polarography. This method is much less prone to interference by accelerators, but has never become widely used. At the time it was developed polarographs were cumbersome and not easy to use. Present-day commercial polarographs are much easier to use and the technique is worthy of reconsideration.

On the other hand, liquid chromatographs (see Chapter 4) are very powerful tools and are becoming increasingly used for the analysis of rubber extracts. Sulphur is one such compounding ingredient which can be so analysed, with high sensitivity and with a high degree of specificity. For an example of the use of this technique for the determination of free sulphur in rubber extracts see the paper by Spagnolo and Malone (1976).

Determination of Sulphide Sulphur
In unvulcanized but compounded rubbers the only source of sulphide sulphur will normally be metallic sulphides present in inorganic fillers such

as lithopone. In vulcanizates, however, the situation is different. As vulcanization proceeds most of the sulphur which was initially added as elemental sulphur becomes bound to the rubber network but some reacts with the zinc oxide present to generate zinc sulphide. At optimum cure the level of sulphur combined with the rubber reaches a *maximum*, decreasing if further heat is applied and crosslink degradation reactions occur. The zinc sulphide level, however, *continues to increase* during these crosslink degradation reactions as well as during the initial vulcanization stage (Bateman *et al.*, 1963). Subsequent to this work, a survey of published data has shown (Tidd, 1975) that for a substantial number of cure systems the ratio zinc sulphide/network combined sulphur ranges between 0·17 and 0·21 at optimum cure. On substantial overcure, this figure can increase to

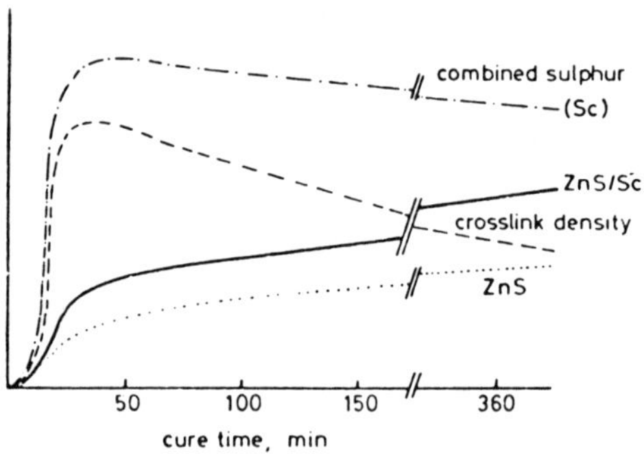

FIG. 6.5. Formation of zinc sulphide during cure (concentrated CBS/S mix). (Courtesy Plastics and Rubber Institute.)

as high as 0·9. At higher cure temperatures, degradation reactions are relatively more important, and the overall levels of zinc sulphide are higher.

Certain cure systems, on the other hand, particularly those containing zinc dithiocarbamates, habitually give very low figures of around 0·02–0·03. Clearly either the actual mechanism of vulcanization changes according to the accelerator being used, or zinc sulphide once formed in the normal manner is consumed by further reaction with the vulcanization additives or their reaction products. Evidence that at least the latter takes place has recently been obtained by Morrison (1982) who has found that after heating zinc sulphide with MBTS (mercaptobenzothiazyl disulphide) only low levels of zinc sulphide are detectable in the product.

The method therefore has limitations for the study of unknown samples. However, under quality control conditions, where the compound used is well known, the technique comes into its own. If the cure temperature is held constant, then increases in the sulphide sulphur level indicate an increasing time of cure. Similarly if the cure time is known, as it is in many situations, an increase in sulphide level must reflect an increase in cure temperature.

The original method for the determination of sulphide sulphur in a vulcanizate was published in BS 903 Part B10. Davey *et al.* (1978), however, significantly improved the procedure in a number of respects, and their method is now shortly to appear as a revision to BS 903 Part B10. The method will in due course appear as an International Standard. At present it is a Draft Proposal (ISO/DP 8054) awaiting issue as a Draft International Standard.

The finely chopped or thinly sheeted rubber is extracted with acetone, and the extracted rubber is dried thoroughly and treated with a mixture of concentrated hydrochloric acid 1 cm^3, water 1 cm^3, and glacial acetic acid 5 cm^3, either in the apparatus described above for the determination of free sulphur by the copper spiral method, or preferably in the improved apparatus produced by Davey *et al.* (1978) (see Fig. 6.6).

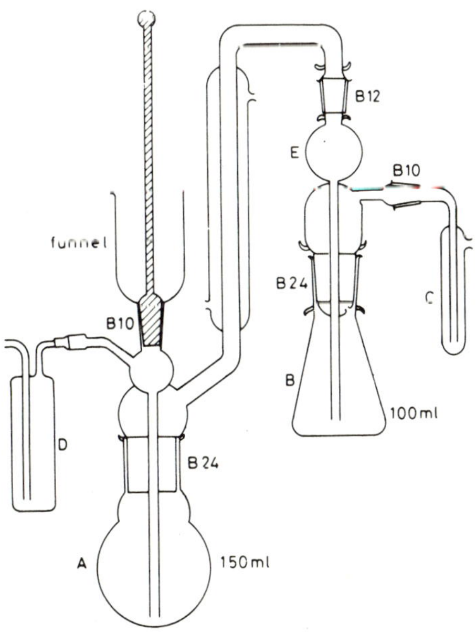

FIG. 6.6. Apparatus for sulphide sulphur determination. (Courtesy Plastics and Rubber Institute.)

All ground glass joints are carefully lubricated with glycerol, and the gas washing bottles contain buffered cadmium acetate. The glacial acetic acid/hydrochloric acid mixture (50 cm^3) is introduced via the funnel into the flask containing the rubber. The contents of the flask are heated gradually to boiling, and boiling is continued for one hour. During the heating, hydrogen sulphide liberated from the sulphide sulphur is swept into the gas absorption flasks containing buffered cadmium acetate solution by a slow stream of nitrogen. The precipitated cadmium sulphide is determined as before by reaction with iodine and back-titration of the excess iodine with sodium thiosulphate solution.

The main features highlighted by Davey *et al.* (1978) are:

(1) The use of hydrochloric/glacial acetic acid. The organic nature of this solvent gives more rapid penetration of the rubber by the acid, and hence a shorter reaction time.
(2) Routine acetone extraction is not necessary in most cases, and even introduces possible errors since any acetone remaining in the sample will distil into the cadmium acetate solution and react with iodine when that is added.
(3) Milling of the sample should be kept to an absolute minimum since this treatment allows oxidation to occur and zinc sulphide to become progressively depleted, although the total sulphur level remains constant.
(4) Milling is more effective than comminution for permitting rapid ingress of the acid mixture, and allows the reaction to be complete in about 1 hour, whereas comminuted samples require up to 5 hours' refluxing.

ION CHROMATOGRAPHY (IC)

Although the technique of ion chromatography is relatively new it has an exciting future and it must merit inclusion somewhere in a book of this type, although its precise location is difficult to decide upon. It is included in this chapter because of its general relevance in the quantification of chloride, bromide, fluoride, phosphate, sulphate etc. in aqueous solution, although its relevance to the examination of aqueous extracts of rubber, latex serum and similar applications should not be overlooked. One specific application is in the analysis of latex serum, by direct injection on

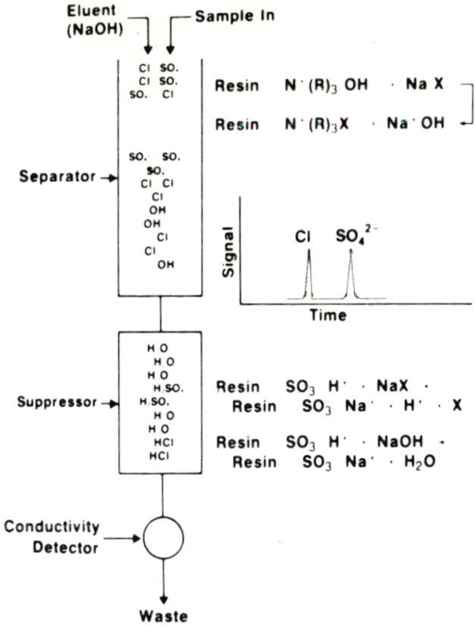

FIG. 6.7. Schematic representation of ion chromatography. (Reproduced with permission of the Dionex Corporation.)

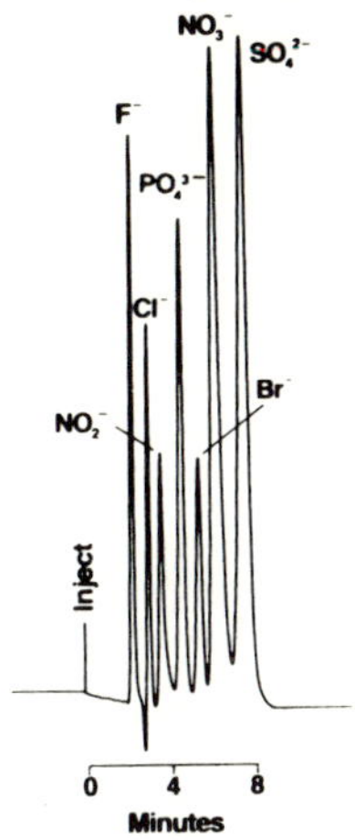

FIG. 6.8. Separation of common anions. (Reproduced with permission of the Dionex Corporation.)

to the column, for both phosphate and sugar phosphates as described by Crafts (1982).

Ion chromatography is a specific form of liquid chromatography and is used for the separation and quantification of both cationic and anionic species, at levels between mg l^{-1} and ng l^{-1}. The procedure is illustrated schematically in Fig. 6.7, and a representative chromatogram is shown in Fig. 6.8.

A liquid sample is introduced at the top of the separator column and as it is pumped through the column the ions move at different speeds, their rates dependent upon their affinity for the resin. The separated ions are then modified to highly conductive acids or alkalis in the suppressor column. As with other forms of chromatography, identification and quantification is based on retention time and peak area respectively.

It will be appreciated that its applications are numerous and, although inevitably expensive, it could effect very substantial savings in many laboratories.

The American Society for Testing and Materials (ASTM) is actually considering this technique on three committees and proposed IC methods were published for comment in 1982.

REFERENCES

Alexander, P. (1904) *Gummi Ztg.* **18**, 729, 1042; *Z. Angew. Chem.* **17**, 1799.
Aluise, V. A., Hall, R. T., Staat, F. C. and Becker, W. W. (1947) *Analyt. Chem.* **19**, 347.
Asboth, A. V. (1895) *Chem. Ztg.* **19**, 2040; *J. Soc. Chem. Ind.* **14**, 1071.
Assn. Offic. Anal. Chem. (1975) *Official Methods of Analysis*, 12th ed., Horowitz, W., Ed., 928.
Auler, H. (1961a) *Gummi Asbest. Kunstst.* **14**, 406.
Auler, H. (1961b) *Gummi Asbest. Kunstst.* **14**, 712.
Baker, C. S. L., Barnard, D. and Porter, M. (1970) *Rubber Chem. Technol.* **43**, 501.
Bartusek, P. (1973) *Textil* **28**, 51.
Bateman, L., Moore, C. G., Porter, M. and Saville, B. (1963) in *Chemistry and Physics of Rubberlike Substances*, Bateman, L., Ed., Maclaren, London.
Bauminger, B. B. (1955) *Kaut. u. Gummi* **8**, WT31.
Bauminger, B. B. (1956a) *Analyst* **81**, 12.
Bauminger, B. B. (1956b) *Trans. IRI* **32**, 218.
Belcher, R. and Godbert, A. L. (1945) *Semi Micro Quantitative Analysis*, Longmans, London.
Bobanski, B. and Sucharda, E. (1936) *Semi-micro Methods for the Elementary Analysis of Organic Compounds*, A. Gallenkamp and Co., London.
Campbell, R. H. and Wise, R. W. (1964) *Rubber Chem. Technol.* **37**, 635, 650.
Chambers, W. T. (1948) *Paper presented at the Rubber Technology Conference*, June, London.
Childs, C. E. and Henner, E. B. (1970) *Microchem. J.* **15**, 590.
Childs, C. E., Cheng, J., Meyers, E. E., Laframboise, E. and Balodis, R. B. (1963) *Microchem. J.* **7**, 266.
Christopher, A. J., Fennell, T. R. F. W. and Webb, J. R. (1964) *Talanta* **11**, 1323.
Corner, M. (1959) *Analyst* **84**, 41.
Crafts, R. C. (1982) Unpublished work at MRPRA.
Craig, D. (1957) *Rubber Chem. Technol.* **30**, 1291.
Cruikshank, S. S. and Rush, C. A. (1962) *Microchem. J., Symp. Ser.* **2**, 467.
Davey, J. E. (1979) Unpublished work at MRPRA.
Davey, J. E. (1981) Unpublished work at MRPRA.
Davey, J. E. and Loadman, M. J. R. (1977) Unpublished work at MRPRA.
Davey, J. E., Edwards, A. D. and Higgins, G. M. C. (1978) *Plastics and Rubb. Mater. and Applic.* 145.
Davies, D. H. (1969) *Talanta* **16**, 1055.
Debal, E. (1972) *Talanta* **19**, 15.
Dionex Corporation (UK Office, Frimley, Camberley, Surrey).
Dunke, M. (1967) *Faserfosch Textiltech* **18**, 123.
Eaton, B. J. and Day, F. W. F. (1917) *J. Soc. Chem. Ind.* **36**, 16; *Agric. Bull. F.M.S.* **6**, 73.
Ehrenberger, F., Gerbach, S. and Mann, W. (1963) *Z. Anal. Chem.* **198**, 242.
Esch, W. (1904) *Chem. Ztg.* **28**, 200.
Fennell, T. R. F. W., Roberts, M. W. and Webb, J. R. (1957) *Analyst* **82**, 639.
Foulk, C. W. and Bawden, A. T. (1926) *J. Am. Chem. Soc.* **48**, 2045.

Freier, H. E., Nippoldt, B. W., Olson, P. B. and Weiblen, D. G. (1955) *Analyt. Chem.* **27**, 146.
Gel'man, N. E. and Kiparenko, L. M. (1965) *Zh. Anal. Khim.* **20**, 229.
Grosse, A. V., Hindin, S. G. and Kirshenbaum, A. D. (1946) *J. Am. Chem. Soc.* **68**, 2119.
Hardmann, A. F. and Barbehenn, H. E. (1935) *Ind. Eng. Chem. Anal. Edn.* **7**, 103.
Haslam, J. and Hall, J. I. (1958) *Analyst* **83**, 196.
Haslam, J. and Soppet, W. W. (1948) *J. Soc. Chem. Ind.* **67**, 33.
Haslam, J. and Whettem, S. M. A. (1952) *J. Appl. Chem. Lond.* **2**, 339.
Haslam, J. and Willis, H. A. (1965) *Identification and Analysis of Plastics*, Iliffe, London.
Haslam, J., Willis, H. A. and Squirrell, D. C. M. (1972) *Identification and Analysis of Plastics*, 2nd ed., Iliffe, London.
Hempel, W. (1892) *Z. Angew. Chem.* **5**, 393.
Henriques, R. (1892) *Chem. Ztg.* **16**, 1595, 1623, 1644.
Huffman, E. W. D. (1962) *Microchem. J. Symp. Ser.* **2**, 811.
Ieki, T. and Daikatsu, K. (1972) *Microchem. J.* **14**, 110.
Jenkins, R. (1974) *An Introduction to X-ray Spectrometry*, Heyden, London.
Johnson, R. N. and Messenger, T. H. (1933) *J. Rubber Res.* **2**, 31.
Kjeldahl, J. (1883) *Z. Anal. Chem.* **22**, 366.
von Konek, F. (1903) *Z. Angew. Chem.* **16**, 516.
Lebedeva, A. I., Nikolaeva, N. I. and Orestova, V. A. (1961) *Izv. Akad. Nauk. SSR, Ctd. Khim. Nauk* 1350.
Light, T. S. and Mannion, R. F. (1969) *Analyt. Chem.* **41**, 107.
Loadman, M. J. R. (1974) Unpublished work at MRPRA.
Ma, T. S. and Gwirtsman, J. (1957) *Analyt. Chem.* **29**, 140.
Ma, T. S. and Rittner, R. C. (1979) *Modern Organic Elemental Analysis*, Marcel Dekker, New York.
Ma, T. S. and Zuazaga, G. (1942) *Ind. Eng. Chem. Anal. Edn.* **14**, 280.
MacDonald, A. M. G. and Turton, G. G. (1968) *Microchem. J.* **13**, 1.
Mackay, J. G. and Avons, C. H. J. (1940) *Trans. IRI* **16**, 117.
McHard, J. A., Servais, P. C. and Clark, H. C. (1948) *Analyt. Chem.* **20**, 325.
ter Meulen, H. (1924) *Rec. Trav. Chim. Pays-Bas* **43**, 463.
Mikl, O. and Pech, J. (1953) *Chem. Listy* **46**, 382.
Moore, C. G. (1964) in *Proc. Nat. Rubber Producers' Res. Assn. Jubilee Conf.*, Mullins, L., Ed., Maclaren, London.
Morrison, N. J. (1982) Unpublished work at MRPRA.
Olsen, E. C. (1971) *Treatise on Analytical Chemistry*, Pt. II, **14**, Kolthoff, M. and Elving, P. J., Eds., Wiley, New York.
Olsen, E. C. and Krivis, A. F. (1960) *Microchem. J.* **4**, 181.
Olson, P. B. (1968) *Microchem. J.* **13**, 75.
Olson, P. B. and Kulver, S. (1970) *Microchim. Acta* 403.
Padowetz, W. (1969) *Microchem. J.* **14**, 110.
Phillips, W. M. (1948) *Plastics (London)* **12**, 587.
Phillips, W. M. (1949) Unpublished work at RAPRA.
Poulton, F. C. J. and Tarrant, L. (1951) *J. Appl. Chem.*, **1**, 29.
Roth, H. (1937) *Quantitative Organic Microanalysis of Fritz Pregl*, 3rd English Ed., translated by Daw, D. E., Churchill, London.

Scheele, W. (1961) *Rubber Chem. Technol.* **34**, 1306.
Schöniger, W. (1955) *Mikrochim. Acta* 123.
Schröder, E. and Waurick, U. (1960) *Plaste. u Kaut.* **7**, 9.
Schültze, M. (1939) *Z. Anal. Chem.* **118**, 241.
Schwarzkopf, O. and Schwarzkopf, F. (1969) *Characterization of Organometallic Compounds*, Tsutsui, M., Ed., Wiley, New York.
Shcherbacheva, M. A. (1957) *Chemical Methods of Analysis of Vulcanized Rubber*, Gosudarst. Nauch.-Tekh. Izdatel. Khim. Lit., Moscow.
Smith, A. J., Meyers, G. Jr. and Shaner, W. C. Jr. (1972) *Microchim. Acta* **2**, 217.
Smith, J. C. B. (1960) *Analyst* **85**, 465.
Spagnolo, F. and Malone, W. M. (1976) *J. Chromatog. Sci.* **14**, 52.
Stern, H. J. and Hinson, D. (1953) *India Rubber J.* **125**, 1010.
Steyermark, A., Lalancette, R. A. and Contreras, E. M. (1972) *J. Assn. Offic. Anal. Chem.* **55**, 680.
Stott, G. (1952) *Research Memorandum R382*, RAPRA.
Sundberg, O. E., Craig, H. C. and Parsons, J. S. (1958) *Analyt. Chem.* **30**, 1842.
Tidd, B. K. (1975) Unpublished work at MRPRA, see (1977) *Plastics and Rubber, Mater. and Applic.* **2**, 100.
Unterzaucher, J. (1940) *Berichte* **73**, 391.
Weber, C. O. (1894) *J. Soc. Chem. Ind.* **13**, 476.
Wetters, J. H. and Smith, R. C. (1969) *Analyt. Chem.* **41**, 379.
Willard, H. H. and Horton, C. A. (1950) *Analyt. Chem.* **22**, 1194.
Zimmermann, W. (1939) *Z. Anal. Chem.* **118**, 258.

Chapter 7

Solution Methods

In the early part of Chapter 3 the terms 'extraction', 'solution' and 'dissolution' were used and 'extraction' was defined as the procedure for removing organic additives from the bulk sample without simultaneously removing significant amounts of the polymer itself. It is now necessary to define and distinguish between the other two terms.

'Solution' is the procedure whereby one or more polymers are removed from the extracted bulk sample by the addition of a suitable solvent, or solvent mixture, in such a way that the polymer(s) may be recovered unchanged from the solution.

'Dissolution' allows a combination of degradation and solvent treatments so that the polymeric phase undergoes chemical reactions which increase its solubility but do not enable it to be recovered unchanged from the resulting solution.

Dissolution is dealt with in a degree of detail in the various chapters wherein its application is relevant, typically in Chapter 8 where a dissolution procedure involving oxidation in a solution of boiling *ortho*-dichlorobenzene (Barnes *et al.*, 1944; Clark and Scott, 1970) is described together with the 'dry' oxidation procedures of LiGotti (1972) and Carlson *et al.* (1970). These represent one end of the dissolution spectrum, with the polymer possibly being only partially solubilized, and suffering relatively little degradation, whilst at the other end is the completely destructive dissolution procedure as described by Kolthoff and Gutmacher (1950) where the object is not to recover the polymer but to free the carbon black and inorganic fillers of all polymeric materials (see Chapter 11). In addition to these methods already covered it is worth noting that Barnard (1956) describes a scheme for the complete removal of natural rubber from grafts with polymethylmethacrylate or polystyrene by the ozonolysis of the natural rubber. This scheme should be applicable to any system containing a mixture of polyunsaturated and saturated polymer and is particularly interesting because, having shown that there was degradation of the remaining polymer after all the natural rubber had been destroyed, he added di-*n*-butyl sulphide to the solution prior to ozonolysis. This, having a reactivity to ozone intermediate between that of natural rubber and the other polymers, provided a 'buffer time zone' so that he could be sure of complete ozonolysis of the former, without degradation of the latter. Here, however, we are concerned with the preparation of solutions and their significance to the analyst *beyond* their obvious purpose of providing a solution for subsequent analysis by one or more of the relevant techniques.

THEORETICAL CONSIDERATIONS

The parameters governing the solubility of a specific polymer in a solvent are numerous and complex and it would be extremely difficult, if at all possible, to predict the behaviour of every polymer with every solvent.

There are, however, several generalizations which may be made. A convenient distinction exists between the dissolving of polymers of high molar mass which are partly crystalline and those which have no crystallinity, i.e. are totally amorphous. In the case of the former, the crystalline regions have first to be melted and thus usually it is necessary to heat the solvent to a temperature near the melting-point of the polymer to effect a rapid solution; hence polyethylene is soluble in boiling benzene (80 °C) or *para*xylene at 100 °C. In the case of the latter, the polymer may be brought into solution at room temperature with relative ease.

In the case where a specific interaction occurs between the polymer and solvent, solution will be obtained. A typical example is the solution of Nylon 6.6 in cold formic acid due to hydrogen bonding between the two.

If we consider an amorphous polymer, and its interaction with a liquid, solubility will occur if the free energy of mixing (ΔG) is negative:

$$\Delta G = \Delta H - T\Delta S \qquad (7.1)$$

As the entropy of mixing (ΔS) is usually large and positive for the organic solvents we are concerned with, the sign of ΔG depends upon the size and sign of ΔH.

If there is a strong interaction between the polymer and the solvent, ΔH

will be negative and thus solution will occur. If, however, there are only dispersive forces involved in the solution we can use the expression of Hildebrand and Scott (1949) to determine the heat of mixing (ΔH):

$$\Delta H = \phi_s \phi_p (\delta_s - \delta_p)^2 \quad (7.2)$$

where ϕ_p and ϕ_s = volume fraction of polymer and solvent respectively; δ_p and δ_s = solubility parameter of polymer and solvent respectively and, δ is defined by

$$\delta = \sqrt{\frac{\Delta E}{V}} \quad (7.3)$$

where ΔE is the heat of vaporization, and $\Delta E/V$ (i.e. ΔE per unit volume) is the *cohesive energy density* (CED).

In this situation ΔH must be positive and therefore should be as small as possible so that ΔG can have as large a negative value as possible (see Eqn (7.1)); this will result if the solubility parameters of the solvent and polymer are equal.

We are now in a position, if we know the solubility parameters of a range of solvents and polymers, to see which are similar and thus probably (or possibly) will allow solution to occur, and which are widely different, making solution unlikely. Note that ΔH is related to the *square* of the difference between solubility parameters, thus there is no advantage in choosing a solvent with a solubility parameter value which is lower than that of the polymer.

Whilst there is intrinsically no difficulty in obtaining a CED value, and hence the solubility parameter, for any *solvent*, it cannot be measured directly for *polymers* as they do not vaporize. Several indirect methods for obtaining the data have been proposed by Gee (1943), Bristow and Watson (1958), and Small (1953) who used methods based on swelling, viscosity and the summation of individual values for small molecular groups, respectively. If we move from an ideal to realistic situation it is necessary to make some allowance for hydrogen bonding as in many polymer–solvent systems this will have a significant effect on solubility. Burrell (1955) divided solvents into three classes:

Poor hydrogen bonding capability
aliphatic/aromatic hydrocarbons, chlorinated hydrocarbons, nitroparaffins.

TABLE 7.1(a)

POLYMER SOLVENTS (ALPHABETICAL ORDER)

	Solubility parameter	*Hydrogen bonding*	*Dipole moment*
Acetone	10·0	9·7	2·9
Acetonitrile	11·9	6·3	3·9
Acetylacetone	10·8	8·4	3·1
n-Butanol	11·4	18·7	1·7
Carbon disulphide	10·0	0	0
Carbon tetrachloride	8·6	0	0
Chlorobenzene	9·5	1·5	1·6
Chloroform	9·3	1·5	1·2
Cyclohexane	8·2	0	0
Cyclohexanol	11·4	18·7	1·7
Cyclohexanone	9·7	11·7	2·7
n-Decane	6·6	0	0
Diacetone alcohol	9·2	13·0	2·5
N,N-Diethylacetamide	9·9	12·3	2·0
Diethyl ether	7·4	13·0	1·2
N,N-Diethyl formamide	10·6	11·7	2·0
N,N-Dimethyl formamide	12·1	11·7	2·0
Dimethyl sulphoxide	12·9	7·7	4·0
Dioxan	9·9	9·7	0
Dipropyl sulphone	11·3	7·7	4·5
Ethyl acetate	9·1	8·4	1·8
Ethyl alcohol	12·7	18·7	1·7
Ethyl benzene	8·8	1·5	0·6
Ethylene glycol	14·2	20·6	2·3
n-Hexane	7·3	0	0
2-Propyl acetate	8·4	8·6	1·9
Methyl alcohol	14·5	18·7	1·7
Methyl ethyl ketone	9·3	7·7	2·7
Methyl chloride	9·7	1·5	1·5
Nitrobenzene	10·0	2·8	4·3
Nitromethane	12·7	2·5	3·4
o-Dichlorobenzene	10·0	0	0
Piperidine	8·7	24·2	2·2
Pyridine	10·7	18·1	2·2
Styrene	9·3	1·5	0
Tetrahydrofuran	9·5	8·6	1·6
Toluene	8·9	4·5	0·4
Trichloroethane (1,1,2-)	9·6	1·5	1·2
Water	23·5	39·0	1·8
Xylene (commercial mixture)	8·8	4·5	0·4

After Mellan (1968).

TABLE 7.1(b)

POLYMER SOLVENTS (IN ORDER OF INCREASING δ)

Polymer	δ	*Polymer*	δ
n-Decane	6·6	Dioxan	9·9
n-Hexane	7·3	Diethyl acetamide	9·9
Diethyl ether	7·4	*o*-Dichlorobenzene	10·0
Cyclohexane	8·2	Acetone	10·0
2-Propyl acetate	8·4	Carbon disulphide	10·0
Carbon tetrachloride	8·6	Nitrobenzene	10·0
Piperidine	8·7	Diethylformamide	10·6
Ethyl benzene	8·8	Pyridine	10·7
Xylene (commercial)	8·8	Acetylacetone	10·8
Toluene	8·9	Dipropyl sulphone	11·3
Ethyl acetate	9·1	*n*-Butanol	11·4
Diacetone alcohol	9·2	Cyclohexanol	11·4
Chloroform	9·3	Acetonitrile	11·9
Methyl ethyl ketone	9·3	Dimethyl formamide	12·1
Styrene	9·3	Ethyl alcohol	12·7
Chlorobenzene	9·5	Nitromethane	12·7
Trichloroethane (1,1,2-)	9·6	Dimethyl sulphoxide	12·9
Tetrahydrofuran	9·5	Ethylene glycol	14·2
Cyclohexanone	9·7	Methyl alcohol	14·5
Methylene chloride	9·7	(Water)	23·5

Moderate hydrogen bonding capability
esters, ketones.
Strong hydrogen bonding capability
alcohols, acids, amines.

He then used these, together with solubility parameters, to select solvents for particular polymers. More recently Crowley *et al.* (1966, 1967) have proposed a three-parameter concept of solubility, based on solubility parameter, hydrogen bonding and dipole forces. Table 7.1 lists a range of common solvents for polymers together with numerical values for these parameters. Hydrogen bonding values of less than 4 can be considered poor, 4–10 moderate, and greater than 10 strong (Mellan, 1968). Table 7.2 provides solubility parameter data for a range of polymers from a variety of sources, including Mellan (1968), Small (1953), Bristow and Watson (1958) and Brandrup and Immergut (1975). The spread includes results obtained by different authors using several methods and allows for

TABLE 7.2

SOLUBILITY PARAMETERS OF POLYMERS

Polymer	*Range of quoted values*	Elastomer	*Range of quoted values*
Polyethylene	7·7–8·2	NR/PI	8·0–8·5
Polypropylene	9·4	SBR (4–40% S)	8·1–8·6
Polystyrene	8·5–10·6	BR	8·0–8·6
Polyacrylonitrile	12·3–12·8	IIR	7·5–8·0
Polyvinylacetate	8·5–9·5	CR	8·1–9·4
Polyvinylchloride	8·5–11·0	NBR (18–30% ACN)	8·7–9·3
Polymethylmethacrylate	8·9–12·7	NBR (40% ACN)	10·4–10·5
Styrene acrylonitrile	10·6–11·2	Silicone(s)	7·0–11·5
Ethylene vinylacetate	7·8–10·6	Chlorinated rubber	9·4
		Thiokol(s)	9·0–10·0
		Polyurethane(s)	9·8–10·3
		EPDM(s)	7·5–8·6

differences due to the different hydrogen bonding properties of the solvents used.

Table 7.3 shows the solubility of a range of polymers and plastics using a scale of 1 to 6 defined in a footnote to the table.

Theta Temperature

It will be apparent from inspection of Tables 7.1 and 7.2 that many solvents *should* dissolve a particular polymer and indeed may well do so. It is then necessary to choose a particular one. Leaving aside specific applications, which will be mentioned later, there are certain theoretical considerations which will influence the choice.

If one considers a polymer dissolved in any solvent, the polymer chain will be extended, to a degree depending upon the polymer–solvent interaction, so that the mean end-to-end chain length (l) is greater than would be predicted for a random coil configuration (d). The ratio l/d is known as the *expansion factor*, α.

As the solution cools the configuration approaches that of a random coil ($\alpha \rightarrow 1$) and at a specific temperature for each polymer–solvent system one reaches the *theta temperature* at which α is unity. This is also the critical miscibility temperature for a polymer of infinitely high molar mass and may

TABLE 7.3

SOLUBILITY OF POLYMERS IN VARIOUS SOLVENTS

Solvent	*A*	*B*	*C*	*D*	*E*	*M*	*N*	*O*	*P*	*Q*
Acetone	6	1	6	6	6	5	1	3	1	3
Acetonitrile	6	5	6	6	6	6	6	4	1	6
n-Butanol	6	6	6	1	5	6	6	6	5	6
Carbon disulphide	1	4	1	1	1	6	5	1	4	1
Carbon tetrachloride	1	5	1	1	1	6	5	1	1	1
Chlorobenzene	1	1	1	1	1	4	1	1	1	1
Chloroform	1	1	1	1	1	5	1	1	1	1
Cyclohexane	1	6	1	1	1	6	6	1	4	6
Cyclohexanol	5	5	5	1	5	6	6	6	1	6
Cyclohexanone	1	1	1	1	1	1	1	1	1	1
Diacetone alcohol	6	1	6	6	6	6	5	5	1	3
o-Dichlorobenzene	1	1	1	1	1	4	1	1	1	1
Diethyl ether	1	6	1	1	1	6	6	3	4	6
Dimethyl formamide	6	1	5	5	6	4	1	1	1	1
Dimethyl sulphoxide	6	5	6	6	6	2	1	6	1	1
Dioxan	1	1	4	5	1	3	1	1	1	1
Ethyl acetate	3	1	5	5	5	6	1	1	1	1
Ethyl alcohol 96%	6	6	6	6	6	6	6	6	1	6
Ethyl benzene	1	1	1	1	1	6	5	1	1	1
Ethylene glycol	6	4	6	6	5	6	6	6	6	6
Hexane	3	6	1	1	1	6	5	3	6	6
Methyl alcohol	6	5	6	6	6	6	6	6	1	6
Methyl ethyl ketone	4	1	4	1	4	3	1	1	1	1
Methylene chloride	1	1	1	1	1	2	1	1	1	1
Nitrobenzene	1	1	1	6	1	2	1	1	1	1
Nitromethane	6	1	6	6	6	6	6	5	1	6
Styrene	1	1	1	1	1	5	2	1	1	1
Tetrahydrofuran	1	1	1	1	1	1	1	1	1	1
Toluene	1	1	1	1	1	5	2	1	1	1
Trichlorethane	1	1	1	1	1	6	5	1	1	1
Pyridine	1	1	5	4	1	2	1	1	1	1
Xylene	1	1	1	1	1	6	4	1	4	1

1 = soluble, 2 = virtually soluble, 3 = strongly swollen, 4 = swollen, 5 = marginally swollen, and 6 = no visible effect.
Polymer Code:

Elastomers	*Plastics*
A—SBR	M—PVC
B—NBR	N—PMMA
C—NR/IR	O—PS
D—IIR	P—PV acetate
E—BR	Q—Chlorinated PP

After Hanson (1967).

be determined by measuring the critical miscibility temperature for a series of polymer fractions, and extrapolating to infinite molecular weight.

It will be apparent that if the theta temperature of a particular polymer–solvent system is near room temperature, the polymer will be near to precipitation and thus the solvent will be a bad one. A good solvent will have its theta temperature substantially below zero.

Mixtures of two solvents may be used to adjust the theta temperature of a particular system and indeed this is the principle of precipitation. If the theta temperature of a solvent is adjusted until it is close to room temperature, the solvent for that particular polymer of that particular molecular weight is called the *theta solvent*.

Flory (1942) and Huggins (1942) introduced the *Flory–Huggins interaction constant* (χ) in order to define numerically the concept of 'goodness of solvent' as described above. The critical value of χ is 0.5 and for a given polymer–solvent system χ must be below 0.5 for solubility to occur.

The practical significance of the constant may be appreciated by considering the determination of molar mass by membrane osmometry (see Chapter 9).

The working equation is given:

$$\frac{h}{c} = \frac{K}{\bar{M}_n} + bc \tag{7.4}$$

where

$$b = (0{\cdot}5 - \chi) \tag{7.5}$$

For a theta solvent (when χ approaches 0.5)

$$\frac{h}{c} = \frac{K}{\bar{M}_n} \tag{7.6}$$

where K is the product of the *gas constant* (R) and the absolute temperature (T). Thus no calibration plot is required and the molar mass ($\bar{M}_n$) may be calculated directly by measuring the osmotic pressure (h) for one solution of concentration (c).

At the other extreme, the 'better' the solvent the smaller will be χ, the steeper the slope of the h/c *vs.* c graph and the greater the error of measuring the intercept. One should therefore choose a solvent which has its theta temperature near to the operating temperature of the osmometer so that the graph is relatively flat. Gee (1940, 1944) used a mixed-solvent system of benzene/methanol to approach these conditions in the analysis of natural rubber, but the use of mixed solvents is a potentially dangerous situation because of differences in diffusion rates and relative volatilities.

Guidelines to Solubility

From the mass of published empirical data on solubility, together with a consideration of the scientific reasons put forward in this chapter, Billmeyer (1962) listed some generalities:

(1) Similarity of chemical and structural makeup of solvent and polymer favours solubility.
(2) Solubility is inversely related to molar mass.
(3) Solubility is inversely related to melting-point.
(4) The presence of polar groups in a polymer will reduce solubility in non-polar solvents (increase in polymer–polymer interaction).
(5) Solubility of copolymers is a function of the relative amounts of each monomer.

These were later expanded by Hanson (1967) to include systems containing two or more polymers:

(1) Polymers must be individually soluble in a solvent for them to be mutually soluble in it.
(2) Individual solubility does not guarantee mutual solubility, particularly if
  (a) both are of high molar mass;
  (b) the solvent is a poor one for one of the polymers;
  (c) the polymers do not have similar and overlapping δ ranges.
(3) As the concentration increases, one polymer in a mutually soluble system can become insoluble.

PRACTICAL CONSIDERATIONS

Perhaps the most important practical consideration for an analyst is that of quickly deciding which solvent is likely to dissolve a particular polymer and therefore Table 7.3 is included, but the choice of which particular solvent to use must obviously be governed by many factors, not least the reason for obtaining the solution. Thus for many elastomers the preparation of a cast film for infrared spectroscopic study is carried out from a chloroform solution but this would be no use if the ultraviolet spectrum of the solution were required. The latter could possibly be obtained from a solution in tetrahydrofuran but, conversely, this is a poor solvent from which to cast a film for infrared examination as it is too volatile and gives bubbles in the polymer film.

Preparation of a Solution

Certain points are worth bearing in mind when the appropriate solvent has at last been chosen. Polymer solutions are often unstable, with the polymer often being prone to oxidation. This is not only true of unsaturated polymers such as natural rubber (Bateman, 1954) but also of saturated polymers whose viscosity has long been known to be time dependent, to a different extent, in different solvents (Mead and Fuoss, 1946; Morrison *et al.*, 1946). The practice of adding stabilizers is widespread but this must obviously depend upon the purpose of obtaining the solution, and a balance must be struck between the time taken for solution to become complete and the time scale over which the polymer molar mass may be considered to be stable.

Most polymers dissolve slowly and, whilst this can be hastened by heating or agitation, care should be taken as Grassie (1956) remarked that there are many references to the effect that 'overenthusiastic' shaking may result in the depolymerization of the polymer. If the sample of polymer is a powder, and this is added to a warm solvent, it may 'lump up', resulting in a very long time to complete solution; it is better to cool the solvent to a little above its freezing-point, add the powder with gentle agitation so that it is completely wetted, and then raise the temperature to speed up the solution processes.

Removal of Solvent

In many applications it is required to remove a solvent after preparing the solution. Two typical examples would be the preparation of a cast film for infrared spectroscopic analysis and the quantitative measurement of the concentration of a solution after filtration to remove carbon black, filler or gel. In either case it is imperative that the solvent be removed completely, and this is by no means simple. Haslam and Willis (1965) illustrated an extremely simple apparatus (Fig. 7.1) which could replace the comparatively expensive vacuum hotplate in many laboratories, but did not comment on this problem.

If the solvent has a high vapour pressure when frozen, the technique of freeze drying may be used to advantage. The solution is rapidly frozen and the solvent removed under high vacuum, without allowing melting to

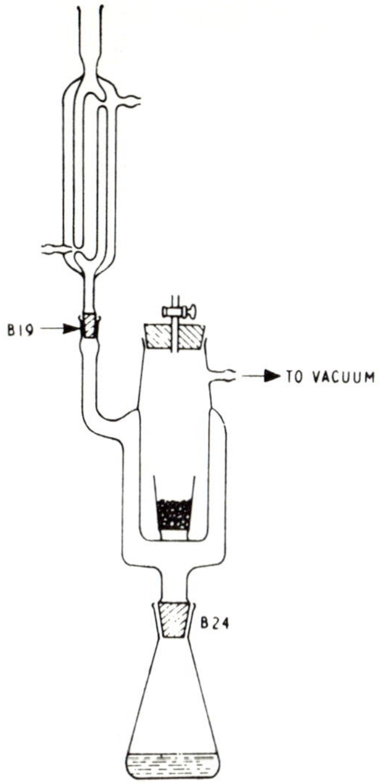

FIG. 7.1. Apparatus for drying washed polymers. (Courtesy Haslam and Willis, 1965.)

occur. This leaves the polymer in a very fine expanded form, the ideal physical state from which to remove final traces of the solvent.

Evans *et al.* (1960) examined in detail the solvent-retaining properties of cyclized natural rubber with a wide range of halogenated solvents, benzene and carbon disulphide, and showed how substantial amounts remain after 'drying'. He also illustrated the effect of altering the level of cyclization on the retention of carbon tetrachloride under fixed conditions. Both sets of data are illustrated, in Table 7.4 and Fig. 7.2. Other particularly noteworthy examples of polymer–solvent pairs which are difficult to separate are polyvinyl chloride–tetrahydrofuran and polystyrene-toluene/benzene.*

As the solvents become higher boiling the situation is exacerbated, and one way round this is to wash the nominally dried samples with a low-

*Benzene is a carcinogen and should never be used if an alternative, such as toluene, is available.

TABLE 7.4

SOLVENT RETENTION OF FILMS OF CYCLIZED NATURAL RUBBER

Solvent	*Molecules solvent retained per 100 moles of C_5H_5 units in polymer*	
	25 °C	*100 °C*
CCl_4[1]	13·42	8·55
$CHCl_3$	9·77	1·64
CH_2Cl_2		0·1
CH_3CHCl_2	7·76	2·89
CH_2ClCH_2Cl	9·75	5·54
$CHCl_2CHCl_2$	13·1	
$CHCl{=}CHCl$	4·31	4·1
$CCl_2{=}CCl_2$	3·8	6·17
CBr_4		2·05
$CHBr_3$	11·7	8·3
CH_2Br_2	10·5	7·3
C_6H_6	2·3	
CS_2	1·84	

[1]After 24 h at room temperature and atmospheric pressure, solvent retention for CCl_4, 19·75; after 3 days at 20 °C, and 10^{-4} mm Hg, 17·75; after 5 h at 155 °C and 10^{-4} mm Hg, 4·56.

The films were cast from various solvents at atmospheric pressure and room temperature and pumped for 3 h at 25 °C and 100 °C, at 0·5 mm Hg. In all cases the films were less than 0·2 mm thick.

boiling liquid which is miscible with the solvent, but which will not dissolve the polymer.

If thermogravimetric analysis (TGA) is available it is well worth while carrying out routine checks on all polymers which are supposed to be 'dry'. Any residual solvent present will be volatized at a relatively low temperature in an atmosphere of nitrogen (<250 °C).

Many published infrared spectra show the polymer to be contaminated with solvents and it is often advisable to cast the films concurrently from different solvents so that one is not misled by an unassigned band. It should also be noted that solvent bands do not always occur in identical positions in different matrices, thus the 14·85 μm band of benzene, as it appears in cyclohexane solution, shifts to 14·75 μm when present as a residual peak in polystyrene.

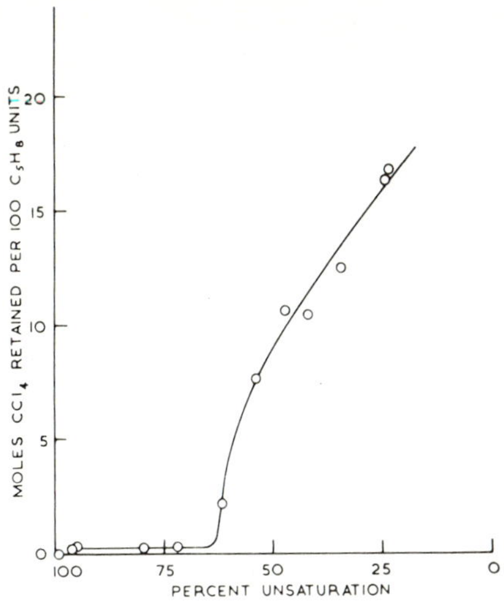

FIG. 7.2. Dependence of solvent retention by films of cyclized rubber on their unsaturation. (Courtesy *J. Appl. Polym. Sci.*)

Selective Solution

There are many occasions when one may wish only to dissolve one polymer from a blend. One is illustrated in Chapter 8, where the natural rubber has been selectively dissolved from a natural rubber–polypropylene blend so that its location becomes visible under the electron microscope. Another is in the examination of laminates, when a knowledge of the total polymer composition and the polymeric constituents of the surface may enable one to dissolve selectively one of the polymers enabling the laminate composition to be obtained in full detail.

Perhaps one of the largest areas where this technique is useful is in the examination of block copolymers or thermoplastics of the rubber–plastic mixture type, when successive extractions with a range of solvents such as methyl alcohol–acetone–ethylene dichloride–tetrahydrofuran followed by infrared spectroscopic analysis of the polymer soluble in each successive solvent will provide a mass of data on the interrelationships between the various monomer species known to be present.

A more sophisticated procedure for multiple extractions has been described by Ceresa (1962) who analysed block and graft copolymer systems such as polyvinyl acetate–polyethylene and natural rubber–polymethylmethacrylate using solvent pairs consisting of a solvent and non-solvent for each polymer. Successive extractions with progressively different mole fractions of the two solvents afforded separation into the two homopolymers and block/graft copolymers of proportionately different compositions.

Selective solution can give misleading results in certain circumstances, the most common one probably being where one of the polymers exists as a discrete phase in a matrix of the other, rather than the two being co-continuous. Attempts to dissolve the discrete phase from the insoluble matrix will often be frustrated due to the exceedingly slow diffusion of solvent and the resulting polymer solution through the sample (see Chapter 3). This can be increased by using a solvent mixture such that bulk swelling occurs, or by using the converse technique of selective precipitation.

Selective Precipitation

This undoubtedly affords the purest materials from any complex blend or mixture of copolymer and polymer and consists of preparing a solution of the total polymeric phase prior to selectively precipitating the individual homopolymers by the addition of the appropriate non-solvent. A dual precipitation will remove both homopolymers, leaving the graft (if present), with its intermediate solution characteristics, in solution.

This process can be extended until it becomes *fractional precipitation*, and it is then used to separate different molecular weight fractions of a particular homopolymer, or graft/block copolymers of different, and progressively graded, proportions of each polymer.

REFERENCES

Barnard, D. (1956) *J. Polym. Sci.* **22**, 213.

Barnes, R. B., Williams, V. Z., Davis, A. R. and Giesecke, P. (1944) *Ind. Eng. Chem. Anal.* **16**, 9.

Bateman, L. (1954) *Quart. Rev. Chem. Soc.* **8**, 147.

Billmeyer, F. W. (1962) *Polymer Science*, Interscience, New York.

Brandrup, J. and Immergut, E. H. (1975) *Polymer Handbook*, 2nd ed., Wiley, New York.

Bristow, G. and Watson, W. F. (1958) *Trans. Faraday Soc.* **54**, 1731, 1742.

Burrell, H. (1955) *Off. Dig. 27* No. 369, 726.

Carlson, D. W., Ransaw, H. C. and Altenau, A. G. (1970) *Analyt. Chem.* **42**, 1248.

Ceresa, R. J. (1962) *Block and Graft Copolymers*, Butterworths, London.

Clark, J. and Scott, R. (1970) *Rubber Chem. Technol.* **43**, 1332.

Crowley, J. D., Teague, G. S. Jr. and Lowe, J. W. Jr. (1966) *J. Paint Technol.* **38** (496), 269.

Crowley, J. D., Teague, G. S. Jr. and Lowe, J. W. Jr. (1967) *J. Paint Technol.* **39** (504), 19.

Evans, M. B., Higgins, G. M. C., Lee, D. F. and Watson, W. F. (1960) *J. Appl. Polym. Sci.* **4**, 367.

Flory, P. J. (1942) *J. Chem. Phys.* **10**, 51.

Gee, G. (1940) *Trans. Faraday Soc.* **36**, 1141.

Gee, G. (1943) *Trans. IRI* **18**, 266.

Gee, G. (1944) *Trans. Faraday Soc.* **40**, 462.

Grassie, N. (1956) *Chemistry of High Polymer Degradation Processes*, Butterworths, London.

Hanson, C. M. (1967) *J. Paint Technol.* **39** (505), 104.

Haslam, J. and Willis, H. A. (1965) *Identification and Analysis of Plastics*, Iliffe, London.

Hildebrand, J. and Scott, R. (1949) *The Solubility of Non-Electrolytes*, 3rd ed., Reinhold, New York.

Huggins, M. L. (1942) *Ann. N.Y. Acad. Sci.* **43**, 1.

Kolthoff, I. M. and Gutmacher, R. G. (1950) *Analyt. Chem.* **22**, 1002.

LiGotti, I. (1972) *Paper presented at 20th meeting, ISO TC45-WG1,* Cologne.

Mead, D. J. and Fuoss, R. M. (1946) *J. Am. Chem. Soc.* **64**, 277.

Mellan, I. (1968) *Compatibility and Solubility*, Noyes Development Corp., London.

Morrison, J. A., Homes, J. M. and McIntosh, R. (1946) *Canad. J. Res.* **24B**, 179.

Small, P. A. (1953) *J. Appl. Chem.* **3**, 71.

Chapter 8

Instrumental Polymer Analysis

Whilst there is no doubt that many of the tests described earlier are extremely useful in the analysis of an unknown vulcanizate, and it would be foolhardy to ignore some of the simpler ones such as the 'burn' test, the analyst with instrumental facilities will find that these rapidly tend to supersede the classical chemical methods in the routine analysis of polymeric materials.

The purpose of the instrumental examinations described in this chapter is to identify the particular elastomer, or blend of elastomers, in a sample and, in the case of a blend, to quantify the component ratio. Although numerous techniques have been covered in the literature of the last forty years two stand out as being of prime importance—infrared spectroscopy (IR) and gas chromatography (GC). More recently thermal techniques such as derivative thermogravimetry (DTG), differential scanning calorimetry (DSC) and differential thermal analysis (DTA), together with nuclear magnetic resonance (NMR), have made increasingly significant contributions in the areas of polymer identification whilst the electron microscope (both scanning and transmitting) has added substantially to an understanding of polymer blend morphology. It should be emphasized that this chapter is not concerned with the microstructure of the elastomer, a subject which will be discussed in Chapter 9, and thus many of the NMR data in the literature will be reserved until then.

INFRARED SPECTROSCOPY (IR)

The infrared examination of a rubber, or rubber-like material, can be separated into three distinct parts:

The presentation of the sample to the instrument.
The running of the spectrum.
The interpretation of the spectral data.

Sample Preparation and Presentation

There are only two ways in which the infrared spectrum of any sample may be obtained: by shining light through it, or by bouncing or reflecting light off its surface. In both cases the spectrometer compares the energy spectrum of the infrared source before and after absorption by the sample and generates a spectrum characteristic of the sample showing the wavelengths at which it absorbs energy. Carbon black interferes by absorbing substantial amounts of energy in the full spectral range whilst inorganic fillers superimpose their spectra on that of the polymer or polymer blend. Ideally, therefore, the polymer should be separated from

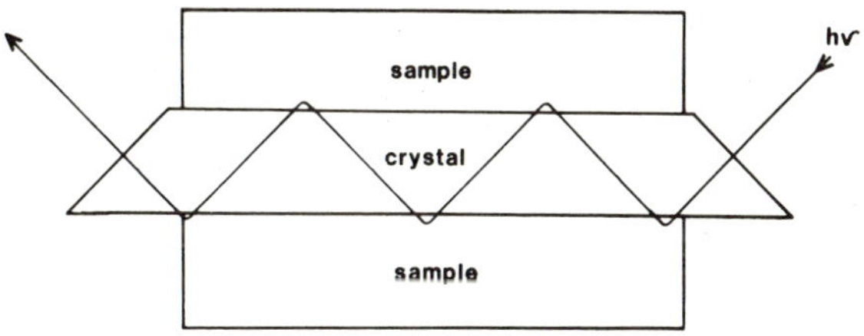

FIG. 8.1. Diagrammatic representation of MIR.

the fillers before its spectrum is obtained, although Corish (1960) has obtained recognizable spectra from microtomed sections of filled vulcanizates. Excepting this, the sample may be offered to the instrument as a film of the polymer itself, as a partially degraded but still polymeric film, or as a liquid film of the severely degraded polymer. The last of these generally produces a spectrum quite different from that of the parent polymer, but nevertheless characteristic of it, and thus perfectly adequate for identification.

If we first consider the two ways of obtaining spectra, that of shining the infrared light through the sample is self descriptive, provided one has a thin film of material some 10–50 μm thick. The second option, of reflecting the light off the surface, was initially proposed by Fahrenfort (1961) and is illustrated diagrammatically in Fig. 8.1. It is correctly described as multiple internal reflectance spectroscopy (MIR); the optional name, attenuated total reflectance (ATR), should really be confined to a crystal giving only one internal reflection.

It is based on the phenomenon of the total internal reflection of light at the interface between media of differing refractive indices. In the system illustrated the rhombohedral crystal is made of a material of high refractive index and, with the elastomer pressed hard against it, produces a spectrum of the elastomer similar to a transmission one. The major difference is that, because in reflectance spectroscopy penetration of the infrared light is wavelength dependent, it is greater at higher wavelengths (lower wavenumbers) and gives a relative increase in spectral intensity as one progresses through the spectrum from 2·5 μm (4000 cm^{-1}) to 25 μm (400 cm^{-1}). The particular advantages of MIR are, first, in examining thin films of elastomeric materials, or packaging, as described by Leukroth (1970), when the presence of laminates or surface treatments can be deduced by comparing reflectance spectra with the more usual transmission ones, and, second, in the examination of pyrolysis products as will be described later.

In all cases of sample preparation, the material under examination should have been extracted prior to spectroscopic analysis. Subsequent treatment will depend upon the type of sample. Coagulums (*ex* latex), raw rubbers and thermoplastic materials can best be examined initially as pressed films obtained by hot pressing at ~180 °C for a few seconds before mounting directly in the instrument.

An alternative method of preparing a sample is as a thin film cast from solution. This can be done either on a rock-salt plate from a suitable solvent, or on to a sheet of glass or mercury, from which it can be lifted on a frame for examination. It is obvious that any chosen solvent must meet certain criteria:

It must dissolve the sample entirely.
It must be inert towards the sample.
It must be completely volatile.
It should leave a smooth film (i.e. must not evaporate too quickly).

Typical solvents are chloroform, dichloromethane, toluene and tetrahydrofuran. The analyst should acquaint himself with the absorption bands of the solvent of choice to be sure that the cast films are solvent free and that the spectrum is not misinterpreted. It should be stated that whilst, as a general principle, complete solution is a prerequisite for a good solvent, selective solution and dissolution, as discussed in the previous chapter, have great significance. If a sufficiently thin film cannot be pressed, and dissolution techniques are inappropriate, then the reflectance procedure can be used.

Vulcanizates are not amenable to direct solution techniques but can be solubilized if a degradation step is carried out first. In one of the earliest papers on the examination of rubbers by infrared, Barnes *et al.* (1944) separated the polymer by dissolution in a high boiling solvent (*ortho*-dichlorobenzene) and removed the carbon black and other fillers by filtration. A similar procedure was described by Dinsmore and Smith (1948) in an extremely detailed paper tabulating solvents of preference for a range of gum and vulcanized polymers. The method is, however, time consuming and Dinsmore estimates 6 man hours per sample on a routine basis! More recently Clark and Scott (1970) improved the rate and extent of dissolution by the addition of Pepton 22 (2,2′-dibenzamidodiphenyl disulphide) to the *ortho*dichlorobenzene before refluxing for 7 hours prior to a complex work-up procedure. In all these cases dissolution is due to oxidative degradation of the polymer. A typical experimental procedure is as follows:

> Samples of milled or chopped extracted vulcanizate are sieved and particles smaller than 0·5 mm used for the decomposition. 2 g of sample, together with 50 cm^3 *ortho*dichlorobenzene (ODCB), are boiled under reflux until dissolution is complete and the mixture is cooled and added to 50 cm^3 toluene in a 250 cm^3 beaker. Mineral fillers are removed by centrifugation but, if carbon black is present, or the solution is still cloudy, a filter aid may be added and the solution filtered. Alternatively the solution may be chromatographed down a short (10 cm × 1 cm) column of silica gel, eluting with toluene. In either case the clear solution is then evaporated to low volume under nitrogen and a thin film of the elastomer cast from this solution.

The spectrum is then recorded with the particular advantage that the structural features of the elastomer are not destroyed and the spectrum will be closely similar to that of the unvulcanized elastomer. The main disadvantage is that one must wait for complete dissolution of the sample to be certain that all components of a blend are solubilized and in some instances the prolonged times which this entails can result in the complete oxidative degradation, and hence loss, of some of them.

An alternative to 'wet' degradation is 'dry' degradation and was described by LiGotti (1972) and Carlson *et al.* (1970). In their procedure the sample is heated in air prior to dissolution according to the following scheme:

> About 2 g of milled, extracted and dried vulcanizate are placed in a

test-tube which is plugged with cotton wool and placed in an oven at 200 °C for 10 min. The sample is then transferred to a beaker containing 50 cm^3 of trichlorethylene and this is heated in a boiling water bath for 30 min. After this the solution is filtered, the clear solution evaporated to low volume under nitrogen and a film cast for infrared examination.

Because the heating time is much shorter than in the wet method, there is less oxidation and the spectrum is even nearer that of the unvulcanized polymer.

The main drawback to this method is again the question of the representative nature of the sample due to the different thermal stabilities of the polymers which could be present in a blend, particularly since, quite deliberately to avoid release of carbon black and fillers, complete dissolution is not aimed at. The advantage is that it is a relatively quick method for a single polymer, or blend of polymers of comparable thermal stabilities.

The most certain way to be sure that a polymer, or blend of polymers, has been completely removed from the filler matrix is to fragment the polymer into volatile species of low molar mass which can be collected and characterized. This procedure, usually carried out either under vacuum or in an inert gas such as nitrogen, is pyrolysis and was originally proposed as a routine method for the identification of vulcanized elastomers by Harms (1953) who heated the vulcanizate, in a horizontal test-tube, in a flame and collected the pyrolysate which condensed in the mouth of the tube for spectroscopic analysis. Although crude, this is still a relatively effective *qualitative* method in the hands of an expert. However, in the same year Kruse and Wallace (1953) recommended an aluminium block, heated to 440–465 °C, with holes drilled for thermometer and test-tube, as a more stable heat source. Cleverley and Herrmann (1960) went a step further and used a temperature gradient up to 200 °C to extract materials of low molar mass thermally before increasing the temperature to 400 °C to pyrolyse the polymer. More recently Gross (1975) has reviewed degradation methods of dissolution and pyrolysis and published the spectra of many polymers obtained by both methods whilst a comprehensive set is illustrated in ISO 4650. (To be published during 1983.)

There appears to be general agreement between authors that variability in the infrared spectra of the pyrolysates of the same polymer or, more particularly, of polymer blends, is due to variation in the temperature of pyrolysis, thus Lerner and Gilbert (1964) note that the pyrolysis of an NR–SBR blend gives different ratios of the 11·0 μm (910 cm^{-1}) and 11·25 μm (888 cm^{-1}) bands (due respectively to pyrolysis products of polybutadiene (*ex* SBR) and polyisoprene) for pyrolysis in the 400–500 °C range from those obtained by pyrolysis at 850–950 °C. In order to obtain quantitative data it is essential that a pyrolysis apparatus capable of reproducible operation is constructed in which the atmosphere, temperature gradient of heating and final temperature are closely controlled. If it is not intended to trap all of the pyrolysate, a standardized trapping system should be devised. The current standard in this field (BS 4181:1967) uses the test-tube/bunsen method and, whilst the draft ISO standard ISO/DIS

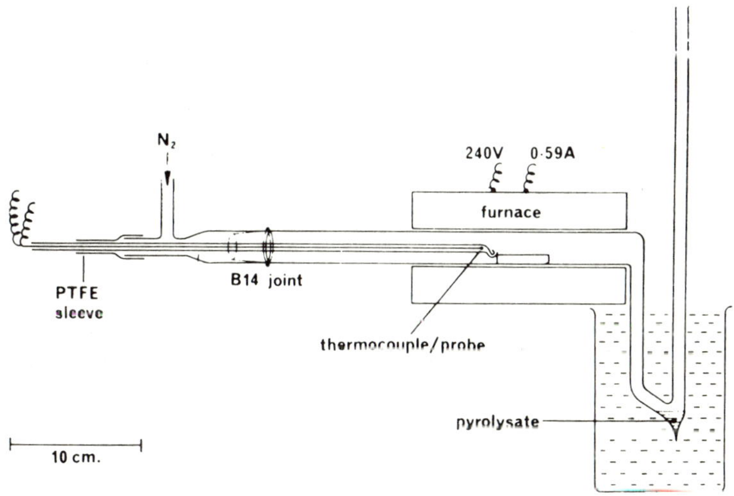

FIG. 8.2. Pyrolysis apparatus (Higgins and Loadman, 1970, 1971). (Courtesy MRPRA.)

4650 also allows this, it offers the option of an electrically heated furnace with the pyrolysate condensed in an open tube. Mackillop (1968) describes an electrically heated furnace, with pyrolysis at up to 390 °C ± 10 °C occurring in a vacuum and collection of all the condensate, whilst Higgins and Loadman (1970, 1971) used a simple system whereby the furnace is maintained at a constant temperature (535 °C) and the sample (*ca.* 100 mg) is always inserted to the same position so that after 15 min pyrolysis the temperature is constant at 515 ± 5 °C. The apparatus is illustrated in Fig. 8.2. The temperature of 515 °C was found to be optimum for this particular system as below this there was incomplete pyrolysis of the SBR, and above

550 °C a reduction in the observed aromatics, presumed to be due to loss of styrene in the volatile, uncondensed fraction.

The variation in spectra of pyrolysates obtained at different temperatures has been used by Dawson and Sewell (1975) to differentiate between natural and synthetic polyisoprene. Black filled vulcanizates of the two elastomers give quite different spectra when pyrolysed at 350 °C, and in the case of blends this effect can be quantified. Analogous differences can be found in thermogravimetric analyses when carried out at a slow heating rate. This will be considered in more detail later in this chapter.

Perhaps the most sophisticated equipment available for infrared analysis of the total pyrolysate is a combined pyrolysis and gas cell, with one side wall of the gas cell an MIR crystal, the temperature of which can be independently controlled. It is thus possible to examine the gas phase products and liquid condensate of the same pyrolysate after one experiment (see Truett, 1977).

The Running of the Spectrum

Most modern infrared spectrophotometers are so pre-standardized, or microprocessor controlled, that there is no real need to discuss this further. The only precautions which should be taken are to make sure that the scan speed is slow enough for very sharp peaks not to be distorted, and to check that no bubbles have materialized in the liquid film whilst the spectrum is being run.

Interpretation of Spectral Data

Spectra will either be of a polymer itself, or of its pyrolysate; in both cases matching to reference spectra gives the quickest means of identification of both blends and single polymer systems.

It is unfortunately contradictory that there will often be small differences between nominally identical pyrolyses of the same polymer, whilst the analyst will be on the look-out for slight differences from 'authentic' spectra as an indication of the presence of a minor component. This can be partly overcome by building up one's own reference spectrum library, under standard conditions, but otherwise it remains a problem which only experience can resolve. Although a 'fresh' analyst will feel overwhelmed by the vast number of published spectra he will soon realize that most of the samples which he analyses fall into a relatively small group, and only occasionally will he have to do a full search.

Undoubtedly one of the best sources of film spectra is the *Atlas* of Hummel and Scholl (1969). Large numbers of spectra are also found in

Haslam *et al.* (1972) whilst, as already mentioned, Gross (1975) illustrates both film and pyrolysate spectra, and BS 4181-1967 pyrolysates. Cleverley (1979) has published the spectra of a wide range of packaging films.

There are, inevitably, problems with the identification of most polymers at levels of less than 20% although some, such as SBR, will be visible at below 5%. Chloroprene sometimes decomposes completely leaving no pyrolysate, whilst halobutyl will be indistinguishable from the butyl rubber

TABLE 8.1

QUANTITATIVE POLYMER BLEND ANALYSES

Polymer system	*Authors*	*References*
SBR–BR	Clark, J. K. and Scott, R. A.	*J. Appl. Polym. Sci.* **14**, 1 (1970)
	Higgins, G. M. C. and Loadman, M. J. R.	*NR Technol.* **10**, 1 (1970)
	Mills, W. and Jordan, M. J.	*J. IRI* **4**, 60 (1970)
	Mackillop, D. A.	*Analyt. Chem.* **40**, 607 (1968)
BR–NR/IR	Higgins, G. M. C. and Loadman, M. J. R.	*NR Technol.* **10**, 1 (1970)
	Mackillop, D. A.	*Analyt. Chem.* **40**, 607 (1968)
	Binder, J. L.	*Appl. Spectroscopy* **23**, 1 (1969)
NR–SBR–BR	Higgins, G. M. C. and Loadman, M. J. R.	*NR Technol.* **10**, 1 (1970)
	Mills, W. and Jordan, M. J.	*J. IRI* **4**, 60 (1970)
	Mackillop, D. A.	*Analyt. Chem.* **40**, 607 (1968)
	Jasper, B. T.	*J. IRI* **3**, 72 (1969)
EPDM	Takeuchi, T., Tsuge, S. and Sugimura, Y.	*Analyt. Chem.* **41**, 184 (1969)
	Brame, E. G. Jr., Barry, J. E. and Toy, F. J. Jr.	*Analyt. Chem.* **44**, 2022 (1972)
	Gardner, I. J., Cozewith, C. and Verstrate, S.	*Rubber Chem. Technol.* **44** (4), 1015 (1971)
(Termonomer)	Altenau, A. G., Headley, L. M., James, C. O. and Ransaw, H. C.	*Analyt. Chem.* **42**, 1280 (1970)
ABR	Scism, A. J.	*Analyt. Chem. Acta* **42**, 177 (1968)
NBR–ABR	Ruzicka, B. and Krotki, E.	*Chem. Anal. (Warsaw)* **16**, 1207 (1971) (Polish)
NR–IR	Dawson, B. and Sewell, P. R.	*Rubber Ind.* **9** (5), 180 (1975)
cis/trans NR/IR (not pyrolysis)	Cunneen, J. I., Higgins, G. M. C. and Watson, W. F.	*J. Polym. Sci.* **15**, 1 (1959)

itself. EP or EPDM could well be missed at levels of up to 30% in NR whilst polybutadiene may be confused with chlorosulphonated polyethylene. For these reasons it is emphasized that it is always advisable to use chemical tests as well as pyrolysis in any polymer identification, and thermogravimetric analysis will also provide further data on the presence of blends.

Quantitative analysis of polymer blends by pyrolysis–infrared spectroscopy constitutes a vast area of research with many hundreds of publications now existing. It is completely beyond the scope of this book to consider these in detail, but a table of references is provided which an analyst might find useful.

PYROLYSIS–GAS CHROMATOGRAPHY (PGC)

When analysing a sample by pyrolysis–infrared spectroscopy the most important problems encountered are in selecting appropriate pyrolysis apparatus and conditions to provide informative and reproducible data. Pyrolysis–gas chromatography not only has these problems but, in addition, the chromatographic conditions have to be optimized to provide worthwhile data on as broad a range of polymers as possible. It seems probable that this wide range of variables, leading to great difficulty in correlating interlaboratory data, has provided one of the reasons why pyrolysis–gas chromatography has not received the level of acceptance accorded to the infrared technique. Nevertheless, many papers have been published since the earliest one of Davison *et al.* (1954).

The pyrolysis stage can be carried out in a sealed system remote from the gas chromatograph and the pyrolysate then injected into the column in a way analogous to the infrared method, but this is rarely used for routine polymer identification although both Dawson and Sewell (1975) and Gelling *et al.* (1979) used low-temperature pyrolysis at 350 °C followed by gas chromatographic investigation to distinguish between natural and synthetic polyisoprenes. A comparison of the ratios of the yields of 1-methyl-4-(1-methylethenyl)cyclohexane and 1-methyl-4-(1-methylethyl) benzene for both raw and black filled vulcanizates enabled Gelling also to distinguish between lithium alkyl- and Ziegler Natta-catalysed synthetic polyisoprenes.

The more usual approach is to pyrolyse the sample, raw or vulcanized, at the head of the GC column and to chromatograph all the volatiles, producing a trace on the recorder known as a pyrogram. In order that good, and reproducible, resolution of the eluted components is obtained there are certain features which must be considered.

(1) The temperature of pyrolysis should be adequate to give a high concentration of volatile components.
(2) The temperature rise should be extremely rapid to reduce the gas chromatographic 'injection time' to a minimum.
(3) The temperature rise profile should be completely reproducible.
(4) Secondary reactions, which complicate the chromatogram, should be kept to a minimum.
(5) The material of the pyrolyser unit should be inert with respect to as many polymers as possible.
(6) The sample size must be small, again to keep the 'injection time' low, and the pyrolysis temperature profile reproducible through the sample.

Most of the development work in the design of pyrolysis units has been towards meeting those objectives, and a number of different designs have been discussed in the literature. Continuous heating units are those consisting of a furnace, operating at a fixed temperature, which one slides over the sample, positioned at the injection port of the chromatograph, as described by Krishen (1972). An alternative is to locate the furnace at the head of the column and drop the sample into it (Ney and Heath, 1968). Both these authors use sample weights of about 1 mg but it will be obvious that the furnace tubes have quite a large head space, thus a degree of spread in the pyrogram is inevitable whilst the heating actually experienced by the sample itself is not necessarily reproducible. Some interesting results, indicating the way in which yields of primary products fall with increasing sample size, are given by Ney and Heath (1968) for NR, SBR and BR (see Fig. 8.3). It is to be expected that this will occur due to the lower temperature gradient experienced by larger samples, and the extended time for secondary reactions to occur; this must indicate a requirement for constant sample sizes for standards as well as samples.

Of equal significance are plots of relative yields of isoprene from the pyrolysis of NR at various temperatures, illustrated by both Ney and Heath (1968) and Krishen (1972) in Figs. 8.4 and 8.5 respectively. It will be noted that the yield found by Ney and Heath peaks at about 625 °C (1 mg) whilst that of Krishen is at 700 °C for a similar sample weight (0·5–0·8 mg). One must therefore conclude that the furnace temperatures were not truly representative of the sample temperatures.

A second method of pyrolysis is to use a platinum coil or spiral which is 'flash' heated by the application of a voltage across it. Such a system is

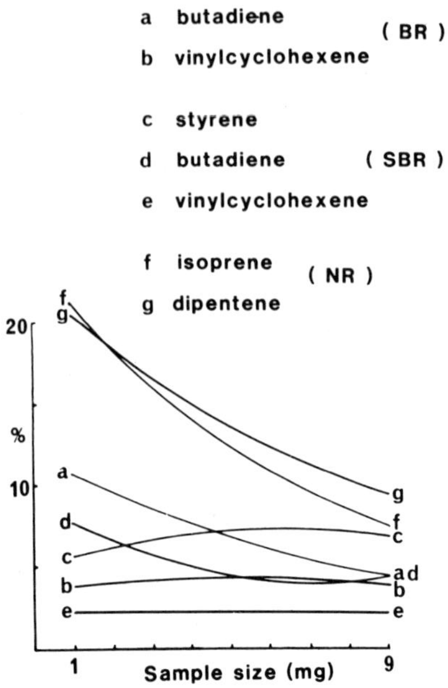

FIG. 8.3. Pyrolysis products as a function of sample size. Pyrolysis temperature 540 °C. (Courtesy *J. IRI.*)

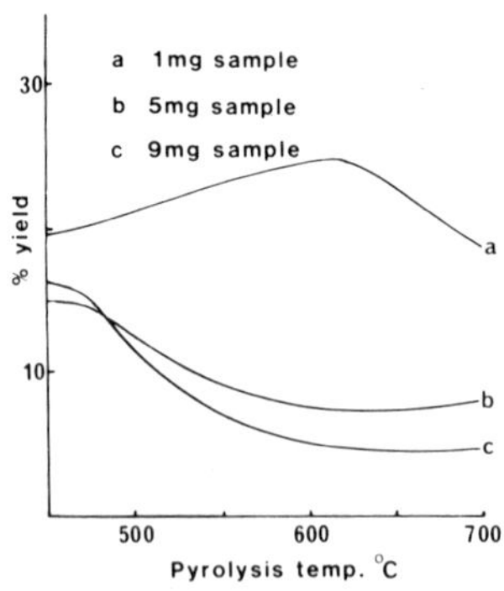

FIG. 8.4. Yield of isoprene v. temperature. (Courtesy *J. IRI.*)

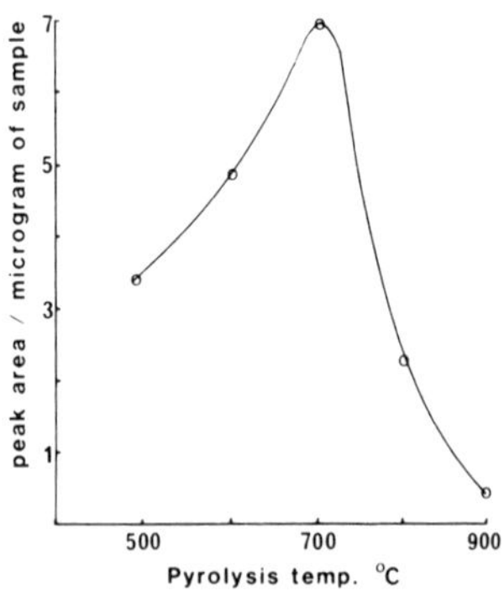

FIG. 8.5. Yield of isoprene v. temperature. (Courtesy *Analyt. Chem.*)

described by Foxton *et al.* (1969) who glue about 0·5 mg of sample to the coil with gum tragacanth and use a 15 s heating cycle at 750 °C. In the long term this system is likely to deteriorate as Hu (1977) comments on the dissolution of carbon in the platinum giving a variation in heating characteristics for a fixed power supply. One way round this, as reported by Hu, is to use a fine quartz tube, containing the sample to be pyrolysed, sleeved by a heating coil.

Trojer and Ericsson (1978) replaced the coil with a piece of platinum foil and used microtomed sections, stamped out to a constant (1·5 mm) OD disc and weighing between 50 and 150 μg in an attempt to standardize the actual heating of the samples, which were stuck to the foil by heating over an infrared light. The reproducibility is much improved but the procedure is time consuming.

The final type of pyrolyser is known as the *Curie point* pyrolysis unit and was first reported by Giacabbo and Simon (1964). This is a metal coil, heated by a high frequency induction current and, when the temperature increases to the Curie point of the metal used, the wire becomes paramagnetic and does not heat further. Typical temperatures are, for a 50/50 iron/nickel alloy, 500 °C; and for iron 100%, 770 °C. As a new filament can be used for each sample, the problems of contamination and reproducibility appear to have been solved.

It is usual, when pyrolysing a vulcanizate for gas chromatographic analysis, to carry out a prior solvent extraction, but Hu (1977) has taken the thermal extraction procedure described earlier (p. 34) a stage further by injecting a small sliver of vulcanizate into the injection port of the chromatograph at 270 °C so that a chromatogram of the volatiles is obtained before the pyrolysis stage is carried out. This provides a reproducible 'fingerprint' for checking that batch formulations are within specification and provides many more data for the absolute identification or matching of two materials for (say) forensic applications.

In spite of the apparent development of the pyrolyser from a relatively crude furnace tube to the Curie point system, there still appears little agreement in the literature as to the best device to use. A recent interlaboratory check (Gough and Jones, 1975) concluded:

> 'Provided that conditions of pyrolysis and chromatography are specified, it is possible to achieve readily identifiable pyrograms from the same polymer, in different matrices, and from one laboratory to another.'

The details specified were that the pyrolyser should be Curie point (700 or 770 °C) with a 5–10 s pulse time.

The most recent standard on pyrolysis–gas chromatography (ISO 5475-1978) allows the use of the furnace/silica tube, platinum coil, or Curie point pyrolyser but makes the point most strongly that fingerprint comparisons, the retention times of particularly indicative peaks, and quantitative peak area comparison must be made against standards obtained on identical equipment, and preferably at the same time.

A range of columns has been used for the analysis of the pyrolysis products, operating both isothermally and under temperature programmed conditions. Typical liquid phases are either non-polar—Apiezon L, SE30, or squalene, or polar—di-2-ethylhexyl sebacate, dibutyltetrachlorophthalate or poly(ethylene glycol adipate). The former provide a

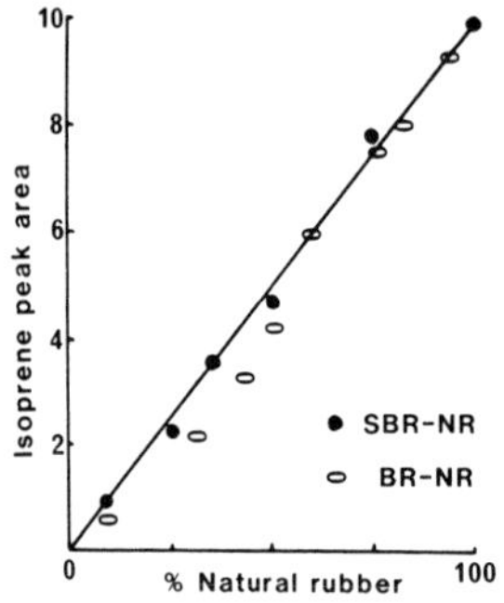

FIG. 8.6. Absolute isoprene peak areas for a range of NR blends. (Courtesy ASTM.)

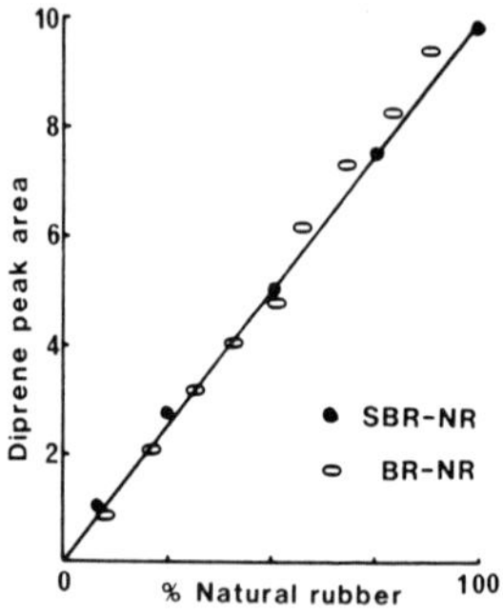

FIG. 8.7. Absolute diprene peak areas for a range of NR blends. (Courtesy ASTM.)

general fingerprint pyrogram whilst the latter should be used if there is a high concentration of rapidly eluted components which require separation for either identification or quantification. Naveau and Dieu (1980) have recently discussed the advantages of capillary columns coated with SE30 in improving resolution and reproducibility.

The mathematical interpretation of the pyrogram data offers, like all other aspects of pyrolysis–gas chromatography, a degree of flexibility! Krishen (1974) recommends a weighed sample and an absolute calculation based on the areas of specific peaks per unit sample weight. Two graphs are reproduced here to illustrate the scatter of the results (Figs. 8.6 and 8.7).

These are compared with the method of calculation more normally used for estimating natural rubber contents—the calculation of area % isoprene and diprene in the total pyrolysate which is illustrated in Fig. 8.8. Cole *et al.*

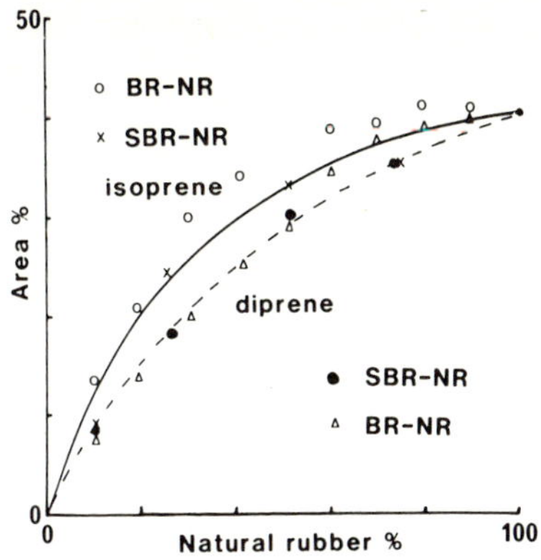

FIG. 8.8. Area % of diprene and isoprene for two series of NR vulcanizates. (Courtesy ASTM.)

(1966) preferred to use peak height ratios and provide a large amount of experimental data on a range of polymer systems.

It is quite obvious that quantification of *good* chromatographic data is not a major problem. It is up to the analyst to make sure that the data are as good as possible by:

(1) Obtaining a quick and reproducible heating to a suitable temperature.
(2) Choosing the best columns for either general or specific analysis.
(3) Optimizing the GC conditions (isothermal/programmed) for good resolution in a reasonable time.
(4) Running standards of a similar composition to the sample immediately before and after it.

Three publications illustrate large numbers of pyrograms, that of Cole *et al.* (1966) giving quantitative data as well. The polymers covered are tabulated to provide some base data for an analyst considering entering this field:

Cole et al. (1966)	*Cianetti and Pecci (1969)*	*Alekseeva (1980)*
NR in NR/BR	NR/IR	NR/IR
IIR in IIR/CR	SBR	SBR
NBR in NBR/CR	NBR	BR
CSM in CSM/CR	BR	Methylstyrene–BR

Cole et al. (1966)	*Cianetti and Pecci (1969)*	*Alekseeva (1980)*
ACN in NBR	IIR	IIR
NBR in NBR/SBR	CR	NCR
EPM and EPDM	EPM	CR
	CSM	EU
	Silicones	AU
	Fluorosilicones	ABR
	CIIR	
	BIIR	

DERIVATIVE THERMOGRAVIMETRY (DTG)

Although thermogravimetry is the oldest of the thermoanalytical techniques, it was only in 1966 that the first derivative thermogravimetric curves were published by Smith (1966*a*, *b*), and it is only in the last ten years or so that modern instrumentation, including the ability to record continuously the first derivative of weight loss, has enabled it to make substantial contributions to the field of polymer analysis. Thermogravimetry will be considered in Chapter 12 but here we are only concerned with the derivative mode, DTG.

The principle of operation is illustrated in Fig. 8.9; the sample, weighing a few milligrams (usually 5–20), is suspended from a microbalance in a

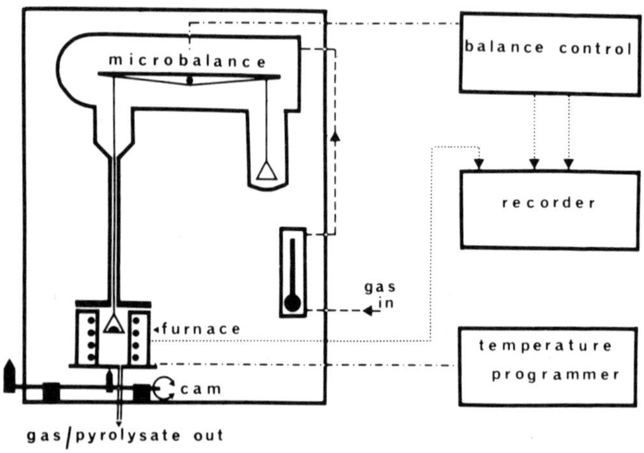

FIG. 8.9. Principle of operation of a thermogravimetric analyser. (Courtesy Stanton Redcroft Ltd.)

small furnace, the whole being enclosed in a glass envelope to enable the atmosphere to be controlled, and the recorder provides a simultaneous reading of temperature and rate of weight change as the temperature programme proceeds. A typical curve, as described by Loadman and McSweeney (1975), is illustrated in Fig. 8.10.

Most elastomers, when heated in an inert atmosphere, undergo thermal degradation in the temperature range 300–600 °C. Figure 8.10 shows the significance of the technique in that the temperature of maximum rate of decomposition, T_{max}, varies with the thermal stability of the polymer

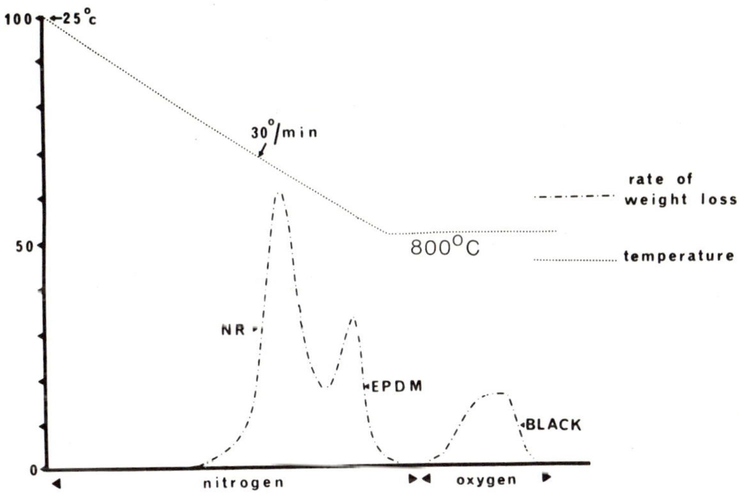

FIG. 8.10. First derivative thermogram, with simultaneous temperature plot.

whilst, provided that the decomposition affords solely volatile products, the area under each curve gives a true and absolute indication of the polymer content. Thus in the case of blends, if good curve separation can be achieved, no calibration is required. Most polymers undergo pyrolysis quantitatively, with the notable exceptions of the chloroprenes and acrylonitrile copolymers, which will be discussed in more detail in Chapter 12.

Polymer Identification

Whilst the use of T_{max} for polymer identification is perfectly valid, it has obvious limitations in that polymers with similar thermal stabilities will decompose at similar temperatures. Table 8.2 lists T_{max} values from three sources, using quite different equipment, but with the *same heating rate* of 10 °C min^{-1} and atmospheres of nitrogen. The results show a good degree of constancy and suggest that, although one would not rely solely on a DTG T_{max} to identify an unknown, the range of possible polymers can be reduced to a very few, whilst the origins of the sample, together with the chemical tests described earlier, could afford a positive identification.

TABLE 8.2

T_{max} FOR VARIOUS RAW ELASTOMERS (MINOR WEIGHT LOSS PEAKS NOT REPORTED)

	Brazier and Nickel (1975)	Sircar and Lamond (1975*a*)	Loadman and Tidd (1976)
Instrument	Dupont 951	(non-commercial)	Stanton Redcroft
Heating rate	10 °C min^{-1}	10 °C min^{-1}	TG 750, 10 °C min^{-1}
Atmosphere	N_2 50 cm^3 min^{-1}	N_2 50 cm^3 min^{-1}	N_2 30 cm^3 min^{-1}
Polymer	T_{max} (°C)	T_{max} (°C)	T_{max} (°C)
NR	373	365	370
IR	373		370
Various BRs	460	465	458
Various SBRs (23·5% St)	445–449	447	447
IIR CIIR BIIR	386		382
EPR EPDM	460		460
Various CRs	375–78/454, 5		367/449
Various CSMs	335–40/465–79		318/464
Various ACMs	370–405		400

It is, however important to realize the effect of altering the heating rate, and this is illustrated in Fig. 8.11.

Brazier and Nickel (1975), Sircar (1977) and Gelling *et al.* (1979) discussed and illustrated how natural and synthetic *cis*polyisoprene may be distinguished by their different DTG curves. The raw polymers show essentially identical curves, the addition of 50 pphr carbon black producing some indication of a high-temperature second peak (Fig. 8.12), but vulcanization (2·5 pphr S, 0·6 pphr CBS) results in quite obvious differences (Fig. 8.13). Sircar postulated that this was due to cyclization of the polyisoprene, possibly catalysed in the case of the synthetics by the polymerization catalyst residues. Strong supporting evidence for this was supplied by Gelling *et al.* who illustrated the DTG curve of cyclized NR and also that of purified Natsyn 2200, with the catalyst residues removed by microfiltration and centrifugation (Fig. 8.14). In this last instance the curve was indistinguishable from that of NR.

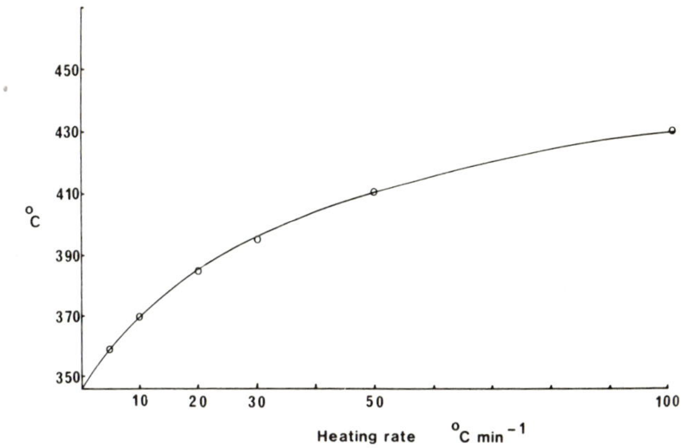

FIG. 8.11. Variation in observed T_{max} for NR with heating rate (after Loadman, 1976).

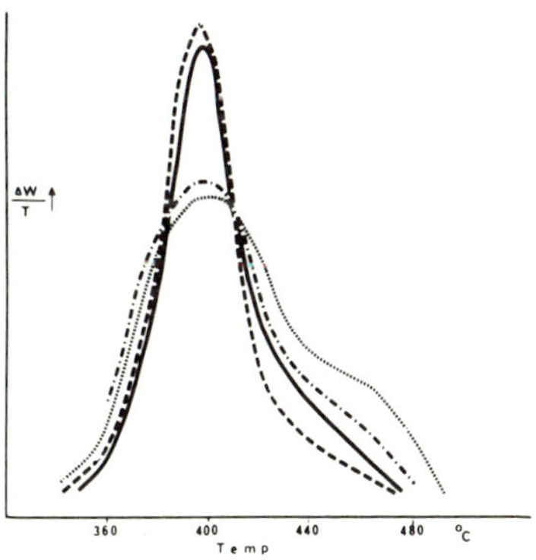

FIG. 8.12 DTG curves of unvulcanized NR and Natsyn 2200 rubbers: NR (–––), Natsyn 2200 (——); NR + 50 pphr HAF black (–·–), Natsyn 2200 + 50 pphr HAF black (······). (Courtesy *J. Polym. Sci.*)

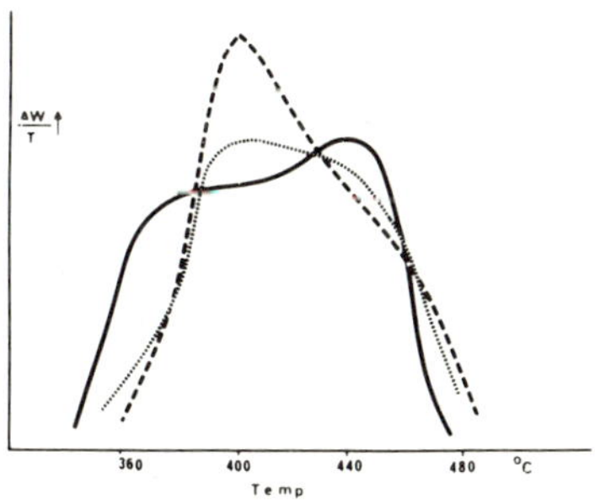

FIG. 8.13. Comparison of the DTG curves of black-filled vulcanized NR (–––), Natsyn 2200 (——), and Cariflex IR 305 or 309 (····). (Courtesy *J. Polym. Sci.*)

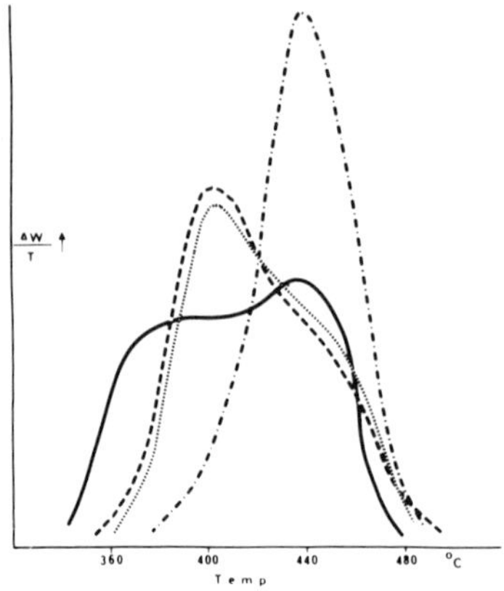

FIG. 8.14. DTG of NR (– – –), Natsyn 2200 (——), and purified Natsyn 2200 (· · · ·), black-filled vulcanizates compared with cyclized NR (–·–). (Courtesy *J. Polym. Sci.*)

Polymer Blend Quantification

Provided that the components of the blend give smooth DTG curves, reasonably resolved from each other and with little carbonaceous residue, there is no difficulty in measuring relative areas under curves, or peak heights, and thus obtaining an adequate estimate of the blend composition; Loadman (1976) used a relative peak height method to analyse blends of natural rubber and polypropylene with a high degree of absolute accuracy ($\pm 1\%$). In instances where one component was too dilute to give a separate T_{max}, measurements of the height of the trace at fixed displacements from the observed T_{max} were taken and, with calibration, levels in the 0–5% range for polypropylene were measured easily and reproducibly. Maurier (1973, 1974) studied NR–EPDM and NR–SBR–EPDM blends whilst Brazier and Nickel (1975) quantified the analysis of NR–BR and NR–SBR–EPDM systems, which have substantial peak overlap, by determining 'response factors', i.e. the peak height per unit of mass degraded for each component elastomer at T_{max} in blends of various compositions. It is important to note that these factors only apply for a given set of experimental conditions. It is, however, particularly advantageous that EPDM and polypropylene are so much more thermally stable, and their levels so easily determined in blends with polyisoprene, as this is one of the most difficult combinations to analyse by the more usual pyrolysis–IR technique. The presence of such a blend is impossible to miss using DTG.

Sircar and Lamond (1975a) illustrate the effect of blending on the observed T_{max} values. Peroxide cured blends of NR–BR were studied over a blend ratio of 80:20 to 20:80 and both elastomers showed a fall in T_{max} as their loadings decreased. In the case of NR this was from 360 °C at 80% to 347 °C at 20% whilst for BR it was 465 °C at 80%, falling to 450 °C at 20%.

DIFFERENTIAL SCANNING CALORIMETRY (DSC)

Differential scanning calorimetry and differential thermal analysis (DTA) provide the same information about a sample, as they both respond to enthalpy (heat content) changes reflecting a chemical or physical change within the sample. DTA is, however, much more difficult to quantify as the observed measurement is a difference in temperature between sample and reference, whilst in the case of DSC it is the difference in energy required to maintain the two at identical temperatures. In the latter case standardization against the melting endotherm of (say) indium allows absolute values of thermal properties to be obtained. Two types of DSC systems are available, described as *power compensated* and *heat flux*, but each gives essentially identical information. In order to illustrate the application of these techniques to polymer analysis, only DSC will be considered, and this relatively briefly, not because of its limited value, but rather because we are now moving towards research rather than 'routine' equipment and this is reflected in the instrument costs. It would be possible to buy an adequate infrared spectrometer, pyrolysis–gas chromatograph and thermogravimetric analyser for the cost of one differential scanning calorimeter!

Figure 8.15 illustrates the events which may be observed in the DSC examination of a sample and these various regions may be analysed to

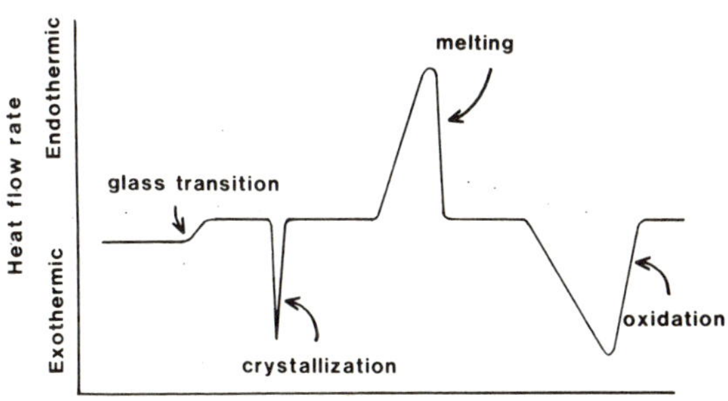

FIG. 8.15. Idealized DSC curve for a polymeric material in an air atmosphere.

provide meaningful data. First, however, the samples must be prepared for analysis and this presents one of the problems of reproducibility of data as heat transfer through the container to the sample must be uniform. Although methods such as melting the polymer in the sample container (Dannis, 1963) have been advocated, it is essential that any such method must be fully reversible and this is not always the case. The authors find that a sliver some 1–1·5 mm thick, stamped to a 6 mm disc with a No. 3 cork borer, provides an ideal sample for encapsulation in that the disc perfectly fits into the aluminium sample container, and is flat for good heat transfer from the small sample heater. A reference capsule containing an equivalent weight of aluminium lids affords a more linear base-line.

Glass Transitions

Glass transitions (T_gs) provide useful *non-destructive* analytical data on polymers, and the T_gs of many elastomers covered in this book may be found in Appendix II. The presence of one broad 'average' T_g or two discrete T_gs for polymer blends affords information on the compatibility of the two phases; NR–SBR vulcanizates show two distinct transitions at −72 °C and −60 °C respectively whilst SBR–BR samples show a gradual progression from that of SBR (−60 °C) to that of the BR used (e.g. −112 °C for 55 NF). The relative displacements of the two T_gs or the overall shapes of the single T_gs enable quantitative blend analysis to be carried out. It is reported by Landi (1972), Chandler and Collins (1969) and Jorgensen *et al.* (1973) that uncured nitrile–butadiene or nitrile–isoprene copolymers exhibit two T_gs when the acrylonitrile content is less than 35–36% of the polymer loading, whilst a similar phenomenon enables SBS block copolymers to be distinguished from randomly polymerized SBR. Fielding-Russell (1972) comments that the polystyrene T_g position varies with different polybutadiene homopolymers due to their plasticization of the polystyrene to different extents. Ikeda *et al.* (1969) have evaluated these observations for the quantification of SBS block copolymers and obtained good results.

Crystallization and Melting

Although the enthalpy of crystalline melting has been used extensively in the analysis of plastics, it has only recently been used to study elastomers. Kim and Mandelkern (1972) investigated NR latex and purified NR. Two melting regions were observed as illustrated in Fig. 8.16 and, whilst it is known that this type of multiplicity may be introduced by the cooling history, Edwards (1975) observed two morphological forms by electron microscopy which he correlated with these events. Loadman and Davey (1978) studied a range of NR crêpes by both density gradient and DSC in order to assess the usefulness of the latter in measuring low levels of crystallinity in sole crêpes. It was found to be as accurate as the former method and much easier to carry out.

Although DSC does not really fall under the heading of polymer blend analysis, it is of technological significance to note that Sircar and Lamond (1973*b*) report the loss of crystallinity of BR when mixed with a second

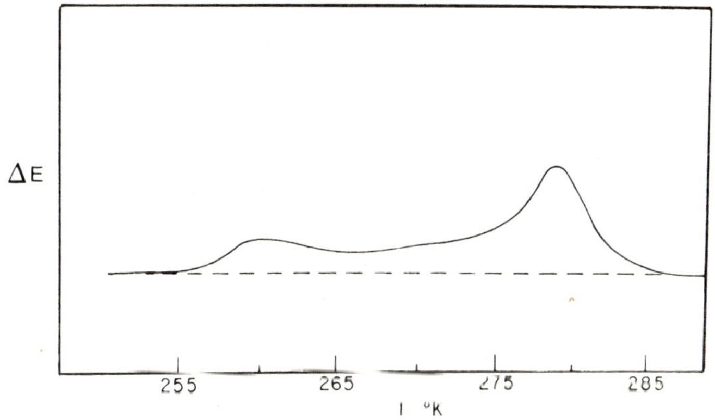

FIG. 8.16. Differential scanning calorimetry fusion curve of purified natural rubber. Sample crystallized at −25 °C for 6 h. Heating rate 5 °C/min. (Courtesy *J. Polym. Sci.*)

elastomer (NR, IR, EPDM, CR–IIR, NBR or CR), both in the presence and absence of carbon black. Lee and Singleton (1979) observed essentially the same behaviour in BR–SBR blends, for which the enthalpy of BR fusion was determined as a function of mixing time by three different procedures. Figure 8.17 illustrates the dependence of the BR fusion enthalpy on mixing time for free mixing. The magnitude of the changes in the enthalpy are consistent with no carbon black transfer between the phases.

The problems of sample placement and reproducibility have been mentioned but dynamic processes such as crystallization can be markedly affected both by the treatment which the sample receives in the DSC itself and by thermal history prior to sampling. This is clearly illustrated in Fig. 8.18 which shows curves for the heating of BR after various cooling procedures.

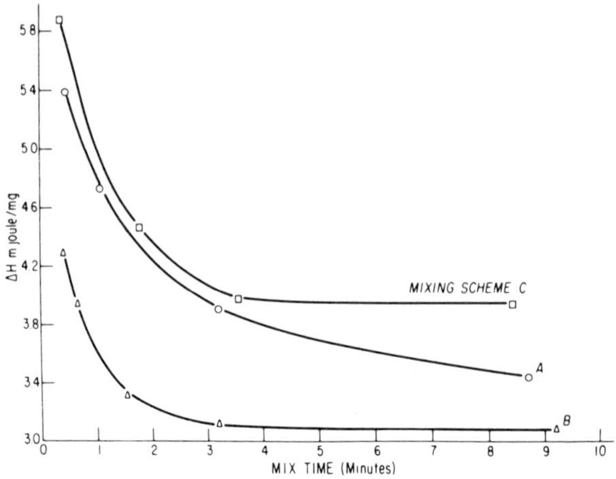

FIG. 8.17. Fusion enthalpy of BR as a function of blending time. Formulation: 20 phr BR, 80 phr SBR, 20 phr carbon black. Mixing schemes: A, free black mixing; B, carbon black premixed in BR; C, carbon black premixed in SBR. (Courtesy *J. Appl. Polym. Sci.*)

Rapid cooling (B) prevents crystallization from being completed before the T_g is passed and then no more crystallization can occur. On heating through the T_g, crystallization continues until a temperature is reached at which melting commences. This crystallization on heating can be completely eliminated if the sample is cooled sufficiently slowly for crystallization to be complete before the T_g is reached (C).

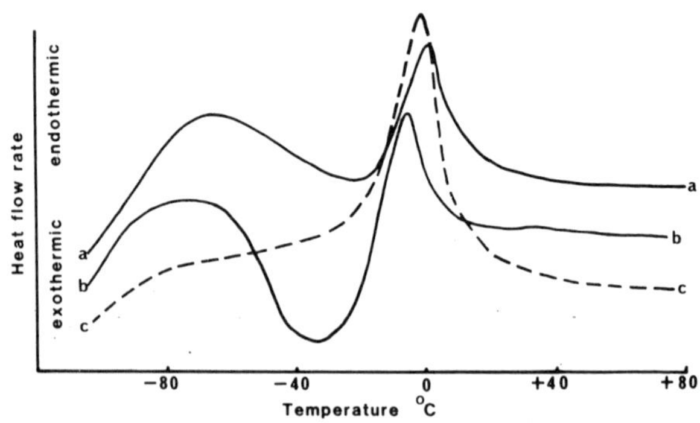

FIG. 8.18. Differential scanning calorimetry curves of BR crystallization; effect of thermal history. (a) Untreated sample; (b) heated and quenched sample; (c) heated sample cooled slowly. (After Sircar and Lamond, 1973*b*.)

High-Temperature Events

In a series of papers covering a vast range of polymers, Sircar and Lamond (1972–77) described how, above vulcanization temperatures, exothermic and endothermic events corresponding to high-temperature reactions in the vulcanizates may be measured. Several elastomers undergo reactions which can be related to the specific chemical structure of the chain. For example, cyclization events in BR, SBR and NBR elastomers were detected quantitatively in the DSC, and the enthalpy associated with the process was used by both Sircar and Lamond (1973*a*) and Sircar and Voet (1970) for analytical purposes.

In an inert atmosphere, the overall thermal degradation pattern results in a characteristic DSC profile which depends upon the type of the elastomer present. The DSC profile is often complex, but the general shape of the profile has found use in the fingerprinting of elastomer samples, and the subsequent identification of elastomer mixtures.

Oxidative degradation usually results in an even more complex DSC profile but, again, can give a characteristic fingerprint for a system. This may also be used for elastomer identification.

Figure 8.19 illustrates a recent series of results by Goh (1980) showing how oxidative degradation can distinguish between natural and synthetic polyisoprenes.

The two examples of thermal analysis—DTG and DSC—covered in this chapter represent the most common, and generally most informative, of a

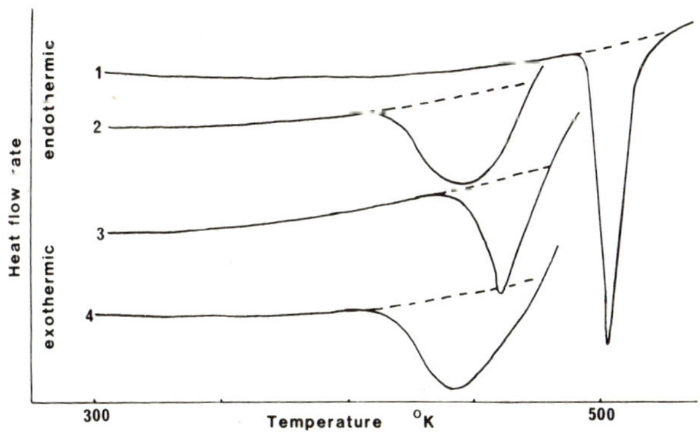

FIG. 8.19. Differential scanning calorimetry curves of various polyisoprenes. Curve 1, natural rubber (sample weight: 0·74 mg); curve 2, Natsyn 2200 (sample weight: 0·62 mg); curve 3, gutta percha (sample weight: 0·68 mg); curve 4, 'Trans-PIP' 100 (sample weight: 0·68 mg). Atmosphere O_2 at 20 ml min^{-1}, scan speed 16 °C min^{-1}. (Courtesy *Thermochim. Acta*.)

TABLE 8.3

CLASSIFICATION OF THERMOANALYTICAL TECHNIQUES*

Measured property	*Derived technique(s)*	*Acceptable abbreviation*
Mass	Thermogravimetry	TG
	Isobaric mass-change determination	
	Evolved gas detection	EGD
	Evolved gas analysis	EGA
	Emanation thermal analysis	
	Thermoparticulate analysis	
Temperature	Heating curve determination[a]	
	Differential thermal analysis	DTA
Enthalpy	Differential scanning calorimetry	DSC
Dimensions	Thermodilatometry	
Mechanical characteristics	Thermomechanical analysis	TMA
	Dynamic thermomechanometry	
Acoustic characteristics	Thermosonimetry	
	Thermoacoustimetry	
Optical characteristics	Thermoptometry	
Electrical characteristics	Thermoelectrometry	
Magnetic characteristics	Thermomagnetometry	

*As defined in *For Better Thermal Analysis*—International Confederation for Thermal Analysis. Instituto di Mineralogia & Petrografia Citta Universitaria, 00185 Rome, Italy, 2nd ed., 1980. (Courtesy Professor G. Lombardi, Vice-President ICTA.)

[a]When the temperature programme is in the cooling mode this becomes 'cooling curve determination'.

whole range of thermoanalytical techniques. The scope is illustrated in Table 8.3 but, because of space considerations, the interested reader will have to pursue these through the specialist publications—*Journal of Thermal Analysis* and *Thermochimica Acta.*

The two final instrumental techniques of nuclear magnetic resonance (NMR) and scanning electron microscopy with X-ray analytical attachment (SEM) to be covered in this chapter differ from those previously illustrated in that they are more usually found in research departments than in the general-purpose laboratory of the rubber-analyst. It is also generally the case that they are operated, and the data interpreted, by a specialist analyst. For this reason the main bulk of these sections will be more concerned with illustrating results from relatively few sources than with experimental details so that the rubber-analyst may appreciate the power of the techniques and the types of information which are potentially available should a particular situation merit their application.

NUCLEAR MAGNETIC RESONANCE (NMR)

The two nuclei of most concern to the rubber-analyst are carbon and hydrogen. NMR experiments may be carried out with each of these. In the case of hydrogen it is the ${}^{1}_{1}H$ isotope which affords a spectrum (^{1}H-NMR) and for carbon the ${}^{13}_{6}C$ isotope (^{13}C-NMR). The principle of both forms of spectroscopy is the same; both types of nuclei are spinning charged particles which generate their own magnetic fields and, when they are placed in a strong external magnetic field, they are forced to precess with their precession aligned either parallel or anti-parallel to the applied field. The former is the lower energy state and hence the more densely populated one. The nuclei can absorb radiofrequency energy resulting in transitions between the two states and giving rise to a net absorption of energy. The frequency at which this occurs is dependent upon the magnetic field experienced by each nucleus and, as each nucleus in a molecule is surrounded by bonding electrons which themselves generate 'mini magnetic fields', the net field experienced by any particular nucleus varies depending upon its chemical environment. Thus a 'spectrum' of radio-frequency absorption is obtained. The modification of the applied field is called 'shielding'. Frequencies of absorption are quoted relative to a standard chemical (tetramethyl silane, TMS) which is more highly shielded, both for ^{13}C and for ^{1}H, than most carbons or protons found in ordinary compounds. This is placed at the right-hand side of the spectrum by convention and 'chemical shifts' for other nuclei quoted relative to this. Shifts may be expressed in hertz (Hz) but shifts then vary with the frequency of the spectrometer used, and thus the latter requires specifying. If, however, the shift is expressed as

$$\frac{\text{Hz from TMS}}{\text{frequency of instrument in MHz}}$$

then the figures obtained, as parts per million, will be independent of the frequency of the NMR spectrometer. The usual frequency-range of modern instruments is 60–200 MHz. Thus a 90 Hz shift with a 90 MHz

NMR corresponds to 1 ppm as does a 200 Hz shift with a 200 MHz machine. The range of a proton spectrum is usually 0–10 ppm (although some protons are shifted by up to 15 ppm) and for a ^{13}C spectrum 200 ppm. For reasons too lengthy to discuss here, ^{1}H-NMR spectra, as usually obtained from solutions of polymers, are not sharp bands, but a series of moderately broad symmetrical peaks, whilst the ^{13}C spectra appear as very sharp bands. It is also of crucial importance to note that the area under each peak in a ^{1}H-NMR spectrum is proportional to the number of that type of hydrogen nuclei present but for ^{13}C as normally shown this is not strictly true but, by the use of special techniques, can be made so. For further study of the principle of ^{1}H- and ^{13}C-NMR spectroscopy the reader is referred to

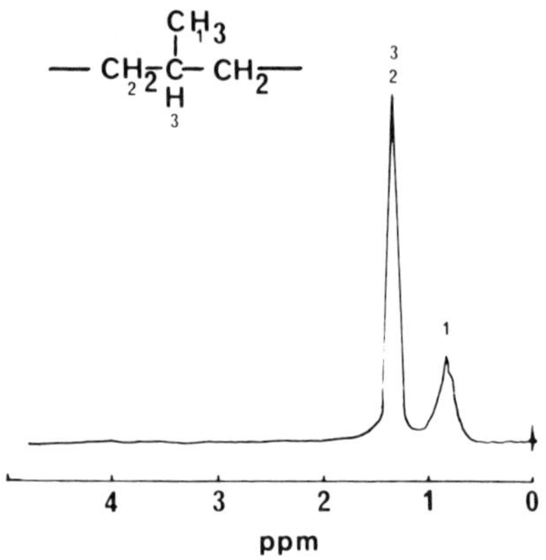

FIG. 8.20. ^{1}H-NMR spectrum of an EPR.

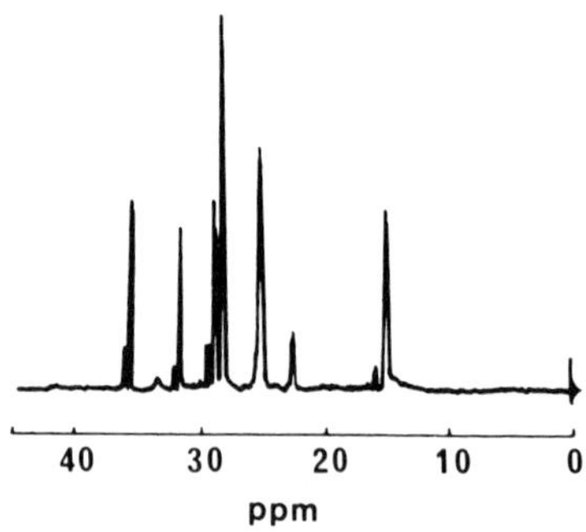

FIG. 8.21. ^{13}C-NMR spectrum of a high-ethylene EPR.

books by Jackman and Sternhell (1969), Stothers (1972) and Abraham and Loftus (1978). Normal 'high resolution NMR' spectra are obtained from solutions, usually in a proton-free solvent (CCl_4) or deuterated solvent ($CDCl_3$, C_6D_6, etc.), and numerous publications exist on the determination of structure and microstructure by these techniques. The latter, however, is the subject of Chapter 9 so here we are only concerned with the former—polymer identification and blend analysis.

A typically illustrative pair of spectra is shown in Figs. 8.20 and 8.21. From the ^{1}H-NMR spectrum we can see bands typical of an ethylene–propylene copolymer. Integration (measurement of the relative peak areas) will give the percentage of protons from the propylene methyl group, and hence by simple calculation the wt. % of polypropylene. The

^{13}C-NMR spectrum shows many more different carbons than the five one would naïvely expect, and this is due to various microstructural features such as sequence distribution etc. as described by Wilkes (1977), Wilkes *et al.* (1973) and Carman and Wilkes (1971) and considered in more detail in the next chapter.

Very early articles by Bovey *et al.* (1959) and Bovey and Tiers (1960) illustrated the use of ^{1}H-NMR for monomer ratio determinations in styrene copolymers and since that date numerous authors have used the technique to examine quantitatively copolymers of styrene with methyl acrylate, methyl methacrylate, propylene, butadiene, isoprene, ethylene, α-methyl styrene, *para*methyl,α-methyl styrene and 2-ethyl hexyl acrylate.

Polyurethanes have also been the subject of several publications on the use of NMR spectroscopy in structural analysis. The reader is particularly referred to those of Yeager and Becker (1977), who use ^{1}H-NMR, and Delides *et al.* (1981), who use ^{13}C-NMR.

It must be pointed out that NMR spectra may be obtained from many nuclei other than those of hydrogen and carbon. In particular fluorine ($^{19}_{9}$F) is important and papers have been published by Ferguson (1960), Naylor and Lasoski (1960), Wilson (1963) and Wilson and Santer (1965). More recently Murasheva *et al.* (1979) and Toy and Stringham (1978) have reported on the analysis of fluorocarbon polymers.

A few $^{15}_{7}$N spectra have also been reported (Kircheldorf, 1978) but these are more relevant to the analysis of plastics than elastomers.

It is obvious without giving further details that provided a solution of the sample can be obtained an NMR spectrum can provide vast amounts of data, particularly since the rules for predicting the chemical shifts of the various nuclei are the same for polymers as for low-molecular-weight materials and are exremely well documented and understood.

The fact that one requires a solution must be considered a limitation although all the methods of solution/dissolution previously discussed can be used, whilst Werstler (1980) describes a modification of the *ortho*-dichlorobenzene dissolution, and tabulates ^{13}C-NMR chemical shift data for a large number of polymers. Over the last decade, however, there has been increasing activity in the generation of ^{1}H-NMR spectra from solid elastomers, cut straight from the sample and placed in the spectrometer. In practice, of course, the situation is very much more complicated than this, and there is still little evidence that this will become a routine procedure within the next decade. Nevertheless, spectra are continually being reported and a recent review by Carman (1979) is worth reading for the interested analyst.

SCANNING ELECTRON MICROSCOPY (SEM)

There are many ways of 'visually' looking at the surface or cross-section of a piece of rubber, with instruments ranging from a simple lens (×10), through a light microscope (×10–×400), scanning electron microscope (×20–×300 000), to a transmission electron microscope (×1000–>×1 000 000). Similarly the different elements present may be identified and, within limits, quantified by such techniques as X-ray fluorescence (XRF), electron spectroscopy for chemical analysis (ESCA), and many others, new acronyrns for which are appearing continually. Whilst all of these have their place, some indeed providing information not obtainable by any other method, the scanning electron microscope, with an integral X-ray analyser, offers a unique combination of advantages which merits its inclusion in this section, as well as in Chapters 10 and 13 dealing with fillers and blooms.

The first true SEM was built by Von Ardenne (1938), but it was not until the mid 1960s that instruments became commercially available and the technique could be considered to have arrived. Most current applications are biomedical, biological or metallurgical and there is still relatively little published on the application of the SEM to rubber analysis.

The principle of operation is simple, and consists of bombarding the sample with a high-energy electron beam, after coating it with a conductive film less than 1 nm thick to prevent build-up of an electric charge, and monitoring the events which result therefrom. The two events of interest here are the ejection of secondary electrons, which are processed to provide a 'visual' TV image, and the generation of X-rays which, in the case of an energy-dispersive X-ray analytical system, are fed to a multi-channel analyser which generates a spectrum of all elements with mass greater than sodium as a histogram on a cathode ray tube. The great depths of field which are experienced with the SEM, coupled with an element spectrum of the same region, thus provide many unique data. It is worth noting that there is another type of X-ray spectrometer available, known as the crystal, or wavelength dispersive, spectrometer to distinguish it from that previously described which is an energy dispersive unit. The crystal spectrometer does not produce an element histogram but operates as a scanning instrument, producing a range of peaks for each element present. It is important because it allows detection and quantification of all elements of atomic mass greater than beryllium and thus allows the study of nitrogen-, oxygen-, or fluorine-containing elastomers. The relevance of these to polymer analysis is twofold, one based on visual interpretation of the data, and the other on the elemental composition.

Visual Aspects of Polymer Analysis

(a) Latex particle size and size distribution are of critical importance in the manufacture of foam rubber goods, as they both reflect on the viscosity of the latex solution. It is a relatively simple matter to see the latex particles, if they are hardened before drying with osmium tetroxide as described by Kato (1966), and to differentiate between NR latex and a synthetic one by the particle size and size distribution. NR shows a range of sizes of 0·4–4 μm in diameter whilst the synthetics are much smaller with few particles above 0·2 μm (as these are too small for good foam manufacture they are generally agglomerated using the process of Talalay (1963), but it is still a simple matter to distinguish the agglomerates from individual latex particles).

(b) An area where much work will probably be carried out using an SEM is that of thermoplastic blends of rubbers and plastics, often called composites, both in the initial manufacture and in subsequent product fabrication. Studies carried out by Loadman and Tinker (1981) are briefly illustrated in photographs P1–4 (see pp. 168–9). The thermoplastics are NR–PP mixtures and the samples prepared by freezing in liquid nitrogen, fracturing with a sharp blow, and then etching out the NR by soaking in chloroform overnight prior to gold coating for SEM examination. At low levels of NR (P1) it exists as a discrete phase within the PP matrix but, as the level increases (P2), the phases become co-continuous, so that the removal of the NR leaves a PP sponge. In both cases the dimensions of the phases can be quantified by the application of stereology using techniques described by Underwood (1969) and James (1977). Thus the effects of varying mixing times, temperatures or types of polymer can be evaluated.

As well as the basic morphological details described, there is additional information to be obtained from the analysis of moulded articles relating to orientation effects. This is illustrated in photographs P3 and P4, taken parallel and at right-angles to the direction of flow respectively. Again these imply that there are substantial numbers of analytical data to be obtained which could enable one to shed light on faults or flaws in a thermoplastic product.

Elemental Analysis

The major application for element identification in polymer analysis is in the examination of halogen-containing compounds to see whether the halogen, easily identified as chlorine or bromine, is present in the bulk of the rubber or as a surface skin. This can be achieved in two ways; as the depth of penetration of the electron beam depends upon its accelerating voltage, typically 5–50 keV, a surface halogenated sample will appear

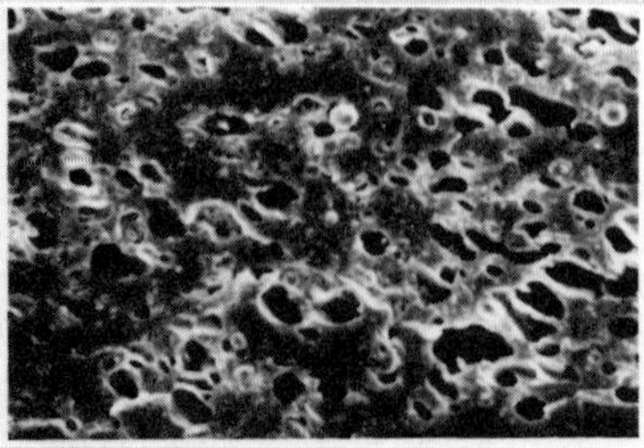

P 1

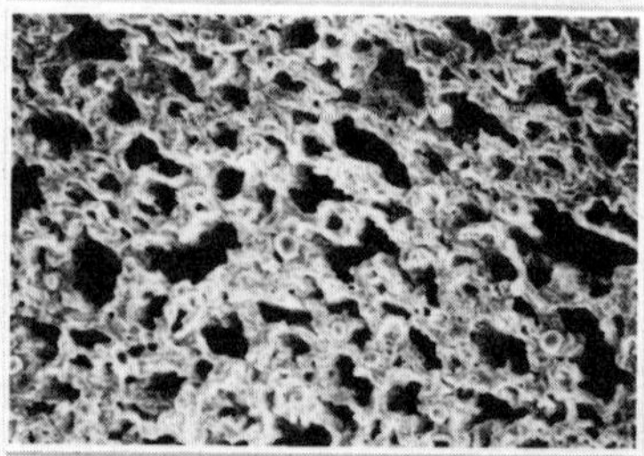

P 2

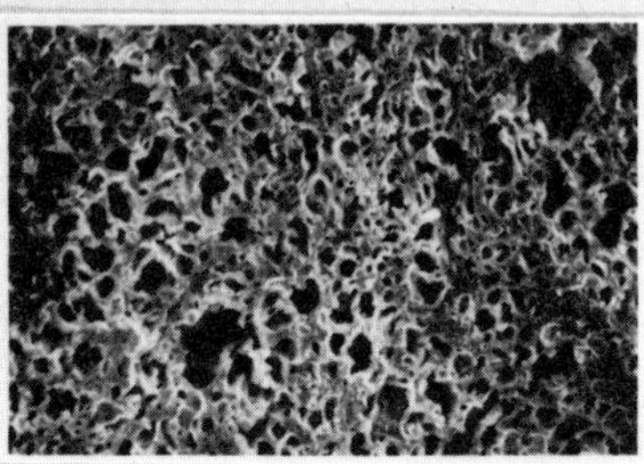

P 3

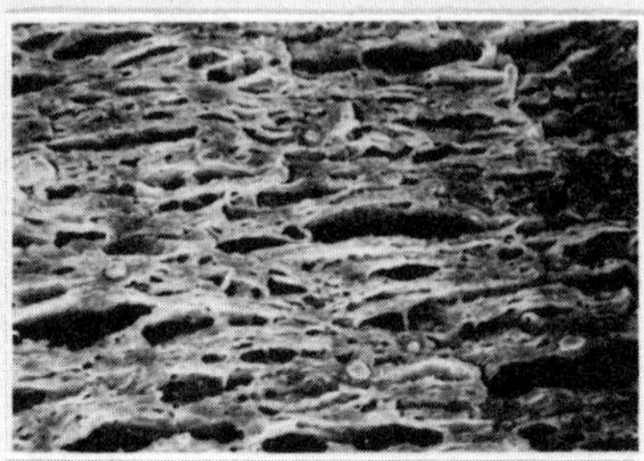

P 4

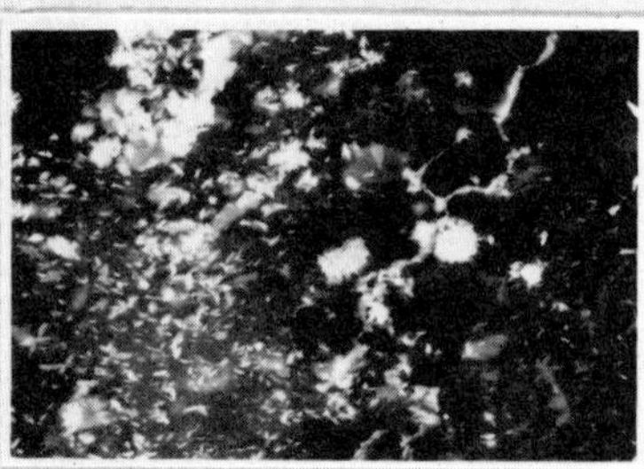

P 5

P 6

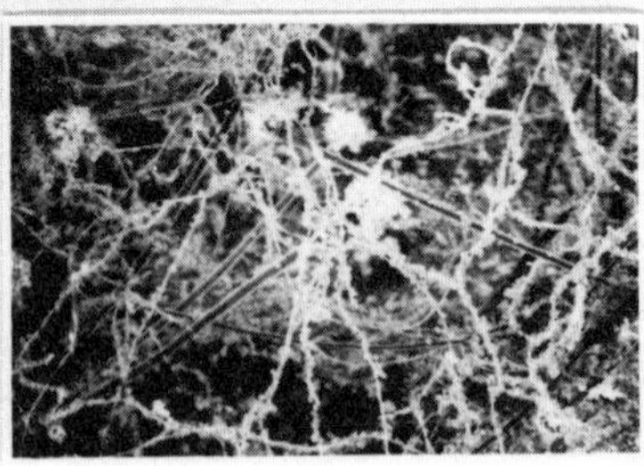

P 7

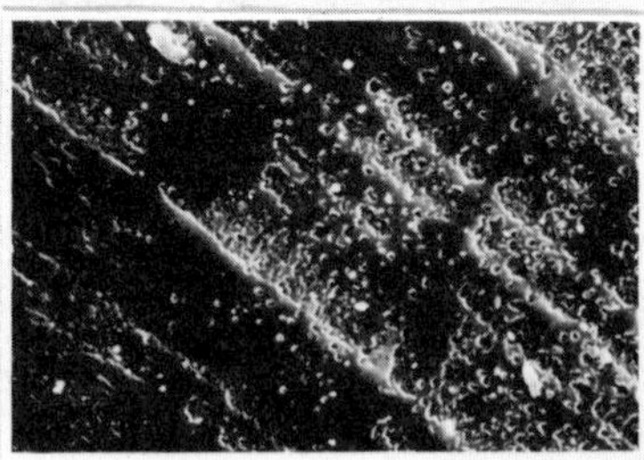

P 8

NR:PP 30:70
Pellets
freeze fractured—
chloroform etched
×1200

P1

NR:PP 70:30
Pellets
freeze fractured—
chloroform etched
×1200

P2

NR:PP 50:50
Injection moulded sheet
freeze fractured ⊥ to
flow; chloroform etched
×1200

P3

NR:PP 50:50
Injection moulded sheet
freeze fractured || to
flow; chloroform etched
×1200

P4

Bloom of sulphur crystals
on partly cured rubber
products
×200

P5

Bloom of sulphur
crystals with large
accelerator crystal
×400

P6

'Bloom'—actually
fungus growth

×120

P7

'Bloom'—actually loss
of surface gloss due to
stress-relaxing oxidation
×120

P8

relatively richer in halogen as the voltage is decreased, and the beam penetrates less into the rubber behind the surface film. Alternatively, the sample may be sectioned and the concentration of halogen measured across the section; a uniform halogen concentration throughout the sample obviously indicates a blend whilst a zero concentration in the bulk of the rubber and a high concentration at the surface indicates surface treatment. A similar procedure may be used with latex-dipped products to analyse for the presence of laminates of different polymers. If the polymers are halogen free it is often possible to observe filler differences which will distinguish laminates from blends. This method of X-ray mapping the concentration of a specific element, and matching the map to the visual display, has also been used to identify crumbed scrap rubber in an article, and to estimate its level in the total product. Such a technique enables one to make sense of bulk polymer or filler analyses, which otherwise may seem unnecessarily complex.

REFERENCES

Abraham, R. J. and Loftus, P. (1978) *Proton and C-13 Spectroscopy, an Integrated Approach*, Heyden, London.

Alekseeva, K. V. (1980) *J. Anal. and Appl. Pyrol.* **2**, 19.

Von Ardenne, M. (1938) *Z. Physik* **109**, 553.

Barnes, R. B., Williams, V. Z., Davis, A. R. and Giesecke, P. (1944) *Ind. Eng. Chem. Anal.* **16**, 9.

Bovey, F. A. and Tiers, G. V. D. (1960) *J. Polym. Sci.* **44**, 173.

Bovey, F. A., Tiers, G. V. D. and Filipovich, G. (1959) *J. Polym. Sci.* **38**, 73.

Brazier, D. W. and Nickel, G. H. (1975) *Rubber Chem. Technol.* **48**, 661.

Carlson, D. W., Ransaw, H. C. and Altenau, A. G. (1970) *Analyt. Chem.* **42**, 1278.

Carman, C. J. (1979) *Am. Chem. Soc. Symp. Ser.* **103**, 97.

Carman, C. J. and Wilkes, C. E. (1971) *Rubber Chem. Technol.* **44**, 481.

Chandler, L. A. and Collins, E. A. (1969) *J. Appl. Polym. Sci.* **13**, 1585.

Cianetti, E. and Pecci, G. (1969) *Ind. Gomma* **13**, 47. See also *Italian Air Force Laboratory Monograph* (English) 1969.

Clark, J. K. and Scott, R. A. (1970) *Rubber Chem. Technol.* **43**, 1332.

Cleverley, B. (1979) *Food Technol. NZ* **14**, 8.

Cleverley, B. and Herrmann, R. (1960) *J. Appl. Chem.* **10**, 192.

Cole, H. M., Petterson, D. L., Sljaka, V. A. and Smith, D. S. (1966) *Rubber Chem. Technol.* **39**, 259.

Corish, P. J. (1960) *J. Appl. Polym. Sci.* **4**, 86.

Dannis, M. L. (1963) *J. Appl. Polym. Sci.* **7**, 231.

Davison, W. H. T., Slaney, S. and Wragg, A. L. (1954) *Chem. Ind.* 1356.

Dawson, B. and Sewell, P. R. (1975) *J. IRI* **9**, 180.

Delides, C., Pethrick, R. A., Cunliffe, A. V. and Klein, P. G. (1981) *Polymer* **22**, 1205.

Dinsmore, H. L. and Smith, D. C. (1948) *Analyt. Chem.* **20**, 11.
Edwards, B. C. (1975) *J. Polym. Sci. Polym. Chem. Edn.* **13**, 1387.
Fahrenfort, J. (1961) *Spectrochim. Acta* **17**, 698.
Ferguson, R. C. (1960) *J. Am. Chem. Soc.* **82**, 2416.
Fielding-Russell, G. S. (1972) *Rubber Chem. Technol.* **45**, 252.
Foxton, A. A., Hillman, D. E. and Mears, P. R. (1969) *J. IRI* **3**, 179.
Gelling, I. R., Loadman, M. J. R. and Sidek, B. D. (1979) *J. Polym. Sci. Polym. Chem. Edn.* **17**, 1383.
Giacabbo, H. and Simon, W. (1964) *Pharm. Acta Helv.* **39**, 162.
Goh, S. H. (1980) *Thermochim. Acta* **39**, 353.
Gough, T. A. and Jones, C. E. R. (1975) *Chromatographia* **8**, 12.
Gross, D. (1975) *Rubber Chem. Technol.* **48**, 289.
Harms, D. S. (1953) *Analyt. Chem.* **25**, 1140.
Haslam, J., Willis, H. A. and Squirrell, D. C. M. (1972) *Identification and Analysis of Plastics*, 2nd ed., Iliffe, London.
Higgins, G. M. C. and Loadman, M. J. R. (1970) *NR Technol.* **10**, 1.
Higgins, G. M. C. and Loadman, M. J. R. (1971) *Ind. Gomma* **15**, 50.
Hu, J. C. (1977) *Analyt. Chem.* **49**, 537.
Hummel, D. O. and Scholl, F. K. (1969) *Infrared Analysis of Polymers, Resins and Additives. An Atlas.* Vol. 1, Carl Hanser Verlag, Munich.
Ikeda, R. M., Wallach, M. L. and Angelo, R. J. (1969) *Block Polymers*, S. L. Aggarwal, Ed., Pergamon Press, New York.
Jackman, L. M. and Sternhell, S. S. (1969) *Applications of Nuclear Magnetic Resonance Spectroscopy in Organic Chemistry*, Pergamon Press, Oxford.
James, N. T. (1977) *Analytical and Quantitative Methods in Microscopy*, Meek, G. A. and Elder, H. Y., Eds., Cambridge University Press, London.
Jorgensen, A. H., Chandler, L. A. and Collins, E. A. (1973) *Rubber Chem. Technol.* **46**, 1087.
Kato, K. (1966) *J. Polym. Sci.* **4B**, 35.
Kim, H. G. and Mandelkern, L. (1972) *J. Polym. Sci. Part A.2* **10**, 1125.
Kircheldorf, H. R. (1978) *Macromol. Chem.* **179**, 2687.
Krishen, A. (1972) *Analyt. Chem.* **44**, 494.
Krishen, A. (1974) *ASTM STP 553* 74.
Kruse, P. F. and Wallace, W. B. (1953) *Analyt. Chem.* **25**, 1156.
Landi, V. R. (1972) *Rubber Chem. Technol.* **45**, 222.
Lee, B. and Singleton, C. (1979) *J. Appl. Polym. Sci.* **24**, 2169.
Lerner, M. and Gilbert, R. C. (1964) *Analyt. Chem.* **36**, 1382.
Leukroth, G. (1970) *Gummi Asbest. Kunstst.* **28**, 1118.
LiGotti, I. (1972) *Paper Presented at 20th Meeting, ISO TC45-WG1*, Cologne.
Loadman, M. J. R. (1976) Unpublished work at MRPRA.
Loadman, M. J. R. and Davey, J. E. (1978) Unpublished work at MRPRA.
Loadman, M. J. R. and McSweeney, G. P. (1975) *Rev. Gen. Caoutch. Plast.* **52**, 805.
Loadman, M. J. R. and Tidd, B. K. (1976) Unpublished work at MRPRA.
Loadman, M. J. R. and Tinker, A. J. (1981) Unpublished work at MRPRA.
Mackillop, D. A. (1968) *Analyt. Chem.* **40**, 607.
Maurier, J. J. (1973) *Polym. Prepr. Am. Chem. Soc. Div. Polym. Chem.* **14**, 518.
Maurier, J. J. (1974) *J. Macromol. Sci. Chem.* **8**, 73.

Murasheva, E. M., Shaskov, A. S. and Galil Ogley, F. A. (1979) *Vysokomol. Soedin. Ser. A* **21**, 882.
Naveau, J. and Dieu, H. (1980) *J. Anal. & Appl. Pyrol.* **2**, 123.
Naylor, R. E. Jr. and Lasoski, S. W. Jr. (1960) *J. Polym. Sci.* **44**, 1.
Ney, E. A. and Heath, A. B. (1968) *J. IRI* **276**.
Sircar, A. K. (1977) *Rubber Chem. Technol.* **50**, 71.
Sircar, A. K. and Lamond, T. G. (1972) *Rubber Chem. Technol.* **45**, 329.
Sircar, A. K. and Lamond, T. G. (1973a) *J. Appl. Polym. Sci.* **17**, 2569.
Sircar, A. K. and Lamond, T. G. (1973b) *Rubber Chem. Technol.* **46**, 178.
Sircar, A. K. and Lamond, T. G. (1973c) *Thermochim. Acta* **7**, 287.
Sircar, A. K. and Lamond, T. G. (1975a) *Rubber Chem. Technol.* **48**, 301.
Sircar, A. K. and Lamond, T. G. (1975b) *Rubber Chem. Technol.* **48**, 631.
Sircar, A. K. and Lamond, T. G. (1975c) *Rubber Chem. Technol.* **48**, 640.
Sircar, A. K. and Lamond, T. G. (1975d) *Rubber Chem. Technol.* **48**, 653.
Sircar, A. K. and Voet, A. (1970) *Rubber Chem. Technol.* **43**, 1327.
Smith, D. A. (1966*a*) *Kaut. u Gummi* **19**, 477.
Smith, D. A. (1966*b*) *J. Polym. Sci. Part B* **4**, 215.
Stanton Redcroft Ltd., Cooper Mill Lane, London.
Stothers, J. B. (1972) *Carbon-13 NMR Spectroscopy,* Academic Press, London.
Talalay, L. (1963) *Rubber Chem. Technol.* **36**, 581.
Toy, M. S. and Stringham, R. S. (1978) *Appl. Polym. Spectrosc.* Vols 1–5, Academic Press, New York.
Trojer, L. and Ericsson, I. (1978) *J. Chromatog. Sci.* **16**, 345.
Truett, W. L. (1977) *Wilks Scientific Application Report No. 9.*
Underwood, E. E. (1969) *Quantitative Stereology*, Addison-Wesley, Reading, Mass.
Werstler, D. D. (1980) *Rubber Chem. Technol.* **53**, 1191.
Wilkes, C. E. (1977) *J. Polym. Sci., Polym. Symp.* **60**, 161.
Wilkes, C. E., Carman, C. J. and Harrington, R. A. (1973) *J. Polym. Sci., Part C.* **43**, 237.
Wilson, C. W. (III) (1963) *J. Polym. Sci.* **A1**, 1305.
Wilson, C. W. (III) and Santer, E. R. Jr. (1965) *J. Polym. Sci., Part C.* **8**, 97.
Yeager, F. W. and Becker, J. W. (1977) *Analyt. Chem.* **49**, 722.

Chapter 9

Polymer Characterization

Two features of polymer characterization are of special interest to the rubber analyst, the first being concerned with the molar mass of the polymer, and the second with details of the microstructure of the polymer chain.

MOLAR MASS

It will be appreciated that a polymer does not consist of a large number of molecules of identical molar masses, and thus there is no such value as *the* molar mass of a polymer. Instead there are most likely to be quoted two values called *the number-average molar mass* ($\bar{M}_n$) and *the weight-average molar mass* ($\bar{M}_w$).

A third value, designated $\bar{M}_z$, may occasionally be met. Described as the *Z-average molar mass* it has no trivial name and is the third term of the power series from which the other two molar masses are derived:

$$\bar{M}_n = \frac{\sum_{i=1}^{\infty} n_i M_i}{\sum_{i=1}^{\infty} n_i}; \quad \bar{M}_w = \frac{\sum_{i=1}^{\infty} n_i M_i^2}{\sum_{i=1}^{\infty} n_i M_i}; \quad \bar{M}_z = \frac{\sum_{i=1}^{\infty} n_i M_i^3}{\sum_{i=1}^{\infty} n_i M_i^2}$$

where n_i is the number of molecules of molar mass M_i. The Z-average molar mass is especially sensitive to high molar mass components in the polymer.

The difference between number- and weight-average molar masses is important. The weight-average value is always greater than the number-average one except in an ideal system of uniform molar mass (monodisperse system) when the two are equal. This leads to the ratio $\bar{M}_w/\bar{M}_n$ being used to describe the *dispersity*, or spread of molar masses in a sample. $\bar{M}_w$ is more sensitive to high molar mass components whilst the converse is true for $\bar{M}_n$. It is a simple matter to illustrate the difference by considering two polymers, one in which all the molecules have a molar mass of 100 000, and the other in which all the molecules are of exactly 10 000 molar mass. In the former case $\bar{M}_n = \bar{M}_w = 100\ 000$, whilst in the latter, $\bar{M}_n = \bar{M}_w = 10\ 000$.

If *equal weights* of each are mixed, the resulting measured values would be

$$\bar{M}_n = 18\ 200 \qquad \bar{M}_w = 55\ 000$$

but if *equal numbers of molecules* of each are taken,

$$\bar{M}_n = 55\ 000 \qquad \bar{M}_w = 92\ 000$$

In subsequent sections dealing with methods of measurement, the type of molar mass so measured will be indicated.

There are many methods for determining some form of average molar mass of a polymer but relatively few for studying the distribution of that molar mass. Both these properties are important in helping to understand variations in processing such as mastication (Baijol, 1972), cold flow characteristics (Purdon and Mati, 1966) and adhesive properties (Koldunovich *et al.*, 1968). A number of the more relevant and successful ones will be discussed in some detail.

End Group Analysis ($\bar{M}_n$)

With certain linear polymers it is possible to estimate the number of end groups by chemical analysis and so derive the number-average molar mass. The *total* number of chain ends is twice the number of polymer molecules but if each polymer molecule contains one group of a particular type at the end of its chain then the number of that end group equates with the number of molecules. It is obviously of fundamental importance that the particular group which is the subject of the analysis is confined only to the end of each chain and also, with branched polymers, that each branch is not terminated with the group of interest. (Obviously this situation would provide data on branching if each branch was so tipped, and a 'true' value of the molar mass was obtained by a different method.) Given these criteria it will be apparent that there are innumerable methods of analysis and it must be up to the analyst to find the most suitable one for any particular end group.

Techniques to be considered include chemical methods, used by Ogg *et al.* as early as 1945 to estimate terminal hydroxyl groups, infrared spectroscopy, first used by Pfann *et al.* a year later, and also nuclear magnetic resonance, pyrolysis–gas chromatography and mass spectrometry. It may also be advantageous to react the functional group chemically prior to analysis. Heacock (1963) estimated carboxyl groups in the presence of carbonyl groups by reacting the former with sulphur tetrafluoride to obtain the thionyl halides which were then quantitatively measured by infrared spectroscopy, whilst Edwards and Loadman (1976) determined the molar mass of hydroxyl tipped polystyrenes by reaction with hexamethyldisilazane and trimethylchlorsilane in pyridine solution followed by measurement of the trimethylsilyl end group relative to a quantitatively added standard using NMR spectroscopy. The replacement of one proton by nine greatly enhances the sensitivity and usefulness of the technique and $\bar{M}_n$ values up to 80 000 have been measured.

Measurement of Colligative Property ($\bar{M}_n$)

A colligative property is one which depends primarily upon the number of molecules in the system and not upon their nature. Perhaps the earliest one observed was the relationship between the depression of freezing point and the concentration of the freezing solution (cryoscopy) enshrined as *Blagdens Law* some two hundred years ago. Others include the elevation of boiling point (ebulliometry), the reduction of osmotic pressure (membrane osmometry, MO) and the reduction of vapour pressure (vapour pressure osmometry, VPO).

Under ideal conditions a general equation defines the calculation of $\bar{M}_n$:

$$\frac{X}{c} = \frac{K}{\bar{M}_n} + bc \qquad (9.1)$$

where X is the colligative property, c is the concentration of the solution, and K and b are 'constants' but different for each technique and for each piece of equipment.

In practice a series of solutions of varying concentrations is prepared and the colligative property measured for each. A graph is then plotted, as illustrated in Fig. 9.1, of X/c against c, the number-average molar mass being the reciprocal of the intercept of the plot, after extrapolation to infinite dilution (c =0), with the abscissa. This method requires a knowledge of the constant K which is normally obtained by calibration against standards of known molar masses.

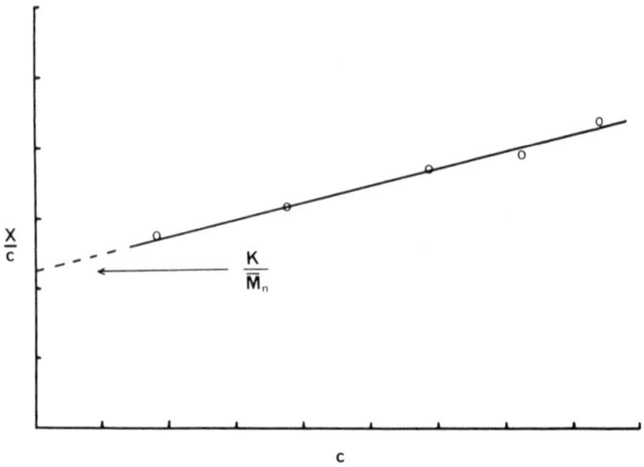

FIG. 9.1. Diagrammatic illustration of the determination of $\bar{M}_n$ from colligative property measurements (X).

In rare instances the solutions do not behave ideally and the plot obtained as above shows a distinct curvature due to the concentration dependence of the colligative property becoming significant. In these cases a plot of $\sqrt{X/c}$ *vs.* c usually provides an acceptably straight line for reliable extrapolation to zero concentration.

Cryoscopy and Ebulliometry

Cryoscopy and ebulliometry can be considered together briefly as they do not feature largely in polymer analysis although a number of publications, such as those of Newitt and Kokle (1966) (cryoscopy) and Ezrin (1968) (ebulliometry), have appeared. These indicate that the techniques are valid up to molar masses of some 30 000, although there are problems with supercooling and frothing respectively. Commercial instrumentation does not, however, appear to be available, probably due to the much greater ease of operation of the vapour pressure osmometer.

Vapour Pressure Osmometry

The technique of indirectly measuring the lowering of vapour pressure of a solvent due to the presence of a dissolved material was first proposed by Pasternack *et al.* (1962). It is based on measurement of the temperature difference between droplets of pure solvent and of polymer solution maintained in an isothermal atmosphere saturated with the solvent vapour. The temperature difference results from the different rates of solvent evaporation from and condensation on to the two droplets.

As with the two previous methods, the colligative property, a temperature difference, is measured electronically and the value of K is obtained by calibration against appropriate standards of known molar masses.

Over the last ten years there have appeared a number of papers which question the validity of an *absolute constant* in the determination by VPO of a molar mass and show how the 'constant' K varies with molar mass; see Bersted (1973), Brzezinski *et al.* (1973), Morris (1977), and Marx-Figini and Figini (1980). Edwards (1977) tabulated K values for a range of molar mass standards as shown in Table 9.1.

TABLE 9.1

RELATIONSHIP BETWEEN $\bar{M}_n$ AND K (SOLVENT, TOLUENE AT 40 °C)

Substance	$\bar{M}_n$	K
8-Hydroxyquinoline	145	10 670
Hexachlorobenzene	285	11 600
Polystyrene standard	970	12 300
Polystyrene standard	2 300	13 300
Polystyrene standard	8 500	14 000
Polystyrene standard	17 500	14 950

However, examination of the literature quoted shows that all these studies were carried out using a Hewlett Packard 302B instrument. When Edwards (1981) repeated this work with a Corona-Wescan 232A vapour pressure osmometer he found that K was indeed a constant, an observation also made by Burge (1979). It would seem, therefore, that each analyst must calibrate his particular instrument for the full molar mass range over which it is likely to be used, since subtle design and manufacturing differences are obviously important.

Membrane Osmometry

Membrane osmometry is the last of the techniques to be considered which relies upon a colligative property of a polymer. The success of this technique depends upon the membrane between the polymer solution and solvent, being permeable to the latter but completely impermeable to the polymer molecules in solution. In practice this means that whereas VPO finds application in the 500–50 000 $\bar{M}_n$ range, MO is restricted to polymers having no components with molar masses less than 15 000; for unfractionated polymers this means an effective minimum $\bar{M}_n$ of 50 000.

The principle of operation is illustrated in Fig. 9.2. The solution of the sample is placed above the membrane and the height of the solvent reservoir is automatically adjusted to keep the bubble in line with the optical detector, and thus equalize the rate of migration of solvent molecules from both sides of the membrane. The colligative property is then directly measured as *h* (the height difference between the membrane and solvent meniscus in cm). In practice the displacement *h* is often converted to a voltage and plotted on a chart recorder. This allows the observation of slow drifts due to diffusion of relatively low molar mass

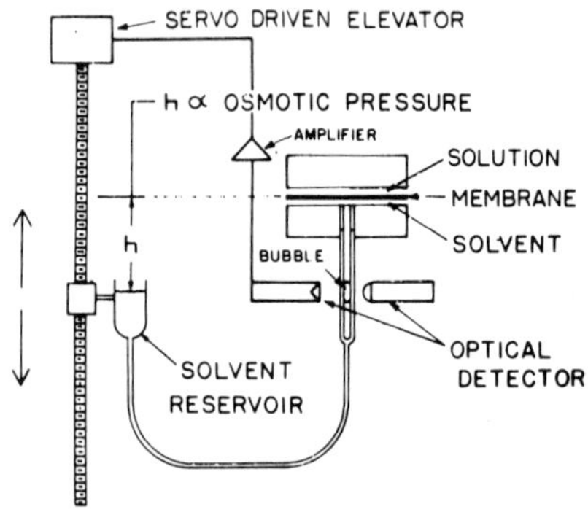

FIG. 9.2. Principle of operation of a membrane osmometer (Mecrolab 500). (Courtesy Hewlett Packard Ltd.)

components, if present, and extrapolation back to zero time to obtain a realistic value for $\bar{M}_n$ in spite of their presence. Elias (1968) reviewed theoretical treatments of this problem and showed that even such extrapolation fails to provide a 'true' figure.

Unlike the previously described techniques, membrane osmometry requires no calibration as the constant $K = RT$ where R is the gas constant and T the absolute temperature.

Viscometry ($\bar{M}_n$ or $\bar{M}_w$)

The use of a viscometer to measure the viscosity (η) of a solution of a polymer, followed by the calculation of intrinsic viscosity ([η]) and hence its molar mass, has been a standard procedure for a number of years, with one of the earliest reports being that of Staudinger and Heuer (1930).

Most analysts will have used a viscometer since schooldays and thus the apparatus requires little description, although ISO 3105 may be consulted for the design of various accepted glass capillary viscometers, the type generally used in measuring polymer solution viscosity. The principle of operation is extremely simple in that the time (t) taken for a solution of known concentration to flow between two marks on a capillary tube is compared with the time taken by the solvent (t_0), and the ratio is a measure of the viscosity of that solution. Full practical details are described in the British Standard BS 5858-1980. Successive dilutions afford a range of concentrations and times (t) from which the intrinsic viscosity $[\eta]$ may be calculated.

The viscosity of the solution (solvent) = $\eta(\eta_0)$ from which η_{sp} (the specific viscosity) may be obtained:

$$\eta_{sp} = \frac{\eta - \eta_0}{\eta_0} = \frac{t - t_0}{t_0} \tag{9.2}$$

and hence $[\eta]$, using either the Huggins equation (1942),

$$\frac{\eta_{sp}}{c} = [\eta] + K_H[\eta]^2 c \tag{9.3}$$

or the Schultz–Blaschke equation (1964),

$$\frac{\eta_{sp}}{c} = [\eta] + K_H[\eta]\eta_{sp} \tag{9.4}$$

Several authors, including Rudkin and Wagner (1975), have described 'one point' methods for determining intrinsic viscosity but Tidd (1976) finds these to have a degree of concentration dependence when applied to natural rubber and does not advocate their use unless sample limitation so demands. Recently Khan and Bhargava (1980) have published a new mathematical approach to the *one point* method for polystyrene and styrene–acrylonitrile copolymers.

Layec-Raphalen and coauthors (1979) raised the problem of the association of macromolecules in dilute solution, together with the effect that this had on viscosity measurement, and showed how association constants can be calculated from viscosity measurements.

Conversion of the intrinsic viscosity $[\eta]$ to a molar mass depends upon the Mark–Houwink–Sakurada expression (eqn (9.5)):

$$[\eta] = KM^{\alpha} \tag{9.5}$$

where K and α are empirical constants. Depending upon the source of these constants, M can be either number-average ($\bar{M}_n$) or weight-average ($\bar{M}_w$). Calculations based on molar mass methods such as membrane osmometry (MO), vapour pressure osmometry (VPO) or end group estimation will give $\bar{M}_n$, whilst those based on light scattering (LS) or sedimentation measurements will give $\bar{M}_w$. Many hundreds of K and α values are tabulated in the *Polymer Handbook*, edited by Brandrup and Immergut (1975), together with solvents, temperatures and techniques used.

Theoretical arguments have been put forward to suggest that eqn (9.5) is not valid over a wide range of molar masses when M refers to $\bar{M}_n$ or $\bar{M}_w$ (see Kurata and Stockmeyer, 1963) but it does seem in practice that valid results for many polymer–solvent systems are obtained if K and α have been determined with reference samples spanning the range of interest. More recently Dondos (1977) has used single and dual solvent systems to study a range of polystyrene samples and showed that, whereas a classical plot of $\log[\eta]$ *vs.* $\log \bar{M}_w$ deviates from linearity at values of $\bar{M}_w < 150\ 000$, a linear relationship exists when $1/[\eta]$ is plotted *vs.* $1/M^{1/2}$. Thus a reliable calibration plot may usefully be constructed for any polymer subject to regular analysis.

Light Scattering ($\bar{M}_w$)

The use of equipment to measure the light scattering behaviour of solutions of polymers and the calculation therefrom of *weight-average molar masses* appears never to have reached the popularity of the techniques described earlier for the determination of *number-average molar mass.* In recent years it has also been eclipsed by gel permeation chromatography (GPC). This situation promises to change with the introduction of low angle laser light scattering (see below).

Much of this earlier neglect has been due to the calculations required together with the lack of reproducibility of results obtained from different workers, and the difficulties experienced in the analysis of standard samples due to problems of sample preparation. Nevertheless the technique can be precise and has been used successfully on samples with molar masses between 10 000 and 10 000 000. The instrumentation is extremely simple, consisting of a monochromatic polarized light source which shines into a solution cell mounted at the centre of the axis of rotation of a prototube which detects the scattered light. Rotation of the prototube is done manually and the angle at which the scattered light is observed is measured.

The problems of sample preparation can be divided into two: choice of a solvent, and clarification of the resulting polymer solution. The former affects the sensitivity and, hence, the accuracy of measurement, with the two major requirements being that the refractive index of the solvent is as different as practicable from that of the polymer, and that the solvent itself is low-scattering. It is the latter, however, which gives the biggest source of variability in results. A general procedure for clarifying solutions of polymers is that of sequential filtration and centrifugation, typically as described by Jennings (1966). Unfortunately when one reads comments such as those of Doty and Bunce (1952), who claim that purification by centrifugation becomes more difficult as the concentration of the solution increases, and Witnauer *et al.* (1955) who reach exactly the opposite conclusion, one realizes that a careful study of each polymer system is required to establish the most viable working conditions. It is also advisable to study an unknown polymer in a selection of solvents to reduce the possibility of artefacts such as polymer–polymer or polymer–solvent interactions.

The basic equation which relates the molar mass to the extent of light scattering is the *Debye equation* (eqn (9.6)) derived by Debye (1944, 1947):

$$K\frac{c}{R_\theta} = \frac{1}{\bar{M}_w P(0)} + 2A_2 c \qquad (9.6)$$

This assumes that the scattering particles are small, compared with the wavelength of the light. If this is invalid, allowance must be made for dissymmetry of scattering throughout the molecule:

$$K\frac{c}{R_\theta} = \frac{1}{\bar{M}_w P(\theta)} + 2A_2 c \qquad (9.7)$$

where K is a calculated constant, requiring a knowledge of, amongst other things, the refractive index of the solution and its variation with polymer concentration (Brice and Halwer, 1951), c is the concentration, R_θ the excess scattering intensity of the solution over the solvent, and $P(\theta)$ the particle scattering function which reduces to unity when the concentration (c) and the scattering angle (θ) are both zero.

Calculation of R_θ from the experimentally obtained data became the subject of a number of papers and is dependent upon the solvent system and cell arrangement used—see Leblanc (1962), Kratohvil (1966), Miyake *et al.* (1970). These data are then treated according to the method of Zimm (1948) to afford the most accurate graphical method for the derivation of $\bar{M}_w$.

A typical Zimm plot is illustrated in Fig. 9.3 for Kc/R_θ *vs.* $\sin^2\theta/2 + kc$ (where k is a convenient arbitrary constant, in this case 100).

Measurements are made at different angles ($\theta_1 - \theta_8$) for a range of solution concentrations (c_1–c_4) and these are plotted as shown (○). Extrapolation of both concentration and angle plots to a common intercept (at $\theta=0$, $c=0$) (●) with the abscissa gives the weight average molar mass ($\bar{M}_w$) as the reciprocal of the intercept, i.e. 1 030 000 in the illustrated example.

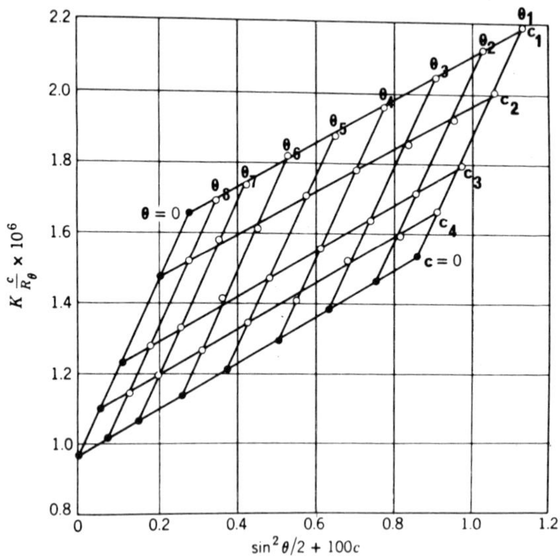

FIG. 9.3. Zimm plot showing the light scattering from a sample of polystyrene in butanone (Billmeyer, 1971). (Courtesy John Wiley & Sons.)

There is a belief that the angular dissymmetry of the scattered light will afford information on the molecular shape of a polymer and whilst this is a reasonable theoretical deduction the practical position appears to be that this may only be so under exceptional circumstances (Benoit, 1968). Carpenter (1966) has concluded that a similar situation prevails with regard to dispersity measurements using this technique. It should also be borne in mind that severe restrictions apply to the interpretation of data obtained from a copolymer solution, although Benoit and Bushuk (1958) developed a theory to encompass copolymers which has been evaluated by a number of workers such as Benoit and Leng (1961), Prud'homme and

Bywater (1971), Shimura *et al.* (1964), Jordan (1968), Spatorico (1974), and Ali (1978); their results, however, show varying degrees of success.

A relatively recent development which overcomes many of the shortcomings of the *classical* light scattering technique and could substantially affect the acceptability of light scattering in the routine rubber-laboratory has been the introduction of low angle laser light scattering (LALLS). The main advantages claimed for this technique are that only one small angle need be used to obtain measurements and that the system can be calibrated absolutely, obviating the complex mathematics of conventional LS. The increased sensitivity provided by the laser light source allows relatively dilute solutions and very low cell volumes (0·1 μl) to be used so that solution clarification is much easier. McConnell (1978) shows how such a system can be combined with a gel permeation chromatograph to obtain very accurate molar mass data on the GPC eluate.

Gel Permeation Chromatography ($\bar{M}_n$, $\bar{M}_w$, $\bar{M}_z$)

Gel permeation chromatography is a particular form of liquid chromatography in which a solution of a polymer is pumped through a series of columns, each packed with a gel of specific pore size, under constant flow conditions, to a suitable detector positioned at the end of the final column. The range of pore sizes is such that it compares with the dimensions of the polymer molecules; thus the largest molecules, which can penetrate few pores, take the shortest route through the column and are eluted first whilst the smaller ones can enter proportionately more pores and take progressively longer to reach the detector.

There is no doubt that, since the early applications of GPC to the characterization of molar mass distributions of polydisperse systems by Vaughan (1960), Brewer (1961) and Moore (1964), the technique has become the one chosen for the routine study of a vast range of polymers (although the calculations carried out to obtain accurate molar mass values rely on calibration data for each system obtained by the methods discussed earlier).

Although conceptually simple, the reproducibility of experimental conditions can generate severe practical problems if care is not taken and the right equipment not chosen. Undoubtedly the biggest potential problem is the pumping system, as the flow of solution through the columns must proceed at an absolutely reproducible and constant rate. Not only does this require a relatively sophisticated pump and associated controls, but also care must be taken in sample preparation using filtration or centrifugation to remove any gel or other insolubles which might progressively block the columns.

Problems with dead space, injection systems and column channelling also occur and, whilst any chromatographer will be aware of these, it is important to note that in the field of GPC they become much more significant as one is rarely concerned just with separating peaks, but more with the absolute retention time (volume) and detailed peak shape.

A wide variety of detectors is available, with the most popular still being the differential refractometer, as used by Moore (1964). Calculations generally assume that the refractive index of a polymer is independent of its molar mass and, whilst this is certainly true when the molar mass exceeds a few thousand, Barrall *et al.* (1968) have shown how a small correction is required for low molar mass polymers. It must also be realized that differential refractometry is not satisfactory for monitoring the fractionation of a mixture of chemically different polymers unless response factors are known for each component.

Spectroscopic detectors, either ultraviolet or infrared, are also used extensively and, if an instrument with a variable wavelength facility is used, there is considerable flexibility in the breadth and selectivity of the data which may be acquired.

Both of these techniques have sensitivities many times greater than that of the differential refractometer but the sensitivity for two polymers may vary by orders of magnitude depending upon their different structures and the specificity of the monitoring wavelength. Ross and Castro (1968) used infrared (2940 cm^{-1}) to obtain data on polyethylene with perchlorethylene as solvent whilst Terry and Rodriguez (1968) monitored methyl methacrylate (1731 cm^{-1}) and styrene (698 cm^{-1}). More recently, Birley *et al.* (1978) have discussed theoretical aspects of quantitative infrared spectroscopic detection. Runyon *et al.* (1969) used an ultraviolet spectrophotometer (260 nm) to detect styrene and, sequentially, a differential refractometer to give 'total detection' for styrene–butadiene copolymers. Spectroscopic types of detectors are particularly useful as the flow can be stopped at any time and a full spectrum obtained. Cooper *et al.* (1969) have shown that there is essentially no loss of resolution with flows interrupted for several hours!

Many other detection systems have been used, including viscometry, thermal conductivity and LALLS as already described by McConnell (1978).

In recent years there has been a tendency to use shorter, narrow bore columns packed with smaller diameter particles in order to reduce the analysis times of GPC runs. This was originally described as high-pressure GPC (HPGPC) but the wording was soon altered to high-performance GPC. Early workers such as Gudzinowicz and Alden (1971) reduced analysis times appreciably but, because they did not have packing materials of sufficiently small particle diameter, their resolution suffered. This problem has now been overcome with the availability of particles less than 10 μm in diameter, compared with the earlier diameters of 50 μm and much has been made of the fact that molar mass distribution measurements can now be carried out in less than ten minutes (see Kato *et al.*, 1974; Kirkland, 1976; Unger and Kern, 1976). Edwards (1981), however, was concerned at differences found in the analysis of natural rubber using both a 'normal' GPC system with column packing particles some 50 μm in diameter and an HPGPC system where the particles were about 10 μm diameter. Whereas the former gave a typical bimodal distribution for the natural rubber (see Subramanium, 1972), the latter showed a steady decrease in apparent $\bar{M}_n$ and a reduction in the relative concentration of those components of higher molar masses with increasing flow rate. The area under each GPC trace was normalized against that for a polystyrene standard of broad molar mass distribution (MWD1 from the NPL).

The results, reported for both natural rubber and a synthetic polyisoprene, are given in Table 9.2 and show that the passage of a dilute solution of polyisoprene of higher molar mass through tightly packed

TABLE 9.2

MOLAR MASSES OBTAINED AT DIFFERENT FLOW RATES USING 10 μm PARTICLES IN HPGPC

Flow rate $cm^3\ min^{-1}$	*Back pressure* *p.s.i*	$\bar{M}_n$	$\bar{M}_w$	$\bar{M}_n/\bar{M}_w$
		Cariflex IR 305		
0·3	150	593 000	1 942 000	3·28
0·7	500	377 000	1 078 000	2·88
2·0	1 200	302 000	797 000	2·64
4·0	1 850	237 000	611 000	2·58
		NR (RRIM 501)		
0·3	150	152 000	828 000	5·45
0·7	500	145 000	664 000	4·57
2·0	1 200	124 000	433 000	3·50
4·0	1 850	112 000	344 000	3·07

microparticulate columns, at high flow rates, produces shear forces sufficient to cause severe degradation of the largest molecules. It is imperative, therefore, that molecular weight determinations are carried out at the slowest flow rates consistent with practical considerations and several runs may be necessary to determine valid conditions. Similar observations have been made for polystyrene (Kirkland, 1976) and polyisobutene (Huber and Lederer, 1980).

Having discussed the problems of obtaining a reliable representative chromatogram it is now necessary to consider how molar mass data can be obtained from the chromatogram.

A primary calibration is usually carried out with polystyrenes of as low a dispersity as can be obtained. The 'most representative' molar mass can be described as $(\bar{M}_n\bar{M}_w)^{1/2}$, so as $\bar{M}_n/\bar{M}_w \rightarrow 1$, $\bar{M}_n = \bar{M}_w = M_{rep}$. This is a reasonable approximation if $\bar{M}_n/\bar{M}_w < 1{\cdot}1$ and a plot of elution volume at peak maximum (V) against log M_{rep} gives the primary calibration. From this, any chromatogram can be expressed in terms of 'molar mass polystyrene equivalent' which, although useful in certain empirical applications, does not provide absolute molar mass data for systems other than linear polystyrene.

If samples of a broader molar mass distribution are used to construct a calibration curve, and the $\bar{M}_n$ and $\bar{M}_w$ values are available, the relationship between molar mass and elution volume, assuming the calibration curve to be linear over the molar mass range of a given fraction, becomes

$$\ln M = AV + D \tag{9.8}$$

where M is the molar mass of the fraction eluted at volume V. A number of methods, such as those by Hamielec *et al.* (1969) and Roy (1976), have been used to evaluate the constants A and D but in essence they are iterative methods based on the assumption that only A significantly influences the calculated $\bar{M}_w/\bar{M}_n$ ratio. An initial value of A is obtained by assuming a log-normal molar mass distribution, when it can be shown that

$$A = \frac{4[\ln \bar{M}_w/\bar{M}_n]^{1/2}}{W} \tag{9.9}$$

where W is the peak width. After an initial calculation, the process is repeated with adjustments to A until calculated and experimental values of $\bar{M}_w/\bar{M}_n$ agree. The availability of computer- or microprocessor-controlled GPC equipment has made this a much less tedious calculation than it used to be.

In 1967 Benoit *et al.* proposed the *Universal Calibration Mark-Houwink Method* and showed that for all polymers a single calibration curve exists for the relationship between the hydrodynamic volume $\log[\eta]M$ and the elution volume V in a given solvent system. In practice there may be slight deviations from the universal curve at its extreme ends, for reasons of molecular shape, as described and discussed by Ambler and McIntyre (1975).

At any given elution volume for polymer 1 and polymer 2 the relationship given in eqn (9.10) holds:

$$[\eta_1]M_1 = [\eta_2]M_2 \tag{9.10}$$

(where $[\eta_1] = K_1 M_1^{\alpha^1}$ as given in eqn (9.4)). From eqn (9.10) is derived the basic universal calibration equation:

$$\log M_1 = (1+\alpha_1)^{-1}\log\frac{K_2}{K_1} + \frac{1+\alpha_2}{1+\alpha_1}\log M_2 \tag{9.11}$$

The Mark-Houwink constants for the two polymers must of course be obtained at the same temperature and in the same solvent.

Equation (9.10) can be rewritten as

$$\frac{[\eta_1]}{[\eta_2]} = \frac{\bar{M}_{n_2}}{\bar{M}_{n_1}} = \frac{\bar{M}_{w_2}}{\bar{M}_{w_1}} = B \tag{9.12}$$

where B is designated the *Benoit factor* and is a constant for the particular polymer being considered.

If a primary calibration has been obtained as described earlier (and as is shown in Fig. 9.4 for monodisperse samples of polystyrene), it is only necessary to have one monodisperse standard sample of another polymer to construct its calibration curve. A knowledge of $\bar{M}_n$ or $\bar{M}_w$ together with the measurement of its elution volume under identical conditions to those used to obtain the polystyrene calibration curve enables the point X to be found, and B calculated after obtaining the *polystyrene equivalent molar mass P* at the same elution volume. It is then a simple matter to construct the full calibration curve for the new polymer.

Using this calibration curve, it is now possible to convert a graphical printout of detector response $h(V)$ at elution volume V into a normalized molar mass plot, showing $w(M)$—the fractional weight of a polymer having molar mass M—plotted against M, via eqn (9.13):

$$w(M) = \frac{1}{M}h(V)\frac{\mathrm{d}V}{\mathrm{d}(\log M)} \tag{9.13}$$

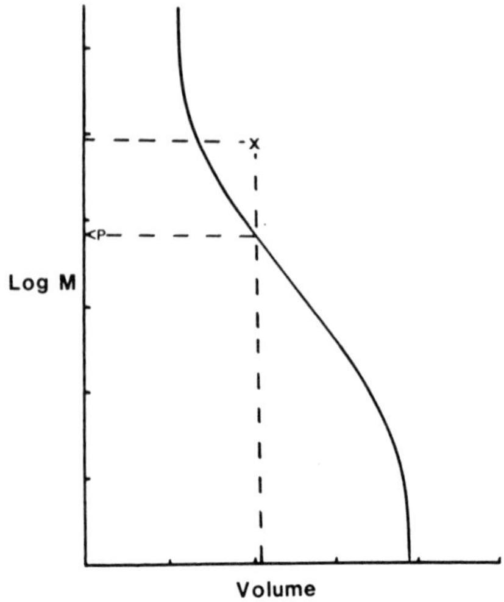

FIG. 9.4. Representative GPC plot of log M v. elution volume.

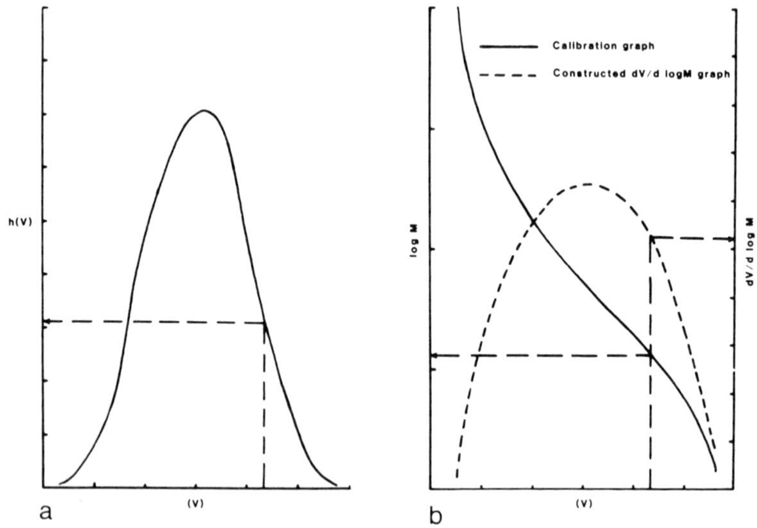

FIG. 9.5. (a) Experimental data recorder response v. elution volume. (b) Calibration graph and calculated relationship dV/d (log M) v. elution volume.

Figure 9.5 shows the experimental data, in the form of a graphical printout, together with the appropriate calibration graph and, superimposed (and calculated from the latter), a graph of $dV/d(\log M)$ v. $\log M$. It is apparent that at any volume V one can read off values for $h(V)$, $dV/d(\log M)$ and $\log M$ (hence M), and calculate $w(M)$.

Implicit in eqn (9.13) is the condition that the detector response is directly related to the weight concentration of the polymer. If this is not valid then a correction must be introduced. It will also be noted that eqn (9.13) contains the reciprocal of the slope of the calibration plot, an important point in converting the V axis to an M axis. This can be ignored if the primary calibration is linear over the whole range of molar masses being studied, but *must* be included if curvature is present—as it almost inevitably is.

Inevitably the situation is never as clear cut as it seems. As already mentioned, Ambler and McIntyre (1975) described how there are deviations from the universal calibration with extremes of spatial configurations of the polymer chains whilst Narasimhan *et al.* (1981) illustrated a concentration effect, and a 'second polymer' effect, on the elution volumes of polystyrenes and polybutadienes.

For a more detailed consideration of many of the problems of converting primary GPC data into absolute molar mass values the reader is referred to papers by Letot *et al.* (1980), Chaplin and Ching (1980), Dawkins (1977), Gilding *et al.* (1981) and Busnel (1982).

No mention has been made so far of peak broadening. The GPC column does not have an infinite resolving power, and thus the observed peak is always broader than its true molar mass distribution warrants. Tung (1966) has shown how this may be corrected for, but as it is generally accepted that the effect is insignificant when $\bar{M}_w/\bar{M}_n > 2$, the usual situation for the rubber-analyst, it is not discussed further.

The techniques described so far in this chapter have a general applicability to the determination of the molar masses of polymers but a number of others have been used for specific applications, three of which are considered briefly.

Differential Scanning Calorimetry/Differential Thermal Analysis ($\bar{M}_n$)

These techniques have been discussed in detail in Chapter 8, but in the context of molar mass measurements we are particularly concerned with the measurement of glass transition temperatures (T_g).

Fox and Flory (1950, 1954) were the first to observe that for a homologous series of fractionated polystyrenes there is a linear relationship between the glass transition temperature and the inverse of the number-average molar mass:

$$T_g = T_{g\infty} - \frac{K}{\bar{M}_n} \qquad (9.14)$$

where T_g is the glass transition temperature for the particular polymer molar mass, and $T_{g\infty}$ the glass transition temperature for the same polymer of infinite molar mass. It will be appreciated that as $\bar{M}_n$ increases, the difference between T_g and $T_{g\infty}$ will become too small for accurate measurement and thus the method is only useful for materials of relatively low molar masses, typically up to 5000 for polyisoprene (Kow *et al.*, 1982), 50 000 for polystyrene (Loadman and Tinker, 1980) and 70 000 for polyacrylonitrile (Keavney and Eberlin, 1960). However, given that reference compounds covering the molar mass range of interest are available, this probably constitutes one of the simplest methods for rapidly checking the molar mass of a homopolymer of relatively low molar mass.

Ultracentrifugation ($\bar{M}_w$)
This is probably the most intricate of existing methods for determining the absolute molar mass of a high polymer but, although it shows a good deal of success in dealing with relatively compact high molar mass materials such as proteins, problems such as entanglements and extended chain effects render the results much less useful for the polymer analyst.

Sedimentation Effects ($\bar{M}_w$ or $\bar{M}_z$)
Many sedimentation experiments have been carried out under different experimental conditions, but they may be generally summarized as being time consuming and of restricted use whilst the results are complicated to evaluate. The interested reader is referred to representative publications on the determination of molar mass distribution by Scholte (1970), and on the use of a mixed solvent system to generate a density gradient, and hence separate polystyrenes of different tacticities, by Morawetz (1965).

MICROSTRUCTURE

Polymer microstructure is most simply defined as the arrangement of the various monomer units which constitute that polymer within the polymer chain. A study of the type and distribution of the monomeric species will give information on isomer specificity, stereospecificity and chain branching whilst, in the case of copolymers, data on the distribution of the chemically different monomers will indicate the randomness or blockiness of the polymerization. It will be apparent that these features are of crucial concern to both the manufacturer and user of synthetic polymers, as relatively small structural changes within the chain can have appreciable effects on the physical properties of the product manufactured therefrom.

If the monomer units are all identical and contain no asymmetric centres, then the structure is completely defined without ambiguity. This would be the situation with an ideal polyethylene $(CH_2)_n$. A monomer such as isoprene, however, could polymerize to give four isomerically different monomer units as illustrated below:

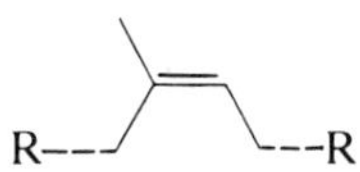

cis 1,4-

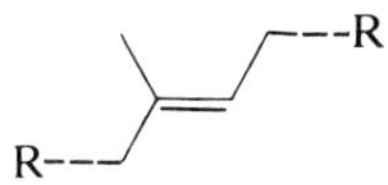

trans 1,4-

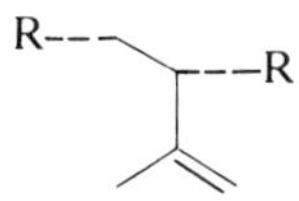

3,4-

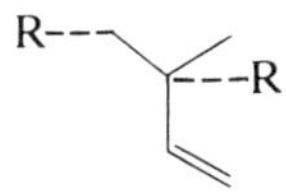

1,2-

It is therefore necessary not only to identify and quantify each of the four possible structures but also to obtain data, if possible, on the *monomer sequence distribution* throughout the chain.

On many occasions the repeat units within the polymer chain will contain asymmetric centres and then we must consider *tacticity*, which describes the relative configurations of these asymmetric centres. Such a monomer is illustrated diagrammatically below. It may polymerize with the configuration of the asymmetric centres identical (*isotactic*), regularly alternating (*syndiotactic*) or random (*heterotactic* or *atactic*).

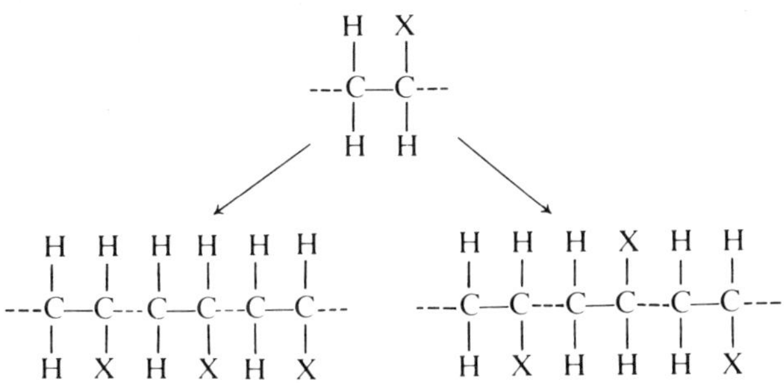

isotactic *syndiotactic*

An alternative nomenclature, which allows one to define in absolute detail larger monomer sequences, considers an existing chain and then adds the next monomer in either a *meso* configuration (m) relative to the last monomer of the chain (i.e. in the same sense) or in a *racemic* configuration (r) (i.e. the opposite sense). Thus m represents a pair of monomers in a like configuration, building to a triad (mm) analogous to the isotactic structure illustrated, whilst rr is analogous to the syndiotactic triad illustrated. The chains can then continue to grow in a specifically defined way:

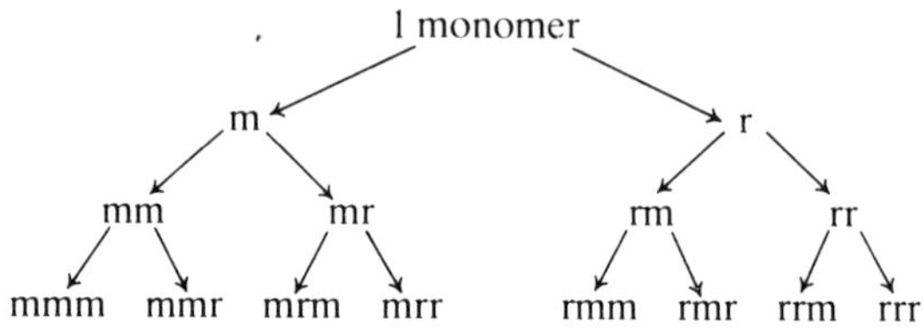

Although this is an area of extreme complexity, much of which is beyond the terms of reference of this book, the rubber-analyst should have some idea of the types of data which may be obtained.

Monomer Type

Exactly the same techniques may be used to study the types of monomer present as have already been described, in Chapter 8, on the instrumental analysis of polymers. Perhaps the most significant difference is that, whereas in the earlier applications we were generally concerned with identifying the major species, much of the microstructural analysis is concerned with the presence of low levels of specific isomeric structures.

As long ago as 1946, Field *et al.* used infrared spectroscopy to examine various polyisoprenes qualitatively and some years later Richardson and Sacher (1953) quantified the analysis in terms of all four possible isomers illustrated earlier. It is interesting to note that whereas one of the modern techniques for analysing a polymer blend by infrared spectroscopy is computerized curve matching, with reference spectra of the pure compounds, Richardson and Sacher confirmed their calculations by manually computing the summed spectra of specific regions from their derived data, and comparing these with the experimentally determined spectra. Cunneen and co-workers (1959) also used infrared spectroscopy to determine the *cis* and *trans* contents of isomerized natural rubber, whilst Kössler and Vodehnal (1964) proposed yet a third infrared method, but the advent of NMR spectroscopy, with the large numbers of data available therefrom, including sequence distribution, tended to restrict the deliberate development of IR in this field, although useful data often appear as a spin-off from other studies.

Pyrolysis–gas chromatography has been described in detail previously (see Chapter 8) and, as with IR, it will be apparent what sort of data are available, particularly if the monomer source of specific peaks can be identified. Recently Tsuge (1981) and Tsuge *et al.* (1980) described a combined pyrolysis–hydrogenation glass capillary gas chromatograph and illustrated how this was used to study the microstructure of polyethylene, polypropylene and copolymers of the two.

The use of DSC in the determination of polymer molar masses has been noted but it may also be used in specific applications to provide microstructural data. For instance Kow *et al.* (1982) point out that the T_g of synthetic polyisoprenes is linearly related to the 1,4 content, whilst inspection of the data given in Appendix II will show many occasions when a relationship between T_g and a microstructural feature exists. It is worth noting that changes in *cis* : *trans* ratio, both for isoprene and butadiene, do not affect the T_g.

Of all the techniques suitable for the determination of microstructure, NMR is undoubtedly the best and many thousands of papers have been published on virtually every aspect of polymer analysis. One example of proton (^{1}H) NMR and one of carbon (^{13}C) NMR will suffice to illustrate the point. Figures 9.6 and 9.7 illustrate the ^{1}H–NMR spectra of natural and a

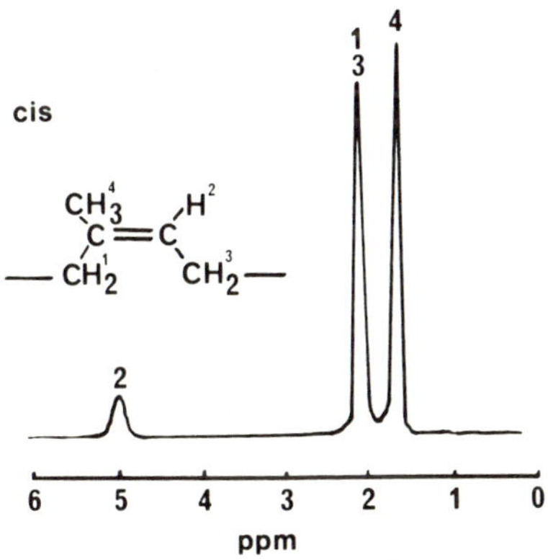

FIG. 9.6. ^{1}H-NMR spectrum of NR.

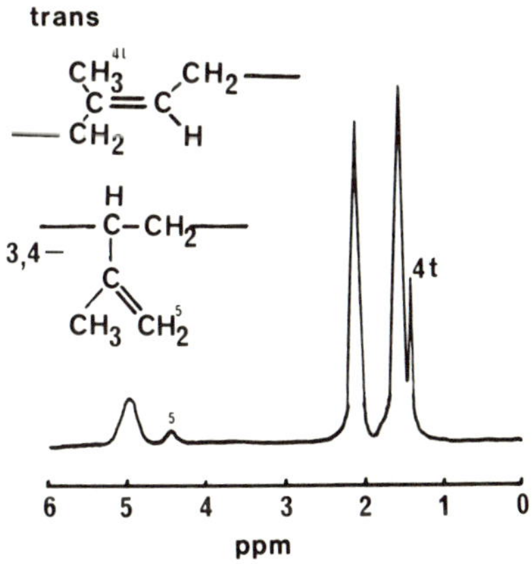

FIG. 9.7. ^{1}H-NMR spectrum of 3,4/ high *trans* PI.

TABLE 9.3

LOW *trans* CONTENTS IN ISOMERIZED NR, MEASURED BY ^{1}H-NMR AND IR

Sample	% trans *content*	
	NMR	*IR*
a	2·9	2·4
b	3·9	3·6
c	4·6	4·2
d	5·5	5·2

synthetic polyisoprene. The *trans* methyl group and 3,4 units are clearly observed and can be estimated by peak area integration, as originally reported by Golub *et al.* (1962). Loadman (1978) used a reference solution of pure natural rubber, with appropriate sensitivity and resolution adjustments, to define exactly the *cis* methyl peak within the *cis–trans* envelope, together with a modification of the infrared method of Cunneen *et al.* (1959), to measure the *trans* content of isomerized natural rubbers. The results in Table 9.3 show that good agreement can be obtained even at low levels of isomerization.

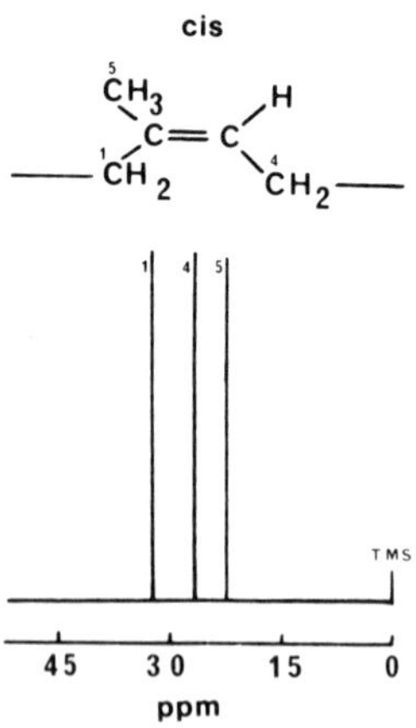

FIG. 9.8. ^{13}C-NMR line spectrum of NR.

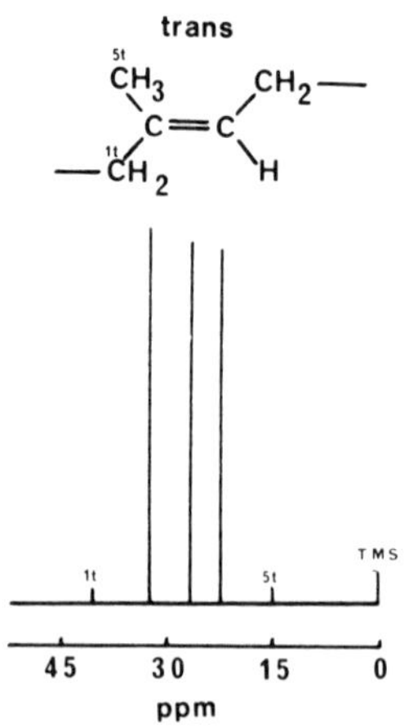

FIG. 9.9 ^{13}C-NMR line spectrum of isomerized NR.

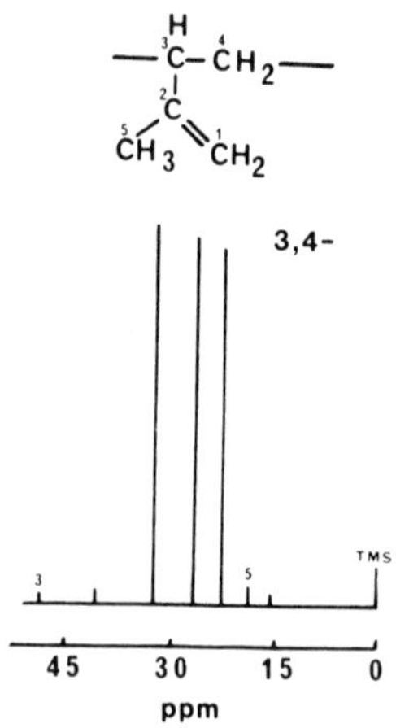

FIG. 9.10. ^{13}C-NMR line spectrum of synthetic (high *cis*) polyisoprene.

The set of spectra illustrated in Figs. 9.8 – 9.10 shows ^{13}C-NMR spectra of the *alkyl regions* of natural rubber, isomerized natural rubber and a synthetic polyisoprene.

As with ^{1}H-NMR the signals due to the *trans* and 3,4 species are clearly visible but with greatly improved resolution and, by using spectral accumulation techniques, very low levels of these may be measured, the limit of detection currently being of the order of 0·05–0·1 moles per 100 isoprene units.

Monomer Distribution

Of those techniques described for the identification of the monomeric species, all can contribute something to the study of monomer sequence but most are of little general significance compared with ^{13}C-NMR.

Infrared, or pyrolysis–infrared spectroscopy, can be used to distinguish between block and random copolymers; for instance the spectrum of a pyrolysate of an SBS block copolymer is the summed spectra of the pyrolysates of polystyrene and polybutadiene, quite different from that of a random SBR copolymer pyrolysate, but beyond this there is little detailed information available.

Tsuge (1981) extended the application of pyrolysis hydrogenation gas chromatography to a study of the high molar mass fragments from the pyrolysis of polyethylene and several polypropylenes of varying tacticity. An examination of those products containing greater than twelve carbons (PP tetramer) showed that information on stereoregularity could be obtained. He does, however, emphasize the points made earlier in the discussion on pyrolysis–gas chromatography of the need for a standardized PGC system to produce universally significant and correlatable data.

Differential scanning calorimetry provides the data already discussed which, in the appropriate situation, can afford information relating to the *blockiness* of a copolymer, and also an indication of the length of the blocks. Roovers and Toporowski (1974) and Kow *et al.* (1982) have studied the T_gs of star-shaped polystyrenes and polyisoprenes respectively. The polystyrene work established a relationship between T_g and the number of chain ends in polystyrene stars with molar masses < 90 000 but the latter authors found no variation with polyisoprenes of molar masses > 10 000. The probable explanation of this is that the polystyrenes had molar masses in the non-asymptotic region of their molar mass *vs.* T_g plot whereas the polystyrenes were all in the asymptotic region.

Perhaps one of the classic illustrations of the use of ^{1}H-NMR spectroscopy in the analysis of microstructural differences is that of polymethylmethacrylate (PMMA), shown in Fig. 9.11 and described by Bovey and Tiers (1960). Three distinct signals are observed due to the central methyls of the isotactic, heterotactic and syndiotactic 'triads' (or blocks of three monomers).

At about the same time, Bovey *et al.* (1959) and Bovey and Tiers (1960) reported the analysis of many styrene copolymers and showed that, as well as obtaining the styrene : butadiene ratio, there was information available on the size of the styrene blocks. Below about eight consecutive styrene units the aromatic peak appears as a singlet, but above eight a

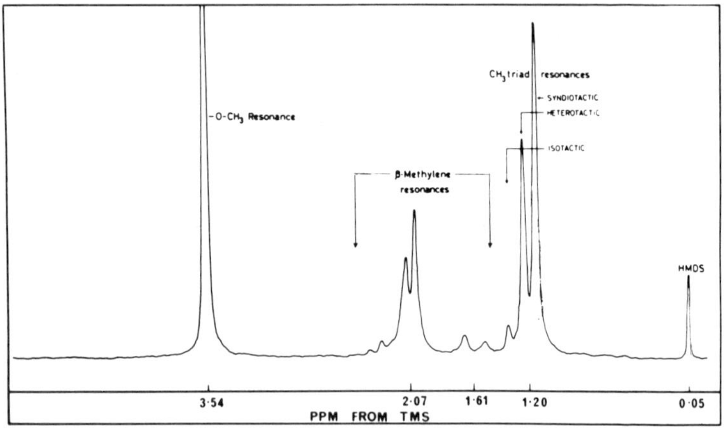

FIG. 9.11. ^{1}H-NMR spectrum of polymethylmethacrylate.

doublet is seen. Random polymerization requires 80% styrene before the average block size reaches eight, hence with most styrene-containing copolymers (if the styrene loading can simultaneously be obtained from the spectrum) this method gives a rapid qualitative criterion to distinguish between the random and block copolymers.

Of all the examples of the application of ^{13}C-NMR to the analysis of microstructure only one relatively simple one is illustrated here to indicate the amount of information available. This is for free radical initiated polyvinylchloride, and uses the data of Carman *et al.* (1971) and Carman (1973).

The ^{13}C-NMR spectrum is illustrated in Fig. 9.12. The spectrum shows five bands assignable to methylene carbons and a further seven (on expanding the spectrum) originating from methine carbons. This large number of signals is due to differences in chemical shifts of the carbons brought about by the variation in configuration of the neighbouring groups. The assignments made by Carman are shown on the spectrum with the rr and mr bands actually separating into two.

The particular advantages of ^{13}C-NMR are that one is able to make extremely precise calculations of a particular chemical shift using established incremental data for carbon atoms several times removed from the one being studied, and intrinsic instrumental resolution allows observation and quantification of these small differences.

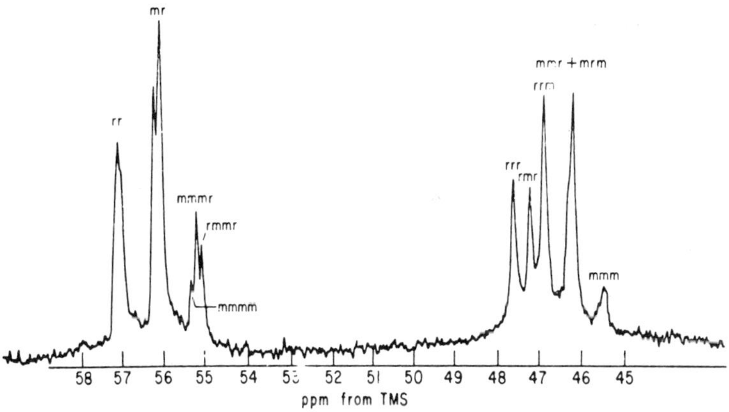

FIG. 9.12. ^{13}C-NMR 'noise decoupled' spectrum of free radical initiated PVC in 1, 2, 4-trichlorobenzene at 120 °C, 25·2 MHz (Carman, 1973). (Courtesy *Macromolecules*.)

The interested reader can obtain a great deal of information from publications by Chen (1968) and Randall (1977, 1979) which, as well as providing background data, discuss in detail many specific polymers. Werstler (1980) discusses the application of [^{13}C]-NMR to the analysis of cured filled elastomers.

METATHESIS

In 1967 Calderon *et al.* described a process which they called *olefin metathesis* whereby vinylic olefins, when treated with a catalyst of tungsten hexachloride, ethyl alcohol and ethyl aluminium dichloride, were transformed according to the scheme

$$\begin{array}{c} R_1CH{=}CHR_2 \\ + \\ R_3CH{=}CHR_4 \end{array} \longrightarrow \begin{array}{c} R_1CH \\ \| \\ R_3CH \end{array} + \begin{array}{c} CHR_2 \\ \| \\ CHR_4 \end{array} + \begin{array}{c} R_1CH \\ \| \\ R_4CH \end{array} + \begin{array}{c} CHR_2 \\ \| \\ CHR_3 \end{array}$$

Since that date a number of papers have been published which refer to the use of this technique in polymer structure elucidation. Thus Michailov and Harwood (1970) treated BR and SBR with 2-butene to obtain 2,6-octadiene from (1,4-) (1,4-) sequences of the BR and both 5-phenyl-2,8-octadiene and 4-phenyl cyclohexene from the (1,4-) (styrene) (1,4-) sequences. The products were identified by gas chromatography. A similar procedure was carried out by Stelzer *et al.* (1977) who used 4-octene as the metathesis olefin whilst Kumar and Hummel (1982) used a catalyst of tungsten hexachloride and tetramethyl tin with 4-octene to study the extent to which a crosslinked 1,4-polybutadiene could be solubilized with varying reaction times. Analysis was by GPC and they concluded that the soluble fraction (particularly the higher molar mass components) passed through a maximum value. More recently Hummel *et al.* (1982) have described how peroxide-cured 1,4-polybutadiene, filled with a wide range of materials, may be broken down with 1-octene and a catalyst of tungsten hexachloride so that the fillers may be quantitatively removed after filtration.

There appears little doubt that more will be published concerning the application of metathesis to the analysis of polymers in the coming years.

REFERENCES

Ali, S. (1978) *Brit. Polym. J.* **10**, 108.

Ambler, M. R. and McIntyre, D. (1975) *J. Polym. Sci., Polym. Letts. Edn.* **13**, 589.

Baijol, M. D. (1972) *Analyt. Chem.* **44**, 1377.

Barrall, E. M. II, Cantow, M. J. R. and Johnson, J. F. (1968) *J. Appl. Polym. Sci.* **12**, 1373.

Benoit, H. (1968) *J. Chem. Phys.* **65**, 23.

Benoit, H. and Bushuk, W. (1958) *Canad. J. Chem.* **36**, 1616.

Benoit, H. and Leng, M. (1961) *J. Chim. Phys.* **58**, 480.

Benoit, H., Grubisic, Z. and Rempp, P. (1967) *J. Polym. Sci.* **B5**, 753.

Bersted, B. H. (1973) *J. Appl. Polym. Sci.* **17**, 1415.

Billmeyer, F. W. Jr. (1971) *Textbook of Polymer Science*, 2nd ed, Wiley, New York.

Birley, A. W., Dawkins, J. V. and Kyriacos, D. (1978) *Polymer* **19**, 350.

Bovey, F. A. and Tiers, G. V. D. (1960) *J. Polym. Sci.* **44**, 173.

Bovey, F. A., Tiers, G. V. D. and Filipovich, G. (1959) *J. Polym. Sci.* **38**, 73.

Brandrup, J. and Immergut, E. H. (1975) *Polymer Handbook*, 2nd ed, Wiley, Chichester.
Brewer, P. I. (1961) *Nature* **190**, 625.
Brice, B. A. and Halwer, M. (1951) *J. Opt. Soc. Am.* **41**, 1033.
Brzezinski, J., Glewala, H. and Komas-Colka, A. (1973) *Eur. Polym. J.* **9**, 1251.
Burge, D. E. (1979) *J. Appl. Polym. Sci.* **24**, 293.
Busnel, J. P. (1982) *Polymer* **23**, 137.
Calderon, N., Chen, H. Y. and Scott, K. W. (1967) *Tet. Letts.* 3327.
Carman, C. J. (1973) *Macromolecules* **6**, 725.
Carman, C. J., Tarpley, A. R. and Goldstein, J. H. (1971) *Macromolecules* **4**, 445.
Carpenter, D. K. (1966) *J. Polym. Sci. A-2* **4**, 923.
Chaplin, R. P. and Ching, W. (1980) *J. Macromol. Sci. Chem.* **14**, 257.
Chen, H. Y. (1968) *Rubber Chem. Technol.* **41**, 47.
Cooper, A. R., Bruzzone, A. R. and Johnson, J. F. (1969) *J. Appl. Polym. Sci.* **13**, 2029.
Cunneen, J. I., Higgins, G. M. C. and Watson, W. F. (1959) *J. Polym. Sci.* **40**, 1.
Dawkins, J. V. (1977) *Eur. Polym. J.* **13**, 837.
Debye, P. (1944) *J. Appl. Phys.* **15**, 338.
Debye, P. (1947) *J. Phys. and Coll. Chem.* **57**, 18.
Dondos, A. (1977) *Polymer* **18**, 1250.
Dondos, A. and Benoit, H. (1967) *Polymer* **18**, 1161.
Doty, P. and Bunce, B. H. (1952) *J. Am. Chem. Soc.* **74**, 5029.
Edwards, A. D. (1977) Unpublished work at MRPRA.
Edwards, A. D. (1981) Unpublished work at MRPRA.
Edwards, A. D. and Loadman, M. J. R. (1976) Unpublished work at MRPRA.
Elias, H. G. (1968) *Characterization of Macromolecular Structure*, McIntyre, M., Ed., National Academy of Science Publication 1573, Washington, D.C.
Ezrin, M. (1968) *Characterization of Macromolecular Structure*, McIntyre, M., Ed., National Academy of Science Publication 1573, Washington, D.C.
Field, J. B., Woodford, D. E. and Gehman, S. D. (1946) *J. Appl. Phys.* **17**, 386.
Fox, T. G. and Flory, P. J. (1950) *J. Appl. Phys.* **21**, 581.
Fox, T. G. and Flory, P. J. (1954) *J. Polym. Sci.* **14**, 315.
Gilding, D. K., Reed, A. M. and Askill, I. N. (1981) *Polymer* **22**, 505.
Golub, M. A., Fuqua, S. A. and Bhacca, N. S. (1962) *J. Am. Chem. Soc.* **84**, 4981.
Gudzinowicz, D. J. and Alden, K. (1971) *J. Chromatog. Sci.* **9**, 65.
Hamielec, A. E., Balke, S. T., Le Clair, B. P. and Pierce, S. L. (1969) *Ind. Eng. Chem. Prod. Res. Dev.* **8**, 54.
Heacock, J. F. (1963) *J. Appl. Polym. Sci.* **7**, 2319.
Huber, C. and Lederer, K. H. (1980) *J. Polym. Sci. Polym. Letts. Edn.* **18**, 54.
Huggins, M. L. (1942) *J. Am. Chem. Soc.* **64**, 2716.
Hummel, K., Groyer, S. and Lechner, H. (1982) *Kaut. u. Gummi Kunst.* **35**, 731.
Jennings, B. R. (1966) *Chem. and Ind. (London)* 463.
Jordan, E. F. (1968) *J. Polym. Sci. A-1* **6**, 2209.
Khan, H. U. and Bhargava, G. S. (1980) *J. Polym. Sci. Polym. Letts. Edn.* **18**, 803.
Kato, Y., Kido, S., Yamamoto, M. and Hashimoto, T. (1974) *J. Polym. Sci. Polym. Phys. Edn.* **12**, 1339.
Keavney, J. J. and Eberlin, E. C. (1960) *J. Appl. Polym. Sci.* **3**, 47.
Kirkland, J. J. (1976) *J. Chromatog.* **125**, 231.

Koldunovich, E. B., Epshtein, V. G., Blyumina, S. B. and Ol'shanskaya, L'A. (1968) *Sov. Rubber Technol.* **6**, 10.
Kössler, I. and Vodehnal, J. (1964) *Coll. Czech. Chem. Commun.* **29**, 2419.
Kow, C., Morton, M. and Fetters, L. J. (1982) *Rubber Chem. Technol.* **55**, 245.
Kratohvil, J. P. (1966) *Analyt. Chem.* **38**, 517R.
Kumar, V. G. and Hummel, K. (1982) *Angew. Makromol. Chem.* **102**, 167.
Kurata, M. and Stockmeyer, W. H. (1963) *Adv. Polym. Sci.* **3**, 196.
Layec-Raphalen, M. N., Silberberg, A., Priel, Z. and Wolff, C. (1979) *Polymer* **20**, 281.
Leblanc, A. (1962) *Mémorial Poudres* **44**, 335.
Letot, L., Lesec, J. and Quivoron, C. (1980) *J. Liquid Chromatog.* **3**, 11, 1637.
Loadman, M. J. R. (1978) Unpublished work at MRPRA.
Loadman, M. J. R. and Tinker, A. J. (1980) Unpublished work at MRPRA.
Marx-Figini, M. and Figini, R. V. (1980) *Makromol. Chem.* **181**, 2407.
McConnell, M. L. (1978) *Internat. Laboratory*, May/June, 89.
Michailov, L. and Harwood, H. J. (1970) *J. Am. Chem. Soc. Org. Coatings and Plast. Chem.* **30**, 129.
Miyake, Y., Nataka, M. and Yamasaki, H. (1970) *J. Polym. Sci., A-2* **8**, 573.
Moore, J. C. (1964) *J. Polym. Sci., Part A* **2**, 835.
Morawetz, H. (1965) *Macromolecules in Solution*, Wiley, New York.
Morris, C. E. M. (1977) *J. Polym. Sci., Polym. Symp. Ser.* **55**, 11.
Narasimhan, V., Huang, R. Y. M. and Burns, C. M. (1981) *J. Appl. Sci.* **26**, 1295.
Newitt, E. J. and Kokle, V. (1966) *J. Polym. Sci., A-2* **4**, 705.
Ogg, C. L., Porter, W. L. and Willits, C. O. (1945) *Ind. Eng. Chem. Anal. Edn.* **17**, 394.
Pasternack, R. A., Brady, P. and Ehrmantraut, H. C. (1962) *Dechema Monograph* **44**, 205.
Pfann, H. F., Williams, V. Z. and Mark, H. (1946) *J. Polym. Sci.* **1**, 14.
Prud'homme, J. and Bywater, S. (1971) *Macromolecules* **4**, 543.
Purdon, J. R. and Mati, R. D. (1966) *Paper presented at Am. Chem. Soc. Div. Rubber. Chem. Soc.*, San Francisco.
Randall, J. C. (1977) *Polymer Sequence Determination, ^{13}C-NMR Method*, Academic Press, New York.
Randall, J. C. (1979) *Amer. Chem. Soc. Symp. Ser.* **103**, 271.
Richardson, W. S. and Sacher, A. (1953) *J. Polym. Sci.* **10**, 353.
Roovers, J. E. L. and Toporowski, P. M. (1974) *J. Appl. Polym. Sci.* **18**, 1685.
Ross, J. A. and Castro, M. E. (1968) *J. Polym. Sci.* **C21**, 143.
Roy, B. R. (1976) *J. Polym. Sci. Polym. Chem. Edn.* **14**, 2321.
Rudkin, A. and Wagner, A. R. (1975) *J. Appl. Polym. Sci.* **19**, 3361.
Runyon, J. R., Barnes, D. E., Rudd, J. F. and Tung, L. H. (1969) *J. Appl. Polym. Sci.* **13**, 169.
Scholte, Th. G. (1970) *Eur. Polym. J.* **6**, 51.
Schultz, G. V. and Blaschke, F. (1964) *J. Prakt. Chem.* **158**, 130.
Shimura, Y., Mita, A. and Kambe, H. (1964) *J. Polym. Sci.* **B2**, 403.
Spatorico, A. L. (1974) *J. Appl. Polym. Sci.* **18**, 1793.
Staudinger, H. and Heuer, W. (1930) *Ber.* **63B**, 222.
Stelzer, F., Thummer, R. and Hummel, K. (1977) *Coll. Polym. Sci.* **225** (7), 664.
Subramanium, A. (1972) *Rubber Chem. Technol.* **45**, 346.

Terry, S. L. and Rodriguez, F. (1968) *J. Polym. Sci.* **C21**, 191.
Tidd, B. K. (1976) Unpublished work at MRPRA.
Tsuge, S. (1981) *Trends in Analyt. Chem.* **1**, 87.
Tsuge, S., Sugimura, Y. and Nagaya, T. (1980) *J. Anal. Appl. Pyrol.* **1**, 221.
Tung, L. H. (1966) *J. Appl. Polym. Sci.* **10**, 371.
Unger, K. and Kern, R. (1976) *J. Chromatog.* **122**, 345.
Vaughan, M. F. (1960) *Nature* **188**, 55.
Werstler, D. D. (1980) *Rubber. Chem. Technol.* **53**, 1191.
Witnauer, L. P., Senti, F. R. and Stern, M. D. (1955) *J. Polym. Sci.* **16**, 1.
Zimm, B. H. (1948) *J. Chem. Phys.* **16**, 1093.

Chapter 10

Inorganic Fillers and Trace Metal Analysis

ASHING

One of the simplest of analytical procedures would seem to be that of combustion of a test portion and the weighing of the residue. In the literal sense this is, of course, true but in many cases merely weighing the residue yields insufficient information, and the ash may be the starting material for further analyses as in the determination of trace metals. In carrying out an ash determination on a compounded polymer the aim is to pyrolyse the polymer, distilling off liquid and vapour without combustion and, only when this is complete, to continue with oxidative heating. In fact, some carbonization of the polymer occurs during the pyrolysis and as this carbon is often very difficult to burn off it is important that flakes of carbon formed in this way should not be lost in air currents. Equally important is the need to avoid setting the polymer alight since material is easily lost in the flame. Some polymers, especially when free of filling powders, liquefy at an early stage of the procedure and froth up over the rim of the crucible; styrene–butadiene copolymers (SBR) are particularly liable to this, but the problem can be avoided by wrapping the test portion in ashless filter paper before placing it in the crucible (Milliken, 1952).

Temperature of Ashing

Owing to the decomposition of certain compounding ingredients simple ashing gives reproducible values only if strict attention is given to an arbitrary scheme whereby the temperature is controlled with respect to time.

ISO 247-1978 offers three different procedures, the first two unsuitable for compounded or vulcanized rubbers containing halogen, whilst the last is not recommended for raw rubbers. The methods need not give the same result and thus the one used must be specified.

Changes in Mineral Constituents During Ashing

The most commonly occurring filler likely to give trouble is whiting. As is well known, calcium carbonate decomposes on heating with the loss of carbon dioxide. Extrapolation from the published data (Johnston, 1910) shows the following figures for the dissociation pressure of calcium carbonate:

°C	mm Hg
300	0·000 016
400	0·003 0
500	0·14
600	2·25

In a closed furnace with unstirred air, carbon dioxide would tend to remain in the crucible with the ash and this would push the equilibrium over and prevent the continuous decomposition which is much more likely in the open air when ashing is carried out over a bunsen burner. Decomposition can be ignored below 520 °C, at which temperature the dissociation pressure of calcium carbonate is about balanced by the partial pressure of carbon dioxide in the atmosphere, but not above 520 °C where decomposition becomes possible. The temperature in ISO 247-1978 of 550 °C ± 25 °C seems therefore a little high whilst the allowed option of 950 °C ± 25 °C will completely decompose the whiting.

The importance of keeping the temperature low is reinforced when silicates are present for zinc oxide forms an insoluble silicate when heated with clay above 700 °C. Poulton (1958) claims that this effect is not found with silica, but Gorsuch (1970) illustrates how, in the presence of inorganic chlorides, reaction can occur with silica crucibles.

As already mentioned, the ISO specification for ashing includes a warning about the presence of chlorine-containing polymers, and Stern and Hinson (1953) published a note demonstrating the magnitude of the effect with zinc oxide and polychloroprene. Some of their results are shown in Table 10.1.

These authors presented evidence in the same note that calcined magnesia could be quantitatively recovered from neoprene by ashing. The question of ashing polychloroprene is discussed later in the section on acid ashing.

Presumably one should also avoid ashing chlorosulphonated poly-

TABLE 10.1

ASHING OF NEOPRENE RUBBER CONTAINING ZINC OXIDE

Composition of mix %		*Ash %*		
ZnO	*Neoprene*	*Bunsen*	*muffle furnace*	*Calculated*
0	100	0·5	0·5	—
5	95	2·9	0·4	5·35
20	80	9·2	5·5	19·85
100	0	97·1	91·5	—

ethylene except by the sulphated ash procedure. For a rather different reason care must be exercised in ashing heavily loaded 'hard rubbers' or ebonites. Smith *et al.* (1959) pointed out that the very high sulphur content can and does react with mineral fillers during ashing. In ebonite containing calcium carbonate both sulphide and sulphate can be found in the residue depending on the conditions and time of heating.

Loss of Trace Elements During Ashing

Apart from the loss of major constituents of the ash there is always the possibility that trace elements present may be lost in one way or another. The element around which interest usually centres is copper and it is with this that most illustrations can be made. Loss by volatilization is avoided by pyrolysing as gently as possible and above all by not letting the test portion catch fire. The most common cause in the past has been the use of porcelain crucibles, the copper or other metal being absorbed by the hot glaze of the crucible. The remedy for this is the use of silica crucibles, but even these become etched after only one or two determinations and while the smooth vitreous surface does not absorb copper an etched surface is said to do so. An alternative is the use of inert material to line the crucible. Magnesium oxide is used for the purpose and has the additional merit of assisting the ashing by helping to introduce oxygen from the air held in the bulk of the powder and also by absorbing liquid on its surface. The use of platinum crucibles is a further option but Zeitlin *et al.* (1958) reported that if a bunsen burner is used to carry out the ashing, unburnt gas in the flame can reduce copper salts to metallic copper which then alloys with the platinum.

Another mode of loss is by reaction with other constituents of the compound. For example, it is very unlikely that any worthwhile estimate can be made of the copper content of a clay-containing rubber compound by a dry ashing procedure. The copper forms a copper silicate which is fused in the clay and cannot be recovered without destruction of the clay by hydrofluoric acid treatment.

Thermogravimetry

A specific application of the dry ashing technique, which completely solves the problem found with whiting, is that of thermogravimetry, the derivative mode of which has been discussed in Chapter 8, and the quantitative aspects of which will be considered in depth in Chapter 12. It must, however, receive mention here because, having removed the polymer by pyrolysis in an inert atmosphere over a temperature range 250–550 °C, heating can be continued up to 800 °C and if any whiting is present carbon dioxide will be evolved quantitatively enabling the whiting content to be calculated. In the same instrument the atmosphere can be changed to air at any specific temperature so that carbonaceous residues can be removed by combustion and a residual ash obtained which, depending upon the conditions chosen, will be identical to one or other of those produced using the two 'dry' methods of ISO 247-1978.

Acid Ashing

A method of overcoming some of the difficulties of dry ashing without introducing all the complications of wet ashing was suggested by van der Bie (1947). We have called this 'acid ashing' because it does not quite fit into the categories of either dry or wet ashing.

A slightly different procedure has now been accepted in ISO 247-1976 and differs from van der Bie's original suggestion by using sulphuric acid instead of nitric acid and is recommended for polychloroprene rubbers. However, whereas the use of nitric acid should, in principle, lead to the same total ash as direct heating, the sulphated ash can give a different figure due to the thermal stability of the sulphates. The final heating of a sulphated ash from polychloroprene is critical as both zinc and magnesium are normally present and it is necessary to ensure the conversion of sulphates to oxides. (N.B. Some of the zinc oxide is fused into the silica crucible used and, although the ash is correctly determined, zinc cannot be estimated.) In the presence of calcium carbonate there is no alternative to weighing the ash as sulphate.

Wet Ashing

Undoubtedly chromic acid systems are the most efficient for obtaining the complete conversion of carbon compounds to carbon dioxide and water

(Houghton, 1945), but they are completely excluded in some trace metal determinations by the certainty of introducing with the reagents just those heavy metals which are to be determined. Nitric and sulphuric acids are not of themselves completely effective and it is usual to follow their use by supplementary oxidizing agents, either hydrogen peroxide or perchloric acid. The latter reagent, in the opinion of the authors, is best avoided. It is important to recall that, relative to the metals to be determined, large

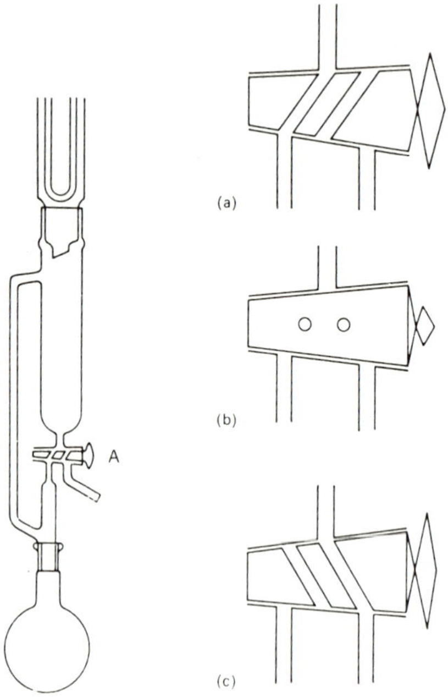

FIG. 10.1. Apparatus for controlled decomposition of organic material. (Courtesy Pergamon Press.)

quantities of reagent are added and evaporated, and concentration of their impurities therefore occurs. For example, the nitric acid used should be of a grade specified for food analysis or a microanalytical reagent. (Reagents with heavy metals such as lead at levels below 1 ppm may still lead to a considerable reagent 'blank'.)

Gorsuch (1970) recommends a sophisticated wet digestion apparatus as illustrated in Fig. 10.1, which has a number of advantages in flexibility of operation depending upon the position of the three-way tap (A).

This allows refluxing (position a), distillation (position b) and removal of some or all of the distillate (position c). Fractions of distillate can be discarded, reserved for further treatment, or examined at leisure in the course of an investigation. In some cases it is convenient to use a two-necked flask instead of the one shown, and to insert a thermometer into the flask. By this means, the lengths of the distillation stages can be controlled by the temperature attained by the reaction solution; this can be of particular value with mixtures containing perchloric acid.

There is one essential difference from dry ashing; the sample size for the latter is 5 g but can be increased to 10 g without inconvenience; with wet combustion, 2 g is about the practical limit. This means that less precision may be achieved if the trace metal content is very low. It may also be noted that, whereas a dry ashing can be performed by an electric furnace, wet ashing is performed by an analyst, i.e. it must be watched and it can take a long time.

It should be apparent that it is impossible to state categorically that one method is superior to another. In general it can be said that for many determinations the advantages and disadvantages can be weighed against each other on the grounds of convenience, but for certain applications there are overriding reasons for choosing one specific one. If there is a possibility of volatilizing the element being analysed then wet oxidation is virtually essential, whilst if extremely low levels of elements are being measured and reagent purity could become a problem, dry or acid ashing would be preferred.

It is probably best to consider each requested analysis separately, using one of the ISO procedures to obtain a *standard* ash value, but using thermogravimetric analysis to obtain a rapid, perfectly valid, ash content, together with whiting loading if any is present, and for subsequent 'bulk' analysis. Separate ashing procedures could be required for each specific quantitative elemental analysis.

BULK FILLER ANALYSIS

Historically the identification of the bulk fillers in a rubber vulcanizate has been made by infrared spectroscopic examination of the dry ash, preferably obtained at *ca.* 500 °C as either a paste, or 'mull', prepared by grinding it with liquid paraffin, or as finely dispersed particles in a matrix of potassium bromide pressed under high pressure to form a thin disc. Comprehensive sets of reference spectra have been published (cf. Scholl,

1981; Corish, 1961) and from these it is a simple matter to identify the inorganic component as a clay, silica itself, whiting, barium sulphate, etc. The main disadvantages of this simple technique are that it does not show zinc oxide, which has no infrared spectrum, nor is it always easy to identify the components of mixed filler systems as the observed absorptions are relatively broad and few, unlike the complex characteristic patterns of organic molecules. Except in exceptional cases it would not be possible to quantify the analysis of filler mixtures by this technique, but the authors' experience is that this is rarely required and, if necessary, can be carried out using standard methods of inorganic analysis which are beyond the scope of the book, or the all-embracing atomic absorption spectrophotometry (AAS) which is equally able to analyse trace metals and bulk fillers—the only differences between the two being in the dilution of the working solution and the correction for matrix effects.

TRACE METALS

A few years ago the only metallic elements, present at trace levels, which the rubber analyst was required to comment on were copper, iron, possibly manganese—all prodegradants—and occasionally lead. Pressures of 'health and safety' have led to a massive expansion of this list, and no doubt it will continue to grow. At the moment ISO Standards only exist for the colorimetric determination of copper (ISO 1654), iron (ISO 1657) and manganese (ISO 1655) in raw rubber, whilst in the case of compounded or vulcanized rubbers ISO 1397 should be consulted. It would also be appropriate to add at this point the ISO Standard for the determination of zinc by the EDTA titrimetric method (ISO 2454). In view of the detailed descriptions given, and the policy of not describing ISO Standards in detail, these are not considered further, and the interested reader is referred to the Standards themselves.

ATOMIC ABSORPTION SPECTROPHOTOMETRY (AAS)

Atomic absorption spectrophotometry is almost certainly the current method of choice for the routine estimation of large numbers of different elements, at concentrations ranging from many percent to less than 1 ppm. The principle is extremely simple and involves the generation of a cloud of atoms of the elements being studied in a flame, such that most contain their electrons in the ground state. One or more electrons can then be raised to higher energy levels by the absorption of energy supplied in the form of photons. Metallic and semi-metallic elements contain valence electrons which are excited by photons of wavelengths in the range 190–800 nm. For each element the difference in energies between any two specific levels is virtually identical for all atoms although a little spread is introduced due to atomic collisions. If radiation of this precise wavelength is passed through the flame and its reduction in intensity measured, this will lead directly to a quantification of that particular element. The ideal energy source for each element is its own specific emission line as found in the emission spectrum generated in a hollow cathode lamp or electrodeless discharge lamp for that particular element. In practice the relationship between concentration and absorption is only linear over a small concentration range, but as samples are bracketed by standards during a routine run this causes few problems, whilst successive dilutions can be used to optimize the concentration range. The particular advantages of this technique on a cost basis are clearly obvious as each additional element may be analysed by the purchase of one extra lamp, and most instruments have multilamp turrets so that one lamp can be changed whilst another is in operation.

The cloud of atoms is usually introduced into the flame by the direct aspiration of a solution of the sample into the flame. The choice of gases for the flame, flame geometry and other operating parameters can easily be adjusted to optimum for each element, so one is only left with making sure that the solution being examined is truly representative of the original sample. Although the problems discussed earlier should not be underestimated, studies at the authors' laboratories (J. E. Davey and R. C. Crafts) would suggest a preferred general approach for a bulk filler analysis to be

(1) dry ash at 500 °C;
(2) repeat additions of HF aq. followed by baking to constant weight (removes SiO_2 quantitatively as SiF_4);
(3) dissolve residue in 1 M HCl (aq.);
(4) dilute as required for aspiration

(if titanium dioxide present (insoluble at 3) repeat (1) and (2), then fuse with sodium hydrogen sulphate prior to dissolution in 5 M HCl).

This should give valid results for silicon, aluminium, potassium, magnesium, calcium, titanium and barium at the bulk level, and copper, iron, manganese, chromium and nickel at the trace level. Cadmium, lead, mercury, arsenic, antimony and bismuth should be 'fixed' using the sulphated ash (in the absence of calcium salts) or nitrated ash method. Zinc levels will be realistically determined, in the absence of halogen, by the dry ash method. In the presence of halogen the sulphated or nitrated ash method may be used although a recent ISO 'round robin' has shown that up to 5% of the zinc present can be lost during sulphated ashing, and recommends the use of wet ashing.

An alternative to dry, sulphated or wet ashing is to use the oxygen flask method as described in Chapter 6. This has been used successfully in the analysis of many elements but there are dangers with lead (Belcher *et al.*,

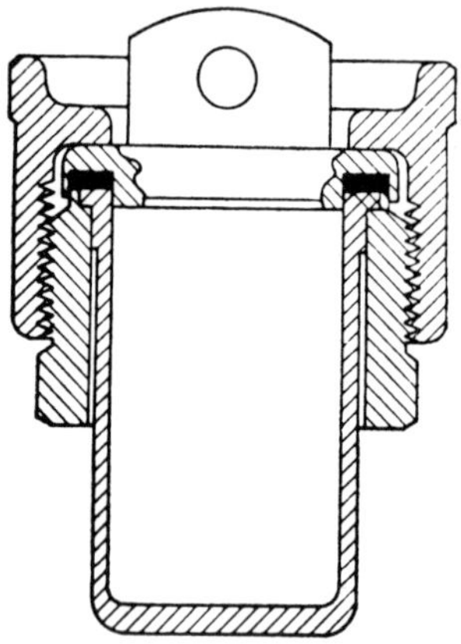

FIG. 10.2. The Parr bomb. (Courtesy Parr Instrument Co.)

1958) and arsenic (Corner, 1959) as each may form an alloy with the platinum foil container. A further option is to use a fusion technique, and this is preferably carried out in a sealed vessel, or 'bomb' (see ISO DIS 6528 and Chapter 6); the classic Parr bomb, made of steel, is illustrated in Fig. 10.2. Modern high-pressure reaction vessels are now available with an inert polytetrafluorethylene (PTFE) lining (Bernas, 1968) and as recently as 1982 Uhrberg described the design of such a bomb for analysing for volatile metals in biological samples.

The history of the Parr bomb, and its application to the analysis of rubber, is discussed in Chapter 6. Although little has been published recently concerning the application of this technique to the preparation of polymeric samples for analysis by atomic absorption spectrophotometry, it appears to have a number of attractions, not least the containment of volatile elements and the speed of the fusion and digestion processes.

For a more detailed discussion of the treatment of specific elements likely to be met by the analyst in the elastomer field the reader is directed to the references suggested in Table 10.2.

TABLE 10.2

SELECTED REFERENCES FOR THE QUANTITATIVE RECOVERY OF INORGANICS FROM ORGANIC MATRICES

Element	*Problem/points covered*	*Authors*
Alkali metals	excessive heat	Joyet (1951)
Aluminium	retention in SiO_2	Sandell (1944)
Antimony	wet oxidation	Gorsuch (1962)
Arsenic	oxygen flask-alloy	Corner (1959)
	various digestions	Banks *et al.* (1948)
	wet oxidation	Gorsuch (1959)
	acetic acid extraction	Moldrai and Petrescu (1965)
Cadmium	wet ashing	Down and Gorsuch (1967)
	acid ashing	Middleton and Stucking (1954)
Chromium	wet *vs.* dry ash	Gorsuch (1960)
Copper	alloy with platinum	Zeitlin *et al.* (1958)
	low-temp. oxidation with excited oxygen	Gleit and Holland (1962)
Iron	loss with chlorides	Gorsuch (1960)
Lead	wet *vs.* sulphated *vs.* oxygen flask	Belcher *et al.* (1958)
	loss with H_2SO_4/Ca salts	Gorsuch (1959)
Magnesium	loss at 740 °C	Davidson (1962)
Manganese	wet *vs.* dry ashing	Heckman (1967)
Mercury	wet ashing $H_2SO_4/KMnO_4$	Gage (1961)
	wet ashing	AMC[a] (1965)
Nickel	wet *vs.* dry ashing	Gorsuch (1970)
Selenium	wet oxidation	Klein (1941)
	loss with H_2SO_4/HNO_3	Gorsuch (1959)
	perchloric acid	Kelleher and Johnson (1961)
	losses on drying	Stanton and MacDonald (1965)
Zinc	losses with chloride	Gorsuch (1970)
	losses with clays	Poulton (1958)
	losses with chloroprene	Stern and Hinson (1953)

[a] Analytical Methods Committee.

TOTAL SAMPLE ELEMENTAL ANALYSIS

In view of the problems associated with the preparation of a truly representative ash and its subsequent dissolution, it is obviously of interest to consider whether one can dispense with these steps, and directly analyse the material with no pre-treatment other than the cutting of a suitably sized piece of sample.

Destructive Elemental Analysis

The techniques which can be used to give a rapid elemental analysis of a total sample of rubber, with the minimum of pre-treatment, may conveniently be divided into two groups, those which result in the destruction of the sample, and those which enable it to be recovered essentially undamaged for subsequent further analyses. The two groups will be considered in this order and it will be apparent that, whilst at the moment the two are complementary rather than in competition, developments in either of the groups could allow one of them to predominate in the future.

Total Sample Atomic Absorption Spectrophotometry (Electrothermal Atomization, ETA)

The problems of ensuring that a solution prepared for atomic absorption spectrophotometric analysis contains the elements to be measured in the same relationship as that in which they exist in the original sample can be overcome by using an electrothermal atomizer attached to an atomic absorption spectrophotometer.

A solid sample weighing about 1 mg—accurately weighed—is heated under reproducibly programmed conditions to remove the polymeric phase, ashed, and then atomized very rapidly. The concentration of the element under investigation can then be determined from the integrated peak area (not the steady state reading as obtained with an aspirating solution) relative to either solids of known composition or solution standards dropped into the atomizer and given the same pretreatment. The atomizer usually consists of a small graphite tube or furnace, or a tantalum cup, which is subjected to a programmed heating cycle optimized to the system under investigation. There are some problems with elements such as titanium which form stable carbides or nitrides but these can be overcome using, for example, coated graphite tubes or an argon purge gas. Other problems relate to the precise reproducibility of positioning of the sample in the furnace but the procedure, still relatively in its infancy although first proposed by L'vov (1961), offers appreciable advantages in total sample analysis, a less obvious one, perhaps, being that one can 'build up' the concentration of a trace component by carrying out a succession of pyrolyses and ashings on several samples before the final atomization step.

The points made previously about the need to be certain that 1 mg is representative of the whole material should always be borne in mind.

In the early stages of the discussion on atomic absorption spectrophotometry the point was made that the flame temperature was such that most atoms of the elements were in their ground state. Some, however, are in excited states and most modern instruments can make use of this by operating in the emission mode, directly measuring the strength of the emission signals as the excited species returns to the ground state. In practice the flame temperature is usually increased somewhat, typically from about 2800 °C to 3300 °C, to increase the population of the excited states but the sensitivity is still relatively low compared with atomic absorption spectrophotometry.

Plasma Spectroscopy

The recent introduction of instruments using a plasma instead of a conventional flame, together with dedicated microprocessors, etc., must lead to a reappraisal of the whole field of elemental analysis. There are two methods currently used for generating the plasma and these give their names to the techniques. One uses inductive coupling (ICPS), the other a DC source (DCPS). The plasma maintains a temperature in the region of 5–10 000 °C, in which range most atoms are in excited states and indeed in many cases the excitation energy is sufficient to ionize them! Each species generates a substantial number of emission lines, as photons are emitted during the return to lower energy states, and because of their extreme sharpness and range of magnitude one has a large choice for 'unobscured analysis'. When compared with AAS the detection limits and quantitative precisions are similar, whilst the concentration:linearity relationship stays linear over a much greater range—several orders of magnitude. In practice the multiplicity of lines is more of a problem than an advantage because interference with the major lines is common and thus weaker lines have to be used, reducing the sensitivity well below that ideally obtainable. The great advantage of this technique over the flame method is its ability to carry out multielement analyses, either using a sequential searching programme when up to 16 elements can be identified and quantified in two minutes or so, or using a multichannel analyser with a number of dedicated data processors analysing up to twenty pre-identified elements simultaneously. The latter will have much smaller solution requirements. Couple this with autosampling capabilities, full microprocessor control to optimize the performance of the instrument, and built-in programmes for background correction, element overlap, etc., and the strength of the technique is obvious. Unfortunately one must pay for this, although it could perhaps be cost effective in a number of applications.

The fact, however, remains that aspiration of the sample into the plasma still requires the sample to be in solution, so the problems discussed earlier still apply, whilst the substantial increase in sensitivity obtainable by total sample analysis using atomic absorption spectrophotometry coupled with electrothermal atomization cannot be matched. For this reason alone it is unlikely that either plasma technique will ever replace atomic absorption spectrophotometry completely. It could be that the latter, using a continuous radiation source rather than hollow cathode or electrodeless discharge lamps, together with a suitable monochromater and 'scanning system' or multichannel analyser for use with electrothermal atomization, offers an alternative way forward.

Non-destructive Elemental Analysis

The second group of techniques to be considered which enables elemental analytical data to be obtained from a sample of rubber, be it raw, compounded or vulcanized, consists of bombarding the sample with one form of energy and monitoring specific induced effects characteristic of each atomic species present. It will be appreciated that this affords *elemental* data, but does not indicate the structural environment of that element; thus, unlike infrared spectroscopic analysis, there will be no distinction between silicon in silica, an inorganic silicate, or silicone rubber. This must be deduced from other data.

One example of such a technique has already been discussed in Chapter 6, when a radioisotope was used to provide X-ray fluorescence identification and quantification of certain elements. There are numerous others as reference to *Analytical Chemistry Reviews: Surface Characterization* shows, but two merit discussion here as they illustrate a complementary pair of effects. In one the bombardment of a sample with X-rays liberates electrons, whilst in the other bombardment with electrons releases characteristic X-rays.

Electron Spectroscopy

The first to be considered is electron spectroscopy, the basis of which is the measurement of the kinetic energy of electrons emitted from a sample in a vacuum following ionization by a monochromatic X-ray source, usually generated by the bombardment of a pure metal such as aluminium with a beam of electrons. This basic technique can be modified in a number of ways to produce subtly different data, but the ones which have the most significance in this application are X-ray photoelectron spectroscopy (XPS) also known as electron spectroscopy for chemical analysis (ESCA) and Auger electron spectrosopy (AES).

A knowledge of the kinetic energy of the ejected core electron together with the irradiating energy enables the binding energy of the core electron to be calculated, and hence its identity to be established. These instruments are, however, very expensive and their use therefore tends to be limited to areas of research such as catalysis, surface structures and electronic properties rather than the routine identification of inorganic constituents of an elastomeric product.

The converse technique would be the bombardment of a sample with an electron beam and the monitoring of the X-rays generated thereby. This merits rather more detailed discussion because it is the principle of X-ray analysis in the electron microscope (as opposed to the more usual 'visual' detection mode which was discussed earlier (Chapter 8) with particular reference to thermoplastics) and it will feature again in Chapter 13 on the analysis of blooms and other surface effects. The many areas which may be investigated using this combination of a scanning electron microscope (SEM) and an X-ray analyser make the very powerful, if expensive, technique worthy of consideration. Even if the analyst is not able to justify purchase of such equipment, contract services exist, and it is thus important that its capability is appreciated.

Wavelength-dispersive X-ray Analysis

There are two distinct and quite different types of X-ray detector which may be fitted to an SEM—those which operate in the wavelength-dispersive mode and those which use energy dispersion to produce a 'spectrum'. The former is achieved by diffraction with a crystal spectrometer, the principles of operation of which are illustrated in Fig. 10.3.

The X-rays leave the sample in all directions but some impinge on the specially shaped crystal which can be fitted so that the angle of incidence (θ) of the X-rays may be altered. Bragg's law states:

$$\frac{2d\sin\theta}{n} = \lambda \qquad (10.1)$$

where λ represents the wavelength of the diffracted X-ray. The range of wavelengths which one crystal can cover is limited so it is usual to have a set, each with different lattice spacings, fitted to the instrument so that for a given range of θ a much larger range of X-rays can be covered. As the wavelength of the X-ray is dependent upon the atomic number of the element, identification of an element merely requires measurement of the angle θ and reference to a table of characteristic angles for each element. The spectrometer can scan through the region of the characteristic wavelength, recording an energy peak, the area of which is proportional to the

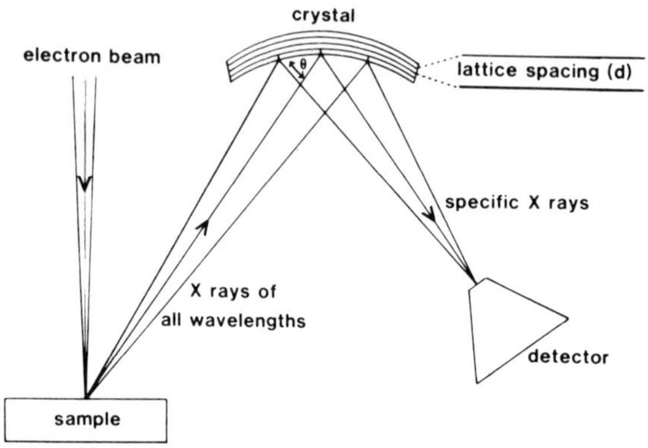

FIG. 10.3. Operation of a crystal spectrometer.

concentration of that element (within the limitations of the matrix effects discussed later).

Energy-dispersive X-ray Analysis

The energy-dispersive system is not mechanical nor does it depend upon the separation of different X-rays. It depends entirely on electronics, the key being the detector which is a lithium drifted silicon crystal, popularly known as a 'Sillie'. This has the property that electron–hole pairs are produced when X-rays fall upon it and these are collected as a current pulse. The number of electron–hole pairs generated by a specific X-ray is dependent upon the energy of that X-ray (i.e. the particular element) and a pulse-height analyser is used to assign each pulse to a particular channel of a multichannel analyser. Each channel is thus dedicated to X-rays of a specific energy and hence to a particular element. As there is no scanning and there is random impingement of all the X-rays on the detector there is an effectively simultaneous accumulation of a full elemental spectrum (although the lowest atomic number element which can be detected in a normal system is sodium unlike the crystal spectrometer which extends down in atomic number to lithium).

This can be displayed on a visual display unit within seconds of beginning the analysis and a typical hard copy printout, obtained after 100 seconds' analysis, is illustrated in Fig. 10.4.

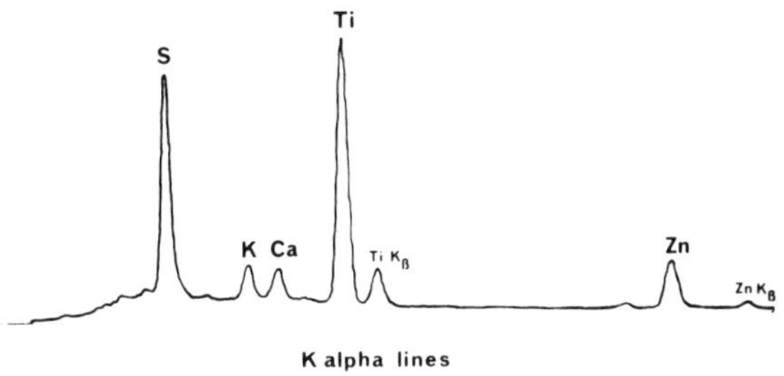

FIG. 10.4. X-ray dispersive spectrum of a rubber vulcanizate.

The major differences between the two techniques, therefore, are that the energy-dispersive system gives a rapid 'total' spectrum but with relatively low resolution whereas the wavelength-dispersive system provides very high resolution peaks for individual elements, but it is extremely tedious if not impracticable to use to obtain a full elemental analysis in spectral form. The latter is, however, very much more sensitive, and allows detection of some species present even at the ppm level, whilst the former is realistically limited to about 0·1% (1000 ppm).

Quantification of X-ray Analytical Data

Unfortunately the quantitative side of X-ray analysis is by no means as clear cut as the qualitative side. Problems can be divided into two areas which may be called *intrinsic* and *specific*. In the *intrinsic*, one must consider background counts in the area being monitored, dead time, when the pulse processor is handling one pulse and rejecting others until the first is dealt with, and interference from other elements. In the *specific* there is absorption of the electron beam by species other than the one of interest, secondary fluorescence where the characteristic radiation of an element is additionally excited by X-rays of a higher energy than the critical energy of the element being monitored, and a diminution of the X-rays being monitored due to their absorption by other elements of the matrix as they pass through the bulk sample on their way to the detector. These are collectively called *matrix effects* and, whilst various methods have been adopted to correct for them (Lucas-Tooth and Price, 1961; Lucas-Tooth and Pyne, 1964), and indeed instrument manufacturers supply 'correction programs' for the computers used with their analysers, the very lack of publications dealing with the quantification of fillers in elastomers, together with the authors' experience at MRPRA, would suggest that the overall problems in this type of matrix are much greater than suggested. It would be fair to add that if accurately known controls are available which bracket, in all elements, the 'unknown' composition, reasonable results can be attained. It seems better, therefore, at the present time to regard the energy-dispersive X-ray spectrum of a bulk sample as an extremely rapid indicator of the elements present (with atomic numbers greater than that of sodium), together with their approximate concentrations which then can be estimated more accurately by other techniques.

It should always be borne in mind that these are *surface* analyses and thus, whilst full use can be made of this together with the magnifying power of the microscope to identify individual particles, or areas of inhomogeneity, care must generally be taken to ensure that the surface under examination is representative of the bulk sample.

REFERENCES

Analytical Methods Committee (1965) *Analyst* **90**, 515.

Banks, C. K., Sultzberger, J. A., Mourina, F. A. and Hamilton, C. S. (1948) *J. Am. Pharm. Assoc.* **37**, 13.

Belcher, R., Macdonald, A. M. S. and West, T. S. (1958) *Talanta* **1**, 408.

Bernas, B. (1968) *Analyt. Chem.* **40**, 1682.

Van der Bie, G. J. (1947) *India-Rubber J.* **113**, 499, 502, 541.

Corner, M. (1959) *Analyst* **84**, 41.

Corish, P. J. (1961) *J. Appl. Polym. Sci.* **5**, 13, 53.

Davey, J. E. and Crafts, R. C. (1981) Unpublished work at MRPRA.

Davidson, J. (1962) *Analyst* **77**, 263.

Down, J. L. and Gorsuch, T. T. (1967) *Analyst* **92**, 398.

Esch, W. (1904) *Chem. Ztg.* **28**, 200.

Gage, J. C. (1961) *Brit. J. Ind. Med.* **18**, 287.

Gleit, C. E. and Holland, W. D. (1962) *Analyt. Chem.* **34**, 1454.

Gorsuch, T. T. (1959) *Analyst* **84**, 135.

Gorsuch, T. T. (1960) PhD Thesis, London.

Gorsuch, T. T. (1962) *Analyst* **87**, 112.

Gorsuch, T. T. (1970) *The Destruction of Organic Matter*, Pergamon Press, Oxford.

Heckman, M. (1967) *J. Assoc. Offic. Anal. Chem.* **50**, 45.

Houghton, A. A. (1945) *Analyst* **70**, 118.

Johnston, J. (1910) *J. Amer. Chem. Soc.* **32**, 938.

Joyet, C. (1951) *Nucleonics* **9b**, 42.

Kelleher, W. J. and Johnson, M. J. (1961) *Analyt. Chem.* **33**, 1429.
Klein, A. K. (1941) *J. Assoc. Offic. Anal. Chem.* **24**, 363.
Lucas-Tooth, H. J. and Price, B. J. (1961) *Metallurgia* **64**, 149.
Lucas-Tooth, H. J. and Pyne, C. (1964) *Advances in X-ray Analysis* **7**, 523.
L'vov, B. V. (1961) *Spectrochim. Acta* **17**, 761.
Middleton, G. and Stucking, R. E. (1954) *Analyst* **79**, 138.
Milliken, L. T. (1952) *Rubber Age (N.Y.)* **71**, 64.
Moldrai, T. and Petrescu, G. (1965) *Industrie Uscara* **12**, 522.
Poulton, F. C. J. (1958) Unpublished work at the Dunlop Rubber Co.
Sandell, E. B. (1944) *Colorimetric Determination of Trace Metals*, Interscience, New York.
Scholl, F. K. (1973) *Infra Red Analysis of Resins and Additives; an Atlas*, Carl Hanser Verlag, Munich.
Scholl, F. K. (1981) *Atlas of Polymer and Plastics Analysis Volume III (Additives and Processing Aids)*, Carl Hanser Verlag, Munich.
Smith, M., Stickland, F. G. and Tarbin, F. G. (1959) *Trans. IRI* **35**, 210.
Stanton, R. E. and MacDonald, A. J. (1965) *Analyst* **90**, 497.
Stern, H. J. and Hinson, D. (1953) *India-Rubber J.* **125**, 1010.
Uhrberg, R. (1982) *Analyt. Chem.* **54**, 1906.
Zeitlin, H., Fredyma, M. M. and Iheda, G. (1958) *Analyt. Chem.* **30**, 1284.

Chapter 11

Carbon Black

Over a period of many years carbon black in vulcanized rubber was estimated by a method which depended on the destruction of the rubber by nitric acid followed by separation of the black by filtration and drying. Finally, the loss in weight on ignition of the dried residue was identified with the carbon black present. This method was first proposed by Jones and Porritt (1914) and is described in one of the earliest books on the analysis of rubber, that by Tuttle (1922). There are now two major methods in use, with the old acid destruction procedure obsolescent. We propose dealing with these fundamentally different methods of estimation separately and to consider how far each is applicable to the range of materials now available and likely to contain carbon black.

DESTRUCTION AND FILTRATION METHODS

It is instructive to follow the steps whereby the original destruction of the rubber by nitric acid has led to the modern recommended standard procedures. Scott and Wilmott (1941) were the first to realize the need for a modified method when dealing with synthetic rubbers. Their modification consisted of swelling the finely divided rubber in hot nitrobenzene, then adding 25% (v/v) nitric acid and heating on a hot plate. Neoprene disintegrates and dissolves in the nitrobenzene in a few minutes. The whole is then heated on a steam-bath for about one hour after which xylene is added and the mixture filtered hot. After filtration the carbon is washed with hot xylene and then with acetone before drying and igniting. The few results they quote for neoprene indicate a standard deviation of about 0·4% on the original compound. This approach to the problem of disintegrating the more resistant synthetic rubbers was followed up by Louth (1948) whose method has been recommended generally for synthetic rubbers. However, in addition to the combination of solvent and acid attack, Louth found it necessary to introduce an ether extraction to deal with the organic material derived from the decomposed rubber.

The method worked out by Galloway and Wake (1946) for estimating the polymer in a compounded and vulcanized butyl rubber leads to the simultaneous estimation of carbon black which can be obtained by drying and igniting the residue on the sintered crucible. The results had a standard deviation of 1·55%, which is rather high and compares unfavourably with results which are obtainable with the modern dissolution method now to be described.

An alternative way of destroying the structure of the rubber is by oxidative dissolution in a solvent, followed by filtration of the rubber. The use of methods for estimating carbon black based on dissolution has not found favour, removal of the black from the solution of the polymer by filtration being tedious since most solvents disperse the black instead of aggregating it and because polymer is tenaciously absorbed on to the black and is difficult to remove completely. Obviously, the more degraded the rubber is, the less viscous the solution and the more easily is the black washed free of it. This leads naturally to the use of oxidation catalysts which are particularly useful for rubbers based on butadiene which show a tendency to crosslinking rather than chain-scission when subjected to oxidative attack by oxygen alone. Kolthoff and Gutmacher (1950) were the first to use *tert*butyl hydroperoxide in the presence of osmium tetroxide to hasten the dissolution of a range of rubbers in boiling *para*dichlorobenzene.

The modification of Kolthoff and Gutmacher's method adopted for BS 903-1964 dropped the use of osmium tetroxide and used only *tert*butyl hydroperoxide as the oxidation catalyst.

The test portion of rubber (0·1 g) is heated in boiling *para*dichlorobenzene (20 g) with the addition of small quantities of *tert*butyl hydroperoxide to a total of 5 cm^3 over a period of 10 min. After boiling for 2 h, 5 cm^3 of mineral oil is added and boiling continued until dissolution is complete. For the highly polar acrylonitrile rubbers, 5 cm^3 of nitrobenzene is added to the 20 g of molten *para*dichlorobenzene at the same time as the test portion is added. After dissolution, the total fillers are filtered off. Filtration is by a silica filter tube 10 cm long with a sintered 10 mm diameter silica disc of porosity 3 on to which is deposited 0·15 g of asbestos* protected from displacement by a 5mm layer of silica wool. After filtration, the dissolution flask and filter are washed with portions of hot diluted nitric acid (3 vol acid to 1 of water) prior to washing with water. This is necessary to destroy residual polymer adsorbed on the black.

Louth (1948) reports estimations of free carbon in butyl, natural, butadiene–styrene and polychloroprene rubbers with an estimated standard deviation independent of the polymer and about 0·15%. Kolthoff and Gutmacher gave results which show a standard deviation of only 0·09% on a carbon black content of 30% although this represents accuracy unlikely to be achieved in the ordinary routine laboratory.

REMOVAL OF POLYMER BY DISTILLATION

In 1949 Bauminger and Poulton described a method for determining polymer loading by a controlled pyrolysis in an inert atmosphere, and pointed out that the residual carbon derived from the polymer is a function of the proportion of material other than hydrocarbon in the polymer.

This effectively constituted the introduction of thermogravimetry (TG) to the rubber-laboratory, and its subsequent development, through instrumentation, to afford a quantitative estimate of the carbon black loading is discussed in detail in the next chapter.

For the purposes of this chapter we will accept that TG can give quantitative data on carbon black loadings and consider what further information may be obtained concerning the carbon black in a vulcanizate.

TYPES OF CARBON BLACK

Carbon blacks used in the rubber industry were initially of relatively few types and were classified according to the properties of the rubber compound and/or vulcanizate derived therefrom. (There are now five basic types, classified according to the method of manufacture, with the furnace and channel types being of most interest to the rubber analyst. The other three are lampblack, acetylene black and thermal black.) It was soon

*Care should be taken in using this material due to its hazardous nature when inhaled.

apparent that the broad classification was inadequate for the materials available and a more detailed system was required. This, in turn, necessitated an understanding of the physical and physicochemical properties of carbon black so let us begin by considering exactly what carbon black is.

Carbon black particles ideally approximate to spheres in shape but are rarely seen individually as they exist in chains or clusters, referred to as aggregates. These in turn tend to cluster together in agglomerates which are believed to break up on mixing with rubber. Aggregates, on the other hand, may occasionally fracture but in essence represent the units of carbon found within a vulcanizate. The type of aggregate indicates the *structure* of the black which may be considered to reflect the ratio of the surface area exposed to the rubber molecules to that hidden from the rubber inside pores or channels too small for the rubber molecules to penetrate; the higher the structure, the greater the number of particles per aggregate. The parameters which *may* require defining are thus:

basic sphere size,
structure (aggregate size and shape),
surface area.

Carbon blacks are currently identified in a variety of ways such as group number, name or symbol, and ASTM designation (ASTM D1765 1979).

The interrelationships between these, together with other data which will be referred to throughout this section, are given in Table 11.1.

Six carbon blacks, N234, N326, N330, N650/683, N660 and N762, are closely defined as standard blacks in ISO/DIS 6809.

Two general points are worth noting; first, as well as the designation 'N' (normal) one may meet 'S' (slow) which serves to distinguish the slow curing channel blacks (or modified furnace blacks) from the normal furnace blacks, and second, whilst the first of the ASTM numbers is identical with the old group number, the last two are arbitrarily assigned and therefore have no scientific significance.

ANALYSIS OF CARBON BLACK TYPE

It was as early as 1920 that Weigand estimated the size of a carbon black particle by light microscopy but since the advent of the electron microscope just prior to 1940 it has been universally used for this purpose. A

TABLE 11.1

ANALYTICAL AND SPECIFICATION DATA ON CARBON BLACKS

ASTM designation	*Type by name*	*Symbol*	*Group No.*	*Mean dia. (nm)*	*Permitted diameter range (nm)*	*BET Nitrogen surface area measured range* ($m^2\ g^{-1}$)	*Typical iodine No.* ($mg\ g^{-1}$)	*DBP No. measured range* ($ml\ 100\ g^{-1}$)
—	Super-abrasion furnace LS	SAF.LS	1	19	11–19	170	—	67–87
N 110	Super-abrasion furnace	SAF	1	19	11–19	125–160	145	110–119
N 219	Intermediate SAF-LS	ISAF.LS	2	22	20–25	105–135	118	70–80
N 220	Intermediate SAF	ISAF	2	23	20–25	115–130	121	108–120
N 242	Intermediate SAF-HS	ISAF–HS	2	21	20–25	110–140	123	120–135
N 326	High abrasion furnace LS	HAF–LS	3	26	26–30	75–105	82	70–80
N 330	High abrasion furnace	HAF	3	29	26–30	70–90	82	95–110
N 347	High abrasion furnace HS	HAF–HS	3	26	26–30	80–100	90	120–130
N 440	Fine furnace	FF	4	33	31–39	43–69	50	58–70
N 539	Fast extruding furnace LS	FEF–LS	5	42	40–48	35–52	42	100–113
N 550	Fast extruding furnace	FEF	5	42	40–48	35–52	43	115–125
N 568	Fast extruding furnace HS	FEF–HS	5	42	40–48	35–52	45	120–132
N 660	General purpose furnace	GPF	6	60	49–60	26–40	36	85–95
N 761	Semi-reinforcing furnace LM	SRF–LM	7	70	61–100	17–32	26	56–73
N 770	Semi-reinforcing furnace HM	SRF–HM	7	62	61–100	17–33	27	65–85
N 880	Fine thermal	FT	8	150	101–200	10–17	—	30–40
N 990	Medium thermal	MT	9	500	201–500	5–10	—	33–36

LS = low structure, LM = low modulus, HS = high structure, HM = high modulus.
Sources: Manufacturers' literature and a wide range of publications.

substantial advance in the carrying out of this procedure was the development of the image analysing computer which permits a fully automated measurement of particle size and distribution and is described in detail by Hess *et al.* (1969). Virgin blacks were examined after separation by ultrasonic techniques, whilst vulcanizates were subjected to ultramicrotomy; thus a comparison gave data on aggregate breakdown during compounding. This equipment is, of course, expensive and, whilst its purchase may be justified in the laboratory of the carbon black manufacturer, it is not likely to be available in the normal rubber-laboratory; simpler and cheaper methods are required which give data which will correlate with the direct measurement of particle size and degree of aggregation.

In earlier times, when the available types of carbon black were fewer, identification of the type could quite simply be made on the black recovered by nitric acid disintegration of the rubber by means of its tinting strength (Dawson *et al.*, 1947; Scott and Wilmott, 1947). The use of tinting strength might still be worth while in the laboratory where more elaborate apparatus is not available and where, perhaps, a distinction is required between a limited number of known blacks of which authentic samples are available. The tinting strength is obtained by grinding 0·1 g of the carbon black with linseed oil with the addition of zinc oxide in small quantities to give a standard shade of grey. The tinting strength is expressed as the ratio of the weight of zinc oxide to that of black and comparison is made with known samples. A valuable ancillary test to this is a *gloss test* reported by Dawson *et al.* (1947) and ascribed by them to D. F. Twiss. In this a small quantity of the black is rubbed out on a filter paper with a metal spatula and the colour and gloss compared with a range of blacks. These are both simple tests but yield quite surprisingly constant results, and their empirical nature and the need for standards should not prejudice the analyst against them for they are valuable, quite rapid, and inexpensive.

Kress and Stevens-Mees (1970) put the gloss test on a quantitative basis when, with the assumption that increasing particle size would give an increase in reflectance, they developed a method for measuring the reflectance of a black filled vulcanizate, without prior isolation of the black or destruction of the polymer, at 540 nm using a spectrophotometer with a reflectance attachment and a gloss black tile as reference material. The results (Table 11.2) show that a reasonable relationship exists between the % reflectance and particle size as judged by an electron microscope, although there is a marked polymer dependency.

They claim that the black loading is not significant in the range 25–35%

TABLE 11.2

RELATIONSHIP BETWEEN PARTICLE SIZE AND % REFLECTANCE FOR A SET OF VULCANIZATES IDENTICAL EXCEPT FOR THE POLYMER TYPE

Type of black (ASTM)	*Particle size (nm)*	*% reflectance in matrix of:* BR	NR	OESBR	IIR
N 110	17	20	24	37	29
N 220	20	26	30	34	38
N 330	27	22	34	39	42
S 300	28	28	28	35	38
N 550	33	40	48	50	54
N 660	55	43	50	57	55
N 761	75	55	58	66	62
N 880	180	76	79	77	81
N 990	470	82	87	91	99

w/w but that white filler, including zinc oxide, should be less than 20% of the black content.

Other workers have also used spectrophotometers to classify carbon blacks but in most cases suspensions have been prepared in a liquid medium prior to examination. Fiorenza (1956) prepared a suspension of carbon black in a benzene solution of natural rubber prior to measuring the absorbance at two wavelengths—430 nm and 750 nm—and calculating a *colour index*:

$$I_c = \frac{\log I_0/I\,(430\text{ nm})}{\log I_0/I\,(750\text{ nm})}$$

This procedure was also used for uncured compounds and vulcanizates but in the case of the latter the rubber was initially destroyed using concentrated nitric acid. Later, thermal degradation, as described by Kolthoff and Gutmacher (1950), was used to render the vulcanizate soluble. It was also noted that for free carbon black a suspension could be prepared in aqueous gum arabic, rather than the rubber solution previously used.

The method for rubber containing carbon black was subsequently investigated by Davies and Kam (1967) who, finding it time consuming and prone to error due to incomplete separation of the black from the rubber, resorted to pyrolysis using a simple furnace ascribed to Chambers (1958) followed by an acid wash before dispersing the black in an aqueous gum acacia solution for measurement of the colour index at wavelengths of 425 nm and 675 nm. The slightly different wavelengths used from those of

Fiorenza result in marginally different values for I_c. Davies and Kam compared the colour index values of specific blacks in their virgin states, and in compounded and vulcanized formulations (Table 11.3) and also showed that, at least for an HAF black, the values were unchanged when obtained from vulcanizates based on NR, SBR, BR and *cis*-BR. A standard deviation of ± 0·01 is claimed.

TABLE 11.3

COLOUR INDEX VALUES OF VARIOUS BLACKS

Type of carbon black	*Colour index (I_c)*		
	Free state	*Recovered 1*	*Recovered 2*
Lamp black	0·95	—	0·95
Fine thermal	0·96	0·98	0·97
Semi-reinforcing furnace	1·03	1·01	1·01
Fine ext. furnace	1·17	1·16	1·16
High modulus furnace	1·21	1·22	1·22
High abrasion furnace	1·35	—	1·36
Super conductive furnace	1·38	—	1·40
Intermediate super-abrasion furnace	1·44	—	1·43
Low modulus furnace	1·48	—	1·47
Medium processing channel	1·55	1·57	1·56

1. Uncured rubber samples. 2. Cured rubber samples. (Courtesy *J. IRI.*)

Davey (1970) investigated the effect that silica had on these values and reported that over the full range of particle sizes a weight of silica equal to that of the black could be tolerated with no alteration of the observed colour index.

Aqueous dispersions of carbon black have also been studied by Ng (1978) who carried out turbidity measurements in the ultraviolet region of the spectrum (200–300 nm) to predict not only the arithmetic mean particle size, but also the size distribution. He used a procedure for removing the rubber from a vulcanizate which consisted of ozonolysis of a comminuted vulcanizate suspended in chloroform at −20 °C.

A second group of analytical procedures for estimating the type or grade of carbon black in a vulcanizate relies on the direct measurement of the surface area of the black. Many authors, however, including Micek *et al.*

(1968) and Kolthoff and Gutmacher (1952), have shown that the reinforcing ability of a black is a function of the *external surface area* of the carbon black, rather than the *total surface area*, which includes the pores in the carbon black aggregates, which are not accessible to the large rubber molecule. Thus the *total surface area* gives an indication of the particle size and the *external surface area* an indication of the structure.

Total Surface Area

Two methods predominate in the determination of total surface area of carbon blacks. These are the 'BET' method and the 'iodine adsorption' method.

BET Method (Nitrogen Adsorption)

In 1938 Brunauer *et al.* published work on the surface adsorption of nitrogen and this has been the standard (BET) method of total surface area measurement ever since. A wide range of instrumentation is now available to do this and, although the original apparatus is described by Barr and Anhorn (1949), its fragility and intricacy have led to the development of more rugged and simpler equipment. The reader is referred to the work of McFearin (1962), Kremens *et al.* (1965) and Atkins (1964). A draft standard (ISO/DIS 4652) is in existence which uses a commercially available instrument, the Ni-Count-1, based on a technique developed by the Phillips Petroleum Company (Krecji and Roland, 1965) whilst the American Society for Testing and Materials describes under ASTM D 3037-1978 four accepted procedures, one of which is that using the Ni-Count-1.

Iodine Adsorption

This provides another method for estimating total surface area of carbon blacks which has the advantage that no equipment other than that found in a normal chemical laboratory is required. It shows a good correlation with the nitrogen surface area results, as indicated in Table 11.1, although if the black is highly acidic the iodine volume can be low. The procedure is given in ASTM D1510-79 but is described in full because of its usefulness and ease of operation.

Procedure for the determination of iodine number (after Schubert et al., *1969).* The weight of carbon to be used is dependent upon the iodine adsorption number. For iodine adsorption numbers up to 130, use 1·000 g of carbon; from 130 to 500 use 0·5000 g; over 500 use 0·2500 g. It may be necessary to run the test with 1·000 g of carbon and repeat the test with a lower weight of sample after a value is obtained. Weigh dried carbon into a 100 cm^3 screw cap bottle or a stoppered test-tube whose length is approximately 200 mm and ID approximately 300 mm. Pipette into the bottle or tube 50 cm^3 of 0·0236 M* iodine solution containing 6·000 g iodine and 57·0 g potassium iodide per litre. Cap or tightly stopper immediately. Shake the iodine–carbon black mixture vigorously for 5 min on a laboratory shaker. The carbon must be intimately mixed with the iodine. If the carbon is difficult to wet, add a few drops of ethanol, correcting for any change in iodine normality. Centrifuge immediately until the carbon settles. Decant the iodine solution into a 100 cm^3 beaker and immediately pipette 20 cm^3 of the solution into a 250 cm^3 Erlenmeyer flask. Titrate with 0·0394 M sodium thiosulphate solution containing 9·7810 g of sodium thiosulphate pentahydrate and 5 cm^3 of 1-pentanol per litre. Treat a reagent blank in the same manner. If a centrifuge is not available, filter into a 250 cm^3 Erlenmeyer flask through a long-stem funnel plugged with fine glass wool. To avoid volatilization of iodine, pass the stem of the funnel through a bored cork which fits the neck of the flask and cover the funnel with a watchglass. Pipette and titrate as above. Calculate the iodine adsorption number, I, in mg of iodine/g of carbon, as follows:

$$I = \left(\frac{B - S}{B}\right)\left(\frac{50}{W}\right)(M)(253{\cdot}82) \tag{11.1}$$

where B = volume of thiosulphate used in the titration of the blank, in cm^3; S = volume of thiosulphate used in the titration of the sample, in cm^3; W = weight of carbon, in g; and M = molarity of the iodine solution.

Both of these techniques were developed for virgin carbon blacks and therefore, if they are to be used on vulcanizates, we must know their validity for recovered blacks. Brown *et al.* (1979) provide a mass of data on the BET nitrogen surface area of blacks recovered by pyrolysis of black filled elastomers (NR, SBR, alone and blended with BR, and IIR) followed by grinding at high speed in an analytical mill (Janke & Kunkel type A10S) before analysis. The original data illustrate black loadings between 20 and 75 pphr but there is little effect with loading and Table 11.4 is for loadings of 40–50 pphr.

*Commercially available solution.

TABLE 11.4

COMPARISON OF BET SURFACE AREAS OF RECOVERED BLACKS FROM DIFFERENT VULCANIZATES

Grade	*Original black*	*BET surface area: m^2/g*			
		ex NR	*ex SBR*	*ex SBR/BR*	*ex IIR*
N-220	120	—	113	—	—
N-234	120	119	—	—	129
N-326	76	—	—	71	—
N-339	94	96	100	—	108
N-347	86	87	88	—	92
N-357	74	79	82	—	83
N-539	42	—	47	46	—
N-550	39	45	45	—	45
N-660	30	—	—	34	—
N-765	23	—	—	—	21

After Brown *et al.* (1979).

A similar table (Table 11.5) obtained by Lamond and Gillingham (1970) confirms the variations from the original total surface areas and suggests that, given a knowledge of the polymer type, recovery by pyrolysis is a valid technique for obtaining the carbon black, the particle size of which can then be related to that of the originally added material.

TABLE 11.5

BET SURFACE AREAS OF BLACKS RECOVERED FROM VULCANIZATES

Grade	*Original black*	*BET surface area, m^2/g*		
		black ex NR	*black ex SBR*	*black ex SBR/BR*
N-110	144.6	132.5	127.5	123.9
N-220	124.2	124.3	112.9	110.0
N-285	103.6	104.0	95.1	81.8
N-330	82.9	86.4	78.2	74.5
N-550	39.7	45.2	45.1	—
N-785	30.5	38.1	33.5	—
N-990	7.4	12.2	10.8	—

Courtesy *Rubber J.*

External Surface Area

The classical test for measuring the external surface area of a carbon black, and thus assessing its structure, is the oil absorption method of Sweitzer and Goodrich (1944) whereby alkali-refined linseed oil is mixed with 1 g of carbon black until just sufficient is added to allow the mix to cohere as a single mass. This can be carried out on either raw or recovered black with the latter again having been ground prior to testing. It will be appreciated that, although this is a valid and quick test in the hands of an experienced operator, it is highly subjective and could benefit from automation. This has now been carried out, and the automated method is covered in detail in various International Standards such as ISO 4656-1978. In this procedure dibutylphthalate (DBP) is automatically added from a burette to a test portion which is kept in motion by rotating paddles. As the DBP is added the powder changes to a semi-liquid mass with an increase in torque. At the limit of adsorbed DBP the torque peaks and this triggers a closing of the automatic burette.

Inspection of Table 11.1, with particular reference to the three blacks in each of the 200, 300 and 500 series, shows that whereas neither the nitrogen nor the iodine values show a difference between the three types of structures present, one is clearly visible when the DBP figures are consulted. It will also be noted that this spread results in an overlap between—say— an FEF-LS, HAF, intermediate SAF and even an SAF, but that the DBP figure, taken together with another estimate such as the iodine number, enables an unequivocal identification to be made.

Lamond and Gillingham (1970) and Lamond and Price (1970) have attempted to reduce these two tests to one by allowing a very much larger molecule than those so far considered to be adsorbed on to the surface of the carbon black. The chemical is Aerosol OT (sodium 2-diethylhexyl-sulphosuccinate) and the procedure is analogous to the iodine test.

Procedure for the determination of Aerosol OT absorption value (after Lamond and Price, 1970). 50 cm^3 of Aerosol OT solution (4 g/l) is shaken for 30 min with 1 g of the carbon black sample and the carbon black is then separated from the OT solution by centrifuging at 24 000 rev/min. 10 cm^3 aliquots of the OT solution are titrated (before and after treatment with carbon black) with cetyltrimethylammonium bromide (CTAB) (concentration 1·125 g/l) using a modification of the method proposed by Barr *et al.* (1948). On addition of CTAB to the OT mixture, the chloroform emulsifies in the aqueous layer and the chloroform–OT–CTAB mixture takes on a milky appearance. On continued addition of the CTAB, de-emulsification occurs and the chloroform layer (containing some emulsified water) separates out. At this point, bubbles are observed in the chloroform layer (probably caused by emulsified water) and the titration is continued until these bubbles disappear instantaneously after allowing the agitated mixture to stand.

The amount of OT adsorbed is given by the following expression:

$$\text{OT adsorbed, mg/g} = \left[\frac{(\text{decrease in CTAB titre}) \times (\text{conc. of CTAB})}{(0{\cdot}1582 \times \text{sample weight})}\right] \qquad (11.2)$$

The procedure is claimed to be extremely quick and accurate (95% confidence limits, ± 1·7 mg/g) but insufficient data are published to afford a full comparison with Table 11.1. There nevertheless appears to be no overlap in the published data between grades for the same polymer base (see Table 11.6).

TABLE 11.6

OT No. OF RECOVERED CARBON BLACKS FROM A RANGE OF VULCANIZATES

Grade	*Original black*	*Black recovered from:*		
		NR	*SBR*	*SBR/BR*
N-110	103.9	98.2	99.1	100.1
N-220	95.9	96.1	92.0	91.0
N-285	77.8	81.7	78.3	81.7
N-347	70.7	61.7	66.7	73.6
N-330	65.0	71.5	62.0	68.9
N-550	33.5	29.4	32.1	—
N-785	26.3	24.1	26.2	—
N-980	6.3	10.0	7.8	—

Courtesy *Rubber J.*

It may be concluded that provided the vulcanizate is properly treated to isolate the carbon black, that is, solvent extracted, followed by pyrolysis under nitrogen or vacuum, a subsequent acid wash, water wash and drying at low temperature (*ca.* 105 °C) followed by a light grinding to break up the agglomerates without damaging the aggregates, the recovered black may be analysed in the same way as a raw black and, if due consideration is given to its history and to the polymeric base from which it was recovered, valid and useful information will be obtained. All of the methods here described are regularly used and thus a choice must depend upon the facilities of the laboratory, and the depth of information required.

Black Type by Thermogravimetry

A new dimension was added to the analysis of carbon black type with a publication by Maurier (1970*a*) in which he claimed that thermogravimetry (TG) could be used, in certain systems, to identify different carbon blacks.

The initial basis for this statement is illustrated in Fig. 11.1 which illustrates a plot of residual sample weight against temperature for a pair of vulcanizates, identical apart from the type of black used.

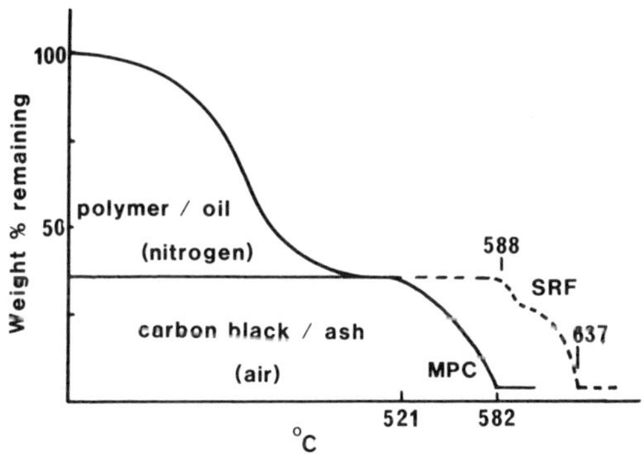

FIG. 11.1. Detection of carbon black differences in 'standard' formulation. (Courtesy *Rubber Age*.)

After pyrolysis in nitrogen, to remove oil and polymer, the sample was cooled to 275 °C, the atmosphere changed to air, and the sample reheated at 15 °C per minute. The different oxidative stabilities of the two blacks, MPC and SRF, are illustrated and this represents a potential method for distinguishing between them. The procedure was improved to allow for the difficulty in determining the exact temperature at which weight loss of the carbon by oxidation commenced, by measuring T_{15} and T_{50}, the temperatures at which 15% and 50% of the carbon black weight were lost. It is well known that a piece of vulcanizate retains its shape after pyrolysis and thus the residue must be a porous matrix through which air can permeate freely. There should therefore be a degree of correlation between the BET surface area and the oxidizability of the black. Indeed this was found to be

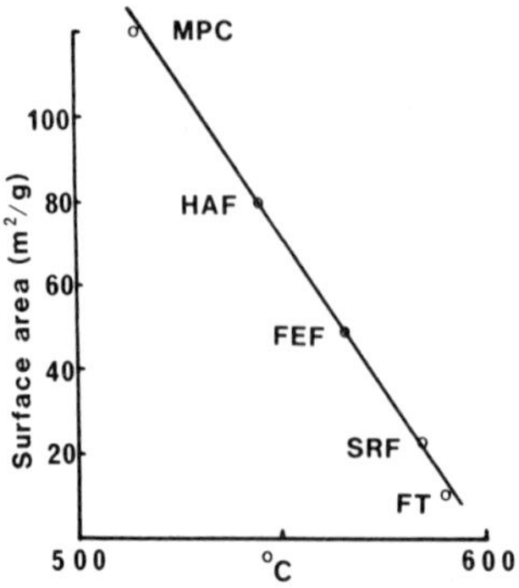

FIG. 11.2. Carbon black in IIR vulcanizate combustion v. surface area. (Courtesy *Rubber Age*.)

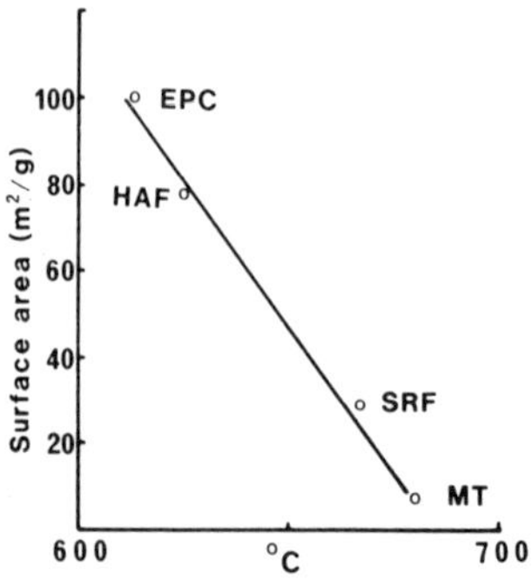

FIG. 11.3. Virgin carbon black combustion v. surface area. (Courtesy *Rubber Chem. Technol.*)

the case as shown in Figs. 11.2 and 11.3 for T_{15}(°C) (see Maurier, 1970*a* and Pautrat *et al.*, 1976). However, the two graphs are by no means superimposable and thus standards will be necessary for any system under consideration. Maurier (1970*b*) showed a further complication when he illustrated the effect of different cure systems on the onset temperature of oxidation of the carbon blacks. This is shown in Table 11.7 for a butyl vulcanizate.

The literature contains other references, however, that suggest that the situation regarding the use of TG for black type identification is far from straightforward; Spacsek *et al.* (1977) were not able to confirm the relationship between T_{15} and surface area whilst Schwartz and Brazier (1978) reported a number of blacks which fell appreciably off a T_{20} v. surface area plot. These latter authors also provide data which suggest that nominally identical blacks manufactured at different locations have different oxidation characteristics.

Attempts to clarify these situations have been numerous and variations such as isothermal oxidation tried (Maurier, 1974), but the current position is realistically summarized by Charsley and Dunn (1981) who studied the following:

The maximum temperature reached during pyrolysis: up to about 600 °C there is little change in the T_{15} but above this there is a relatively linear increase of about 10 °C in T_{15} for each 50 °C rise in final temperature of pyrolysis.

TABLE 11.7

EFFECTS OF CURE SYSTEM AND BLACK TYPE ON DECOMPOSITION OF BUTYL RUBBER VULCANIZATES

	Average values			*Polymer decomposition (°C)*		*Carbon black decomposition (°C)*	
Cure system	*% polymer*	*% black*	*% ash*	*Temp. for 50% loss*	*Terminal temp.*	*Onset temp.*	*Terminal temp.*
A	65.0	31.0	4.0	420	456	497	553
	65.2	30.8	4.0	418	455	529	582
	64.9	31.4	3.8	416	452	550	596
	64.8	31.8	3.3	412	454	551	605
B	65.0	32.0	3.3	418	453	513	558
	66.0	30.3	3.7	417	455	543	588
	65.1	31.3	3.6	415	451	567	608
	65.0	32.0	3.0	411	450	581	623
C	65.3	31.8	3.0	421	457	528	561
	65.8	31.3	2.8	421	464	601	612
	64.8	32.5	2.8	420	455	597	617
	64.0	33.0	3.3	416	452	595	628
D	63.3	30.0	6.7	417	464	427	461
	63.0	30.0	7.0	413	448	468	479
	62.8	30.4	6.9	414	453	464	497
	62.5	31.0	6.8	407	442	461	529

Cure systems (pphr):
A—Altax (1·0), Tellurac (1·5), sulphur (1·0), ZnO (5·0), stearic acid (2·0). B—Sulfasan R (2·0), Tuads (2·0), ZnO (5·0), stearic acid (2·0). C—SP-1055 (12·0), ZnO (5·0), stearic acid (2·0). D—Altax (4·0), GMF (1·5), red lead (5·0), ZnO (5·0), stearic acid (2·0).
Courtesy NBS.

Isothermal hold during pyrolysis: again 600 °C appears significant; there is no observable increase in T_{15} if the hold is up to 30 min at 540 °C but there is a gradual increase if the hold is at 655 °C. This amounts to about 1 °C per 2 min hold.

Effect of air-flow rate over the sample: there is a regular decrease in T_{15} as the air-flow increases; a T_{15} value of 597 °C at a flow rate of 5 $cm^3\ min^{-1}$ falls to 563 °C at a 50 $cm^3\ min^{-1}$ flow rate.

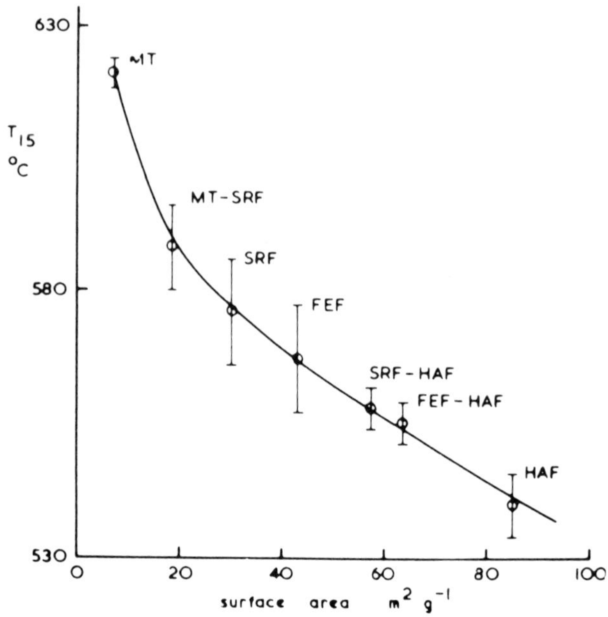

FIG. 11.4. T_{15} valves for carbon blacks in NR rubber formulations. Pyrolysis temperature increase: 100 °C min^{-1}; nitrogen flow: 25 cm^3 min^{-1}; maximum temperature of pyrolysis: 540 °C, start temperature for air run: 400 °C; air flow rate: 25 cm^3 min^{-1}; heating rate: 15 °C min^{-1}.

Having optimized and standardized all these conditions, the authors produced the results shown graphically in Fig. 11.4 where the error bars represent the range for six nominally identical analyses of each sample. They also comment that these values are for a set of results determined on one day and that repeat analyses at a later date gave the same degree of spread but with the curve displaced by a few degrees.

Effect of cure: although the full range of cure conditions as used by Maurier (1970*a*) (Table 11.7) was not examined, it was shown that appreciably different T_{15} values were obtained for sulphur and peroxide cured vulcanizates of EPDM and SBR.

From these data Charsley and Dunn conclude:

> The experimental variables which are found to affect significantly the measured T_{15} value for compounded carbon blacks are: (a) the maximum temperature achieved during the pyrolysis step, and (b) the flow rate of air and the heating rate used during the oxidation step.

There is a definite correlation between the T_{15} value and the surface area of a carbon black, both in its free form and when compounded in a rubber.

The T_{15} value is dependent on the cure method of the rubber and has also recently been reported to depend on the manufacturing source of the carbon black. This technique, therefore, cannot be recommended as suitable for the identification of a carbon black type in an unknown formulation. It can be used, however, as a routine quality control check on batch rubbers.

These comments, together with the observation that the technique appears valid for checking the consistency of a specific grade of virgin black from a particular source, describe the current position in the application of TG to black identification.

(Data and conclusions of Charsley and Dunn reprinted with permission of *Rubber Chem. Technol.*)

REFERENCES

Atkins, J. H. (1964) *Analyt. Chem.* **36**, 579.

Barr, T., Oliver, J. and Stubbings, W. V. (1948) *J. Soc. Chem. Ind.* **67**, 45.

Barr, W. and Anhorn, V. (1949) *Scientific and Industrial Glassblowing and Laboratory Techniques*, Instrument Publishing Co., Pittsburgh.

Bauminger, B. B. and Poulton, F. C. J. (1949) *Analyst* **74**, 351.

Brown, W. A., Schleifer, D. E. and Patel, A. C. (1979) *Paper to Rubber Div. Am. Chem. Soc. Meeting*, Atlanta.

Brunauer, S., Emmett, R. H. and Teller, E. (1938) *J. Am. Chem. Soc.* **60**, 309.

Chambers, W. T. (1958) Unpublished work at MRPRA.

Charsley, E. L. and Dunn, J. G. (1981) *Plast. and Rubber Process. Appln.* **1**, 3. (See also (1982) *Rubber Chem. Technol.* **55**, 382.)

Davey, J. E. (1970) Unpublished work at MRPRA.

Davies, J. R. and Kam, F. W. (1967) *J. IRI* **1**, 231.

Dawson, T. R., Porritt, B. D. and Scott, J. R. (1947) *J. Rubber Res.* **16**, 199.

Fiorenza, A. (1956) *Rubber Age*, **80**, 69.

Galloway, P. D. and Wake, W. C. (1946) *Analyst* **71**, 505.

Hess, W. M., Ban, L. L. and McDonald, G. (1969) *Paper to Rubber Div. Am. Chem. Soc. Meeting*, Los Angeles.

Jones, H. W. and Porritt, B. D. (1914) *Rubber Ind. London*. 199.

Kolthoff, I. M. and Gutmacher, R. G. (1950) *Analyt. Chem.* **22**, 1002.

Kolthoff, I. M. and Gutmacher, R. G. (1952) *J. Phys. Chem.* **56**, 740.

Krecji, J. C. and Roland, C. H. (1965) *Paper to Rubber Div. Am. Chem. Soc. Meeting*, Cleveland.

Kremens, J., Lagarius, J. S. and Deitz, V. R. (1965) *Paper to Pittsburgh Conf. on Analyt. Chem. and Appl. Spectrosc.*
Kress, K. E. and Stevens-Mees, F. (1970) *Rubber Age* **102**, 49.
Lamond, T. G. and Gillingham, C. R. (1970) *Rubber J.* **152**, 65.
Lamond, T. G. and Price, C. R. (1970) *Rubber J.* **152**, 49.
Louth, G. D. (1948) *Analyt. Chem.* **20**, 717.
Maurier, J. J. (1970*a*) *Rubber Age* **102**, 47.
Maurier, J. J. (1970*b*) *NBS Spec. Publ. (US)* **338**, 165.
Maurier, J. J. (1974) *J. Macromol. Sci. Chem.* **178**, 73.
McFearin, T. C. (1962) *Rubber Age* **91**, 611.
Micek, E., Lyon, F. and Hess, W. M. (1968) *Rubber Chem. Technol.* **41**, 1271.
Ng, T. S. (1978) *Prog. Coll. and Polym. Sci.* **65**, 271.
Pautrat, R., Metivier, B. and Marteau, J. (1976) *Rubber Chem. Technol.* **49**, 1060.
Schubert, B., Ford, E. P. and Lyon, F. (1969) *Encyclopedia of Industrial Chemical Analysis* **8**, 179, Wiley, New York.
Schwartz, N. V. and Brazier, D. W. (1978) *Thermochim. Acta* **26**, 349.
Scott, J. R. and Wilmott, W. H. (1941) *India-Rubber J.* **101**, 177.
Scott, J. R. and Wilmott, W. H. (1947) *J. Rubber Res.* **16**, 204.
Spacsek, K., Somolo, A. and Soos, I. (1977) *J. Thermal Anal.* **11**, 211.
Sweitzer, C. W. and Goodrich, W. C. (1944) *Rubber Age* **55**, 469.
Tuttle, J. B. (1922) *Analysis of Rubber*, Chem. Catalog Co., New York.
Weigand, W. B. (1920) *Canad. J. Chem.* **4**, 160.

Chapter 12

Formulation Derivation and Calculation

In the preceding chapters, methods and instrumentation have been discussed whereby specific quantitative and qualitative analyses may be carried out. The major omission to date is the overall determination of polymer content. The purpose of this chapter is to fill this gap and to illustrate how the primary analytical data—that is, the results actually obtained—may be manipulated and intercorrelated to provide the best estimate of the formulation actually used in the manufacture of the article.

POLYMER CONTENT

Various methods for determining specific polymer loadings will have become apparent to the reader of the earlier chapters—typically the estimation of chloroprene rubber content based on chlorine analysis—but what we require initially is a basic analytical scheme which can subsequently be developed as more information on the sample becomes available.

The possibility of removing the organic part of a compound, be it rubber or plastic, by simple heating, leaving the inorganic fillers behind is very attractive. If the material is first extracted with a suitable solvent, thus removing plasticizers and cure residues, quantitative pyrolysis of the remainder should provide a figure for the total polymer content and also a value for the combined carbon black and inorganic fillers contents. It will also provide material in the form of a pyrolysate which can be used, as described in Chapter 8, for identifying the polymer. If complete pyrolysis were always possible, an analytical scheme could be built around the process without further ado. Unfortunately this is not the case but nevertheless the procedure is sufficiently valid in so many cases that it does provide our 'starting-point'.

When a long-chain molecule is heated it can behave in more than one way. The possibilities are:

(1) Fracture of the chain at points randomly disposed along it leading to a steady fall in molar mass.
(2) Fracture of the chain at certain weak points along it. This also will lead to a drop in molar mass but the molar mass distribution will differ from that obtained by random fracture.
(3) Depolymerization by 'peeling off' of monomeric units. This should, ideally, cause the molar mass of the residue to fall much more slowly than the other two processes but there will also be monomer distilled off.
(4) Polymerization of material formed by depolymerization. This can become evident as crosslinking or chain branching and may lead to a hard gel being formed which remains stable at a given temperature, a rise in degradation temperature being required before it too breaks down.
(5) Decomposition of the molecule either while still having a high molar mass or else decomposition of the monomer formed by depolymerization. This will become apparent by the evolution of, for example, hydrogen chloride or water, or by carbonization.

All degradation processes involve free radicals and will thus be modified or suppressed by suitably active molecules if they remain in the system up to a significant temperature. Although the authors are not aware of publications dealing with this topic from the point of view of vulcanizate analysis, a number of papers have been written, and will be referred to later in this chapter, which show that the carbon black loadings, determined by pyrolysis of unextracted vulcanizates, are generally higher than would be expected from the formulations. This could indicate a modification of the decomposition route by the extractable ingredients present.

Some of the earliest work on the quantitative pyrolysis of polymers was carried out by Bauminger and Poulton (1949). This was incidental to their main purpose of studying the carbon black (see Chapter 11) but they published some data on the residual carbon contents of a number of polymers. These data are tabulated in Table 12.1. More recently Brazier (1980) and Sircar and Lamond (1978) have published data which are reproduced in Table 12.2.

TABLE 12.1

RESIDUAL CARBON FROM PYROLYSIS OF RAW POLYMERS AT 600 °C

Polymer	*Residual carbon* %
Crêpe rubber	0·27
Smoked sheet rubber	0·23
Butadiene–styrene rubber	0·23
Butyl rubber	0·10
Butadiene–acrylonitrile rubber (I)	1·5
Butadiene–acrylonitrile rubber (II)	1·9
Butadiene–acrylonitrile rubber (III)	3·7
Ethylene disulphide rubber	2·9
PVA	2·7
PVC	5·9
Neoprene	12·3, 15·4

After Bauminger and Poulton (1949).

TABLE 12.2

CARBONACEOUS RESIDUES (550 °C) FOR VARIOUS RAW POLYMERS HEATED IN NITROGEN

Polymer type/name	*% Carbonaceous residue at 550 °C*	*Polymer type/name*	*% Carbonaceous residue at 550 °C*
All polymers containing only C/H	<1	FKM Viton A	4·0
Silicones	<1	Viton B	3·0
CIIR/BIIR	<1	Viton C10	7·0
CR Neoprene W	21·0	Viton E60	3·7
GT	22·0	NBR (ACN content measured)	
AJ	23·0	18·5	2·1
CSM Hypalon 20	2·0	25	2·7
40	3·5	28·8	2·9
45	2·0	32·2	5·2
PU/AU urethanes	1–5	34·1	5·5
	(higher value the	38·5	6·1
	more aromatics	47·0	11·6
	present)	47·5	12·5
		ACN 94·8	44·0

It is interesting to note the difference in values obtained for Neoprene by Bauminger and the more recent authors. The major difference appears to be in the heating rate, as the earlier pair of workers inserted the samples directly into a hot furnace whilst the more recent ones increased the temperature of the samples slowly from ambient to 550 °C. Work in the authors' laboratory using heating rates of 30 °C per minute (Loadman, 1975) has also given a value of about 20% for Neoprene W, together with 3% for Viton B and 5% for Hypalon 30.

It can be seen that the vast majority of elastomers leave a carbonaceous residue of less than 1%, whilst those leaving 1–5% can be reasonably corrected for. It is only with the higher values that there can be problems which require more than a simple mathematical correction.

In 1958 Chambers described a very simple furnace system to pyrolyse quantitatively hydrocarbon polymers, as illustrated in Fig. 12.1.

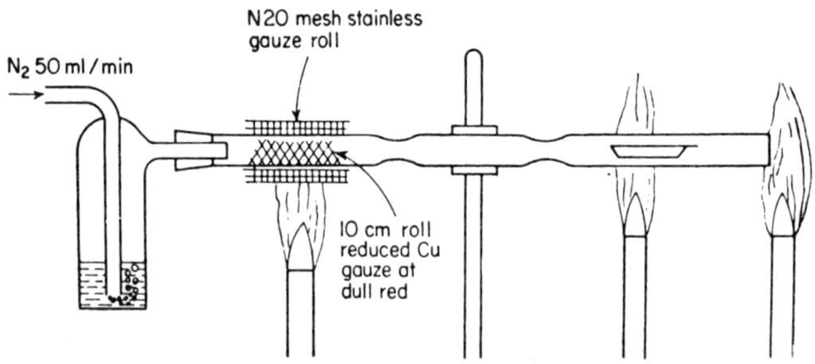

FIG. 12.1. Pyrolysis apparatus. (After Chambers, 1958.)

This has given very satisfactory results even in the presence of fillers, such as clay, if proper correction is made for the moisture loss. Such correction involves knowledge of the actual clay or other material used in the compound, and Frey (1960) gives the following weight losses for several inorganic fillers, when they are heated at 550 °C and 800 °C: kaolin, 5% and 7%; Aerosil, 3% and 8%; Durosil F, 11% and 13%; Keiselguhr 69, 10% and 12%; hydrated alumina, 40% and 53%; lithopone, 3% and 7%; and chalk, <1% and 42%.

Subsequently Higgins and Loadman (1970, 1971) replaced the bunsen burner with a controlled temperature furnace, put a 'U' tube receiver on the outlet from the furnace to collect the pyrolysate for IR examination, and found that modern white spot nitrogen, or the boil-off from a bulk liquid nitrogen storage system, was so low in oxygen that the heated copper gauze was no longer required. This apparatus has already been illustrated (Fig. 8.2).

There does not, at the moment, exist an ISO Standard for the quantitative determination of polymer content by a pyrolytic method as such although ISO 1408 uses the furnace tube to determine black loading quantitatively on selected polymers whilst BS 903. Part B.11 and ASTM D 297 allow the polymer to be determined as (100 – ash – black – extract) %, with the first two being determined by pyrolysis/combustion.

In the last decade the availability of thermogravimetric analysers and an increased awareness of the amounts of information available by obtaining a continuous plot of weight against temperature with a reproducible heating cycle in a controlled atmosphere have led to the publication of a number of papers discussing quantitative aspects of the technique. The International Standards Organization is currently carrying out an interlaboratory study into the reproducibility of thermogravimetric analysis (TGA).

Several types of thermogravimetric analyser are available commercially, one of which has been illustrated previously (Fig. 8.9). A first derivative plot for a mixed polymer system was also illustrated (Fig. 8.10). By using a three-pen recorder the temperature, weight change and first derivative of weight change can all be monitored simultaneously.

It is interesting, and perhaps instructive, to note that whereas many techniques which give a graphical presentation on to a chart recorder, such as gas chromatography, are moving towards a numerical print-out of data only, the converse is true for thermogravimetry. The original procedure of a sample weighing after each stage (extraction, pyrolysis, combustion) has been replaced with a full graphical presentation of the manner in which the weight changes with temperature and atmosphere, and much more useful data are thus obtained.

The first point to be considered is whether the sample should be extracted prior to thermogravimetric analysis or not. To a large extent this depends upon the information required. If a full thin layer chromatographic examination is to be carried out, in addition to thermogravimetric analysis, there is no advantage in not using the extracted sample. If, on the other hand, the study is being carried out on a quality control basis, then the matching of integral and derivative curves with those of a standard material will give the necessary data without extraction. An excellent review was published in 1969 by Maurier in which he considers the many options open to the rubber-analyst when carrying out an investigation on a vulcanizate using thermogravimetric analysis. He discusses the advantages

TABLE 12.3

RAPID TG ANALYSIS FOR QUALITY CONTROL OF MASTERBATCH SYNTHETIC RUBBERS

Sample[1]:	*A*	*B*	*C*	*D*
% Carbon expected:	*25·0*	*30·0*	*34·5*	*35*
% Carbon determined				
1	24·6	30·0	34·8	34·7
2	25·2	30·2	34·8	35·0
3	25·2	30·2	34·5	35·1
4	24·8	30·2	34·5	34·8
5	24·9	30·0	34·3	34·9
6	25·2	30·8	34·8	34·9
Mean	24·98	30·23	34·62	34·90
Std. dev.	0·25	0·29	0·21	0·14

[1]A: Oil-extended SBR + HAF, laboratory prepared. B: Oil-extended BR + HAF, laboratory prepared. C: BR + HAF, laboratory prepared. D: Oil-extended BR + HAF, commercial production. Courtesy *Rubber Chem. Technol.*

TABLE 12.4

DETERMINATION OF CARBON BLACKS IN VARIOUS ELASTOMERS

Black	*Elastomer*	*Quantity of black (mg)* *added*	*found*	*Difference mg*
HAF	EPDM	30·4	31·3	+0·9
	IIR	28·2	30·0	+1·8
	NR	31·5	33·7	+2·2
	SBR	30·0	31·0	+1·0
SRF	EPDM	41·7	43·4	+1·7
	IIR	33·2	34·4	+1·2
	NR	31·5	33·0	+1·5
MT	EPDM	48·5	51·6	+3·1
	IIR	34·0	35·5	+1·2
	NR	31·5	33·5	+2·0

Courtesy *Rev. Gen. Caoutch. Plast.*

of solvent extraction and of thermal extraction, together with the effect of sample size and heating rate on the reproducibility of the data obtained.

Most published data refer to the analysis of unextracted vulcanizates or masterbatches, and it is of relevance to note that whereas Harris (1977) determined the carbon black content of a range of BR and SBR masterbatches, with and without oil extension, with a high degree of precision as shown in Table 12.3, Pautrat *et al.* (1975) found black levels consistently higher than calculated for vulcanizates of several polymers with a range of carbon blacks, as illustrated in Table 12.4.

Jaroszynska *et al.* (1977) studied vulcanizates of the same four elastomers, including oil extended BR and SBR, and found black loadings

TABLE 12.5

TYPICAL TG/DTG ANALYSIS OF PRODUCTION COMPOUNDS

Batch No.	*% NR*[1]	*% BR*[1]	*Oil*[2]	*Carbon black*[3]	*% Ash*[4]	*% Sulphur*
1	31·3	21·4	8·6	31·7	2·1	0·82
2	32·8	23·1	8·0	31·1	2·3	0·80
3	31·7	21·0	8·9	32·2	2·2	0·71
4	31·6	21·4	8·8	31·6	2·4	0·79
5	32·9	21·9	8·9	31·8	2·2	0·95
6	31·1	21·9	8·5	32·5	2·4	0·89
7	30·7	21·4	8·1	31·9	2·5	0·81
8	30·8	22·4	8·9	32·3	2·1	0·86
9	31·4	22·5	8·5	32·2	2·2	0·82
10	31·6	22·4	8·7	32·2	2·3	0·88
11	32·4	22·7	9·4	32·1	2·1	0·81
12	31·6	22·0	8·9	32·2	2·2	0·84
13	33·2	22·4	9·3	31·9	2·2	0·81
14	32·2	22·5	8·5	31·8	2·4	0·90
15	33·2	23·3	8·6	32·5	2·5	0·76
16	31·0	22·3	8·6	31·7	2·2	0·82
17	32·0	23·4	8·3	32·5	2·1	0·74
18	31·9	22·6	8·4	32·5	2·1	0·71
19	33·2	22·2	8·4	32·0	2·3	0·80
20	32·0	23·3	8·5	32·1	2·4	0·85
Nominal	33·0	22·0	8·8	30·3	1·65	0·82

[1] ±0·5%. [2] Wt. % loss at 300 °C. [3] Wt. % loss at 550 °C in nitrogen-less ash. [4] Residual weight at 575 °C in oxygen.

Courtesy *Rubber Chem. Technol.*

0·6–1·8% higher than compounded (with the exception of one SBR sample). Brazier and Nickel (1975) list a series of production compounds (Table 12.5) and again find consistently high black loadings.

On the basis of these data there seems little doubt that, for polymers which pyrolyse completely, and contain carbon black and/or oil, a prior extraction in the case of a vulcanizate enables a true elastomer content to be obtained by TGA whereas the similar treatment of an unextracted vulcanizate gives a small amount (0·5–2%) of carbonaceous residue, possibly due to the interactions discussed at the beginning of the chapter. The problems of quantifying the polymer content of an unextracted vulcanizate or masterbatch in the presence of oils, or indeed other volatiles, have been considered in detail by Swarin and Wims (1974). Figure 12.2 shows a thermogram for the pyrolysis and combustion of a sample based on an EPDM formulation whilst Fig. 12.3 shows the three possible methods of obtaining the 'intercept point'. A relies on the superimposition of the pure 'standard' polymer curve on that of the sample to identify the boundary between the weight loss of the plasticizer and that of the polymer, B uses the intercept of the two measurable tangents, whilst C measures the weight loss at the minimum of the derivative plot.

The quantitative results shown in Table 12.6 suggest that the *reference method* is the most accurate, but the authors prefer the derivative method as the accuracy is acceptable, no reference compound is necessary, and it is easier to measure.

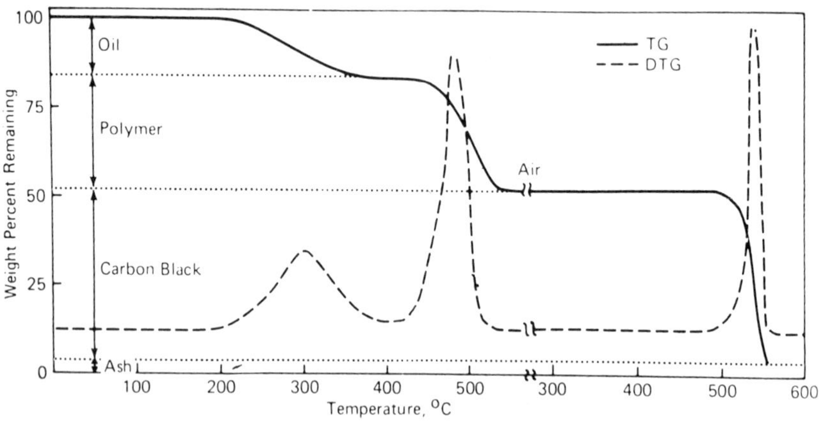

FIG. 12.2. Thermogravimetric curves of EPDM rubber formulation showing weight losses for extender oil, polymer and carbon black. (Courtesy *Rubber Chem. Technol.*)

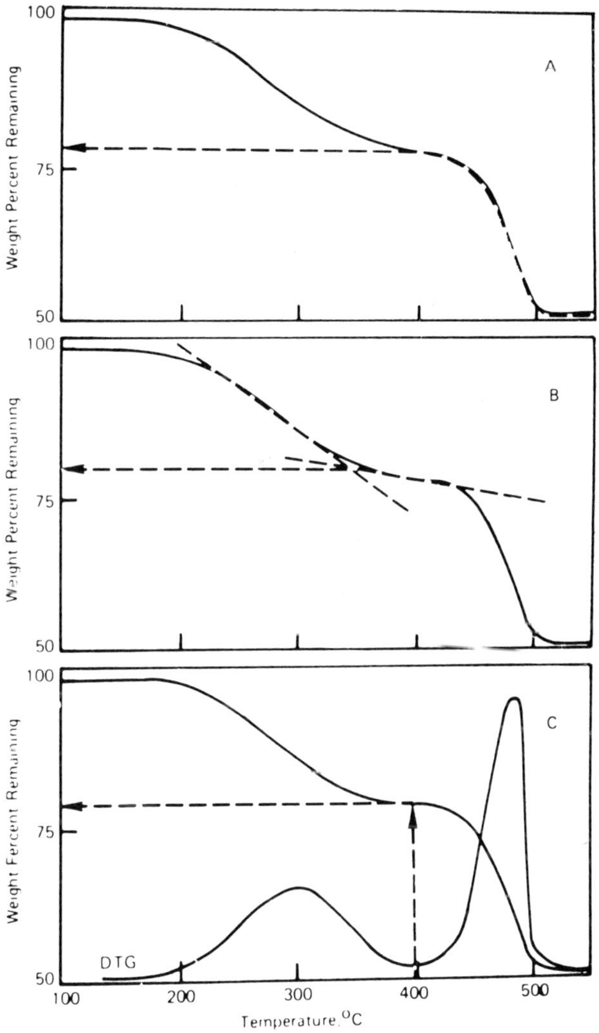

FIG. 12.3. Methods for the determination of oil and polymer in EPDM rubber by thermogravimetry. A, reference overlay method; B, extrapolation method; C, derivative method. (Courtesy *Rubber Chem. Technol.*)

TABLE 12.6

COMPARISON OF METHODS FOR THE DETERMINATION OF OIL AND POLYMER CONTENTS OF EPDM FORMULATIONS

	Composition, %[1]							
	Expected		*Reference polymer method*		*Extrapolation method*		*Derivative method*	
	Oil	Polymer	Oil	Polymer	Oil	Polymer	Oil	Polymer
Sample A	21·6	47·5	21·6	47·0	20·7	47·9	22·1	46·5
Sample B	27·3	33·5	27·3	33·0	25·1	35·2	27·5	32·8

[1]Each value is the average of three determinations.
Courtesy *Rubber Chem. Technol.*

The second problem is that of the polymers which leave a carbonaceous residue on pyrolysis. Of these the chloroprenes (CR) and nitrile–butadiene copolymers (NBR) are by far the most important of the elastomers which are found in black filled products, although urethane rubbers should also be borne in mind. Polyvinyl chloride (PVC), either plasticized or with a copolymer, is probably the most common polymer of the light-coloured halogen-containing products.

The usual procedure for the analysis of halogenated polymers (CR and PVC) is to extract quantitatively and then determine the combined pyrolysate and combustible levels of the extracted polymers by thermogravimetry. A chlorine content of the extracted material is also obtained and then calculation of the polymer level based on the experimentally found halogen contents for the pure polymers (40% CR; 56·7% PVC) gives the required data. If the polymeric phase has been found to consist of more than one polymer, or in the case of PVC a copolymer, a complete elemental analysis of the acetone extracted sample to give absolute values for carbon, hydrogen, oxygen and chlorine levels, when compared to the total pyrolysate and combustible figure, usually enables the polymer system to be fully quantified as described in Chapter 6.

Nitrile–butadiene copolymers and polyurethanes pose more of a problem due to the variability of their nitrogen contents. Swarin and Wims (1974), using an unextracted NBR vulcanizate, illustrate (Fig. 12.4) how there is an initial pyrolysis loss of plasticizer and cure residues, etc., followed by most of the elastomer. When the weight loss stabilizes (550 °C) the heating is stopped, the furnace cooled to 300 °C, the inert gas (N_2) flow

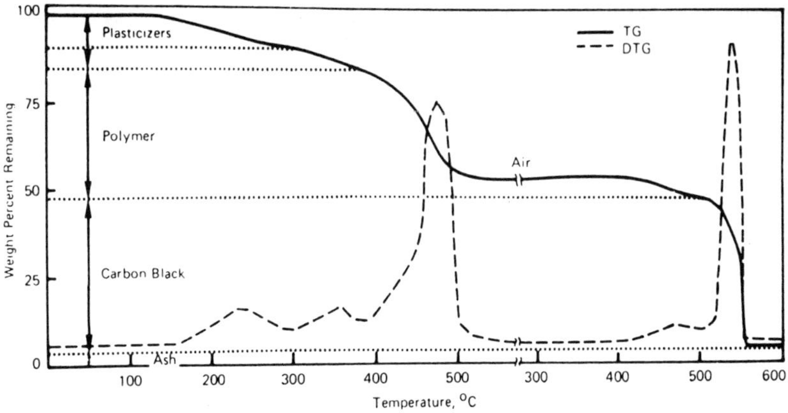

FIG. 12.4. Thermogravimetric curves of NBR rubber formulation showing weight losses for plasticizers, rubber and carbon black. (Courtesy *Rubber Chem. Technol.*)

changed to air, and the heating recommenced at 10 °C per minute. There is an initial gain in weight as oxidation of the residual elastomer occurs followed by a weight loss in the 400–500 °C region shown to be due to its subsequent volatilization. This is followed by the 'true carbon' weight loss.

The quantitative data given in Table 12.7 would suggest that the method is valid, but unfortunately Swarin and Wims do not define the NBR used

TABLE 12.7

TG ANALYSIS OF PREPARED NBR FORMULATIONS[1]

		A	*B*	*C*
Plasticizer, %	Expected	15·6	15·6	11·3
	Determined	16·0	16·0	11·3
	std. dev.	0·1	0·1	0·2
Polymer, %	Expected	57·5	57·5	56·8
	Determined	57·7	57·6	56·0
	std. dev.	0·3	0·2	0·2
Carbon black, %	Expected	23·3	23·3	28·3
	Determined	22·3	22·2	27·8
	std. dev.	0·3	0·1	0·3
Ash, %	Expected	3·6	3·6	3·6
	Determined	4·0	4·2	4·9
	std. dev.	0·1	0·2	0·1

[1]Four determinations on each sample.
Courtesy *Rubber Chem. Technol.*

whilst both Sircar and Lamond (1978) and Pautrat *et al.* (1975) describe problems in differentiating between residual carbon from some nitrile–butadiene rubbers and several carbon blacks. Because of this, Sircar and Lamond (1978) prefer to use glass transition measurements (T_g) and the relationship illustrated in Fig. 12.5 to quantify the acrylonitrile loading of the nitrile–butadiene copolymer, and then to obtain a value for the carbonaceous residue of that copolymer from the data in Table 12.2. Obviously any sample being subjected to a study to determine its glass transition temperature should be extracted prior to that study so that the possible plasticizing effect of an extending oil is removed. The possibility of

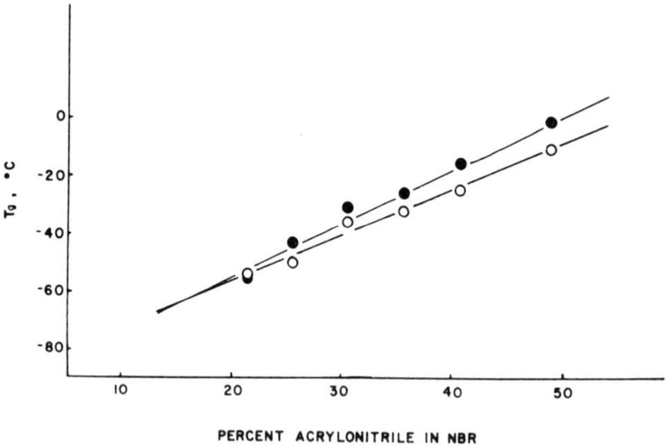

FIG. 12.5. Variation of glass transition temperature of NBR with nitrile content. ●, Raw elastomer; ○, vulcanizate. (Courtesy *Rubber Chem. Technol.*)

the analytical result being modified by an oil or other additive must always be borne in mind, and the decision whether to extract or not considered for each analysis.

This method is equally valid for nitrile–butadiene/polyolefin blends as a separate T_g is still observed, enabling the particular NBR to be identified and the total polymer loading calculated as above. The problem is less easy for polyurethanes and there would appear to be little option other than the identification of the elastomeric constituents as described in Chapter 5 (Dawson *et al.*, 1970) or an estimate of the carbonaceous residue based on the type of polyurethane as identified by pyrolysis–infrared or other technique. In almost every case this latter must be the quicker method and will probably be of sufficient accuracy for most purposes.

Copolymers
Somewhere between physical blends and random copolymers lie the block copolymers, their exact position depending upon their degree of 'blockiness'. It will bc apparent that if the thermal stabilities of the two monomers, as homopolymers, are sufficiently far apart it will be possible to distinguish between the alternatives of block and random copolymerization whilst possibly more detailed interpretation of the data might provide information on the extent of the *blockiness* itself. Studies in this field have been reported by Chiu (1966) and Baer (1964).

Latex
ISO standards do exist for the quantitative determination of natural rubber, or butadiene homo- or co-polymer loadings in latex (ISO 126-1974; ISO 2028-1974). However these methods are time consuming and Loadman (1975) has used thermogravimetric analysis to obtain the data in a few minutes. Perhaps more significantly, Loadman and Tidd (1976) showed how this could be used to obtain the polymer and whiting loadings of a mixed latex carpet backing compound, with a precision quite adequate for quality control purposes, thus enabling a manufacturer to check in ten minutes that settling had not occurred during the standing of the compound.

FORMULATION DERIVATION

By the time that we are interested in deriving the formulation of an elastomeric sample, be it a vulcanizate or thermoplastic elastomer, we will have accumulated an appreciable amount of primary analytical data, which will require further manipulation before it can give a realistic estimate of the formulation.

It is doubtful whether any analyst would claim to be able to describe a complete formulation with an absolute quantitative certainty. It must be considered to be the role of the analyst to assist the technologist by supplying a good approximation to the formulation so that the latter has a valid base from which to start, if a formulation match is the aim, or to provide sufficiently reliable and comprehensive data to enable a formulation to be checked for both correctness and consistency if the purpose is quality control.

The discussion on consistency of compounding is better suited to inclusion in Chapter 14, whilst formulation derivation will be considered here in the form of the reconstruction of some vulcanizate formulations, probably the most complex situation.

Relatively little has been published on the quantitative aspects of formulation derivation although Loadman and McSweeney (1975) showed how a microsample of 10 mg of vulcanizate could be subdivided and analysed to provide the maximum amount of data (Table 12.8), and how the results therefrom are converted to parts per hundred rubber (pphr) and compared with the mixed formulation (Table 12.9).

TABLE 12.8

BREAKDOWN OF 10 mg SAMPLE FOR FORMULATION ANALYSIS

	mg	
% Total sulphur	0	
% Total zinc		
....	2	
% Sulphide sulphur		
....	3	....
Deproteinized NR?		% Acetone extract:
....	4	Natural or synthetic polyisoprene
Carbon black type		Antioxidant
		Antiozonant
....	6	Cure residues
% Unextractable sulphur		Free sulphur
		Oil (extenders)
....	8	Zinc
% Polymer and type(s)		
% Black		
% Inorganic filler and type(s)		
% Zinc oxide	10	

Courtesy Plastics and Rubber Institute.

They do not, however, provide details of the calculations used, but two recent papers, by Putman *et al.* (1979) and Leyden and Rabb (1979) do, the former using a scheme analogous to thermogravimetry on extracted sample, whilst the latter quantified the polymer loading by chemical methods as described in Chapter 5. Putman and colleagues, however, make many more assumptions, particularly on sulphur loadings which they do not determine. Since much importance is placed on sulphur analysis in this book with regard not only to conventional *vs.* efficient cure systems but also to the state of cure of a vulcanizate, examples are considered from the

TABLE 12.9

RESULTS OBTAINED WITH A 5 MG SAMPLE

Experimental results		*Results as parts per hundred rubber*		
Extracted sample	%		as found	as mixed
Polymer (type)	55·2 (79 NR; 21 BR)			
Black (type)	40·1 (HAF)			
Ash	4·7			
(Ash not ZnO)	1·1	Polymer	NR 79	NR 80
			BR21	BR20
Total sample				
ZnO	2·9	Black	71	73
Sulphur	1·46	ZnO	6·3	6·7
Acetone extract	19·0	Other ash	2·0	0
		Sulphur	*ca.* 2	2·0
Acetone extract		CBS	*ca.* 2·5	2·3
β-sitosterol *ex* NR		Nonox ZA	*ca.* 2·5	2·7
Nonox ZA	*ca.* 1	Oil	*ca.* 30	33
MBT, Cyclo-hexylamine } *ex* CBS	high	Zinc stearate		3·6
Oil	aromatic			
Zinc stearate				

After Loadman and McSweeney, 1975. Courtesy Plastics and Rubber Institute.

data of Leyden and Rabb but results have also been calculated taking correction factors used in the authors' laboratory to illustrate the extent of the adjustments which they introduce.

We will consider each piece of the primary analytical data in turn and show the type of corrections which need to be made in order to convert this to a technologically meaningful value.

Extract

Extraction may be carried out with a solvent, or thermally using a thermogravimetric analyser or some other apparatus involving a heated inert atmosphere. For most analytical schemes the thermal methods afford values nominally identical with the solvent ones and, whilst a variety of solvents may be used, acetone extraction is taken as the standard procedure for an unknown system. The acetone soluble content of different polymers is of course a variable, but the same is also true of nominally identical ones, due, for instance, to variations in the soap contents of synthetic rubbers, and to inter-clonal differences in the case of natural rubber. The biggest range of values is found with styrene–butadiene copolymers, which can contain up to 8% extractables when emulsion polymerized, but only about 2% when solution polymerized. In general, values of 1–4% cover the vast majority of rubbers and rubber-like materials. Putman takes 2·5% for a natural rubber/polybutadiene (50/50) blend and 1% for an ethylene–propylene terpolymer whilst Leyden takes 3·7% for a natural/styrene–butadiene/butadiene (50/25/25) vulcanizate, zero for nitrile (Hycar 1042) and 8% for a styrene–butadiene copolymer formulation. This last figure indicates that he is allowing 2·3% for the NR/BR in the first example. Leyden also makes a deduction of 1·5% from his measured acetone extract value due to 'miscellaneous materials' and this is not restored elsewhere in the calculation. It could indicate the level of self polymerization found for the acetone under his extraction procedure.

Polymer

As has been indicated already, for most practical purposes the polymer, black and ash loadings are determined by thermogravimetric analysis directly, or by calculating the percentage polymer as (100 − %black − %ash) (cf. Putman *et al.*, 1979). In these cases the only corrections required are to the polymer loadings thus obtained for any carbonaceous residues which they may generate, and to add the raw polymer extract levels which have been deducted from the extract itself. If a chemical method for determining the polymer content is used, the situation is quite different because a further correction must now be introduced to allow for the unextracted non-rubbers. Thus in a quantitative determination of NR via (e.g.) the Kuhn–Roth method (see Chapter 5) it is the rubber hydrocarbon which is determined and this probably accounts for only some 97% of the extracted sample. In the case of the quantification of styrene–butadiene rubber by the determination of styrene content, all minor corrections must become insignificant relative to the assumption of the styrene content in the copolymer.

Black

Black loadings, as determined by thermogravimetric analysis, with a subsequent correction for any carbonaceous residue from the polymer, represent the best approximation to the true levels present, although it is again emphasized that care must be taken if whiting is suspected of being present. For this reason a thermogravimetric analyser used with a nitrogen atmosphere up to 800 °C, prior to the introduction of air or oxygen, is again recommended. With the graphical presentation not only can the carbon dioxide be measured, but one can gain additional or confirmatory data from a study of the weight-loss and derivative curves. Modern blacks have ash and volatile levels well below 1% at manufacture although, being hygroscopic, they can pick up moisture on standing prior to compounding. This becomes such a variable that it is quite unrealistic to attempt any further correction at the analytical stage; indeed Wake (1969), in discussing a series of interlaboratory trials prior to the introduction of BS 903-1964, expresses surprise at the level of agreement between the added carbon black loadings and the values found by the modified solution method of Kolthoff and Gutmacher (1950) adapted for BS 903-1964. He suggests that oxidation of the black, giving an increase in weight, could counteract the loss of volatiles and indeed all through these manipulations many minor corrections could be made both adding to and subtracting from each piece of primary data. For practical purposes those not considered here can be safely ignored and normally assumed to 'cancel out'.

Ash

Ash contents again appear usually to require little correction beyond the inevitable one if whiting is present. This can be made using either the carbon dioxide weight loss or the measured calcium content although the latter is suspect as calcium oxide or other calcium-containing compounds could have been added, typically 'Caloxol'. The problems associated with ash determination by thermogravimetric analysis are covered in Chapter 10, in the discussion on dry ashing, and should be borne in mind, particularly if a chlorinated elastomer or high sulphur vulcanizate is being analysed.

Sulphur

The intricacies of sulphur analysis have been discussed in detail in Chapter 6. Here we are solely concerned with the corrections to and implications of the total sulphur value. In the absence of a sulphur-containing polymer, or inorganic fillers such as barium sulphide, barium sulphate or calcium sulphate, the biggest correction that will be required is for the sulphur content of the black, whilst there could also be a further contribution from extending oils. Leyden assumes that 0·5% of the sum of the acetone extract level (after the corrections described earlier) and the carbon black loading is sulphur, whilst in the authors' laboratory it is more usual to ignore the oil initially because of the possible variability of its sulphur content and allow 0·8% of the black loading as sulphur (this does not require a correction of the black content as this sulphur is not extracted nor does it appear to be removed by pyrolysis under conditions used to volatilize the polymer). Should the corrected sulphur value still appear too high for the application, the type and level of oil present can then be considered and possibly a further correction made. This corrected sulphur content is not, of course, due only to the added elemental sulphur but includes the sulphur present in the accelerator. Table 12.10 gives values for the sulphur contents of a number of accelerators and it will be appreciated how important this correction is, particularly if the cure system is a semi-efficient or efficient one.

TABLE 12.10

S% IN SOME COMMON ACCELERATORS (TO NEAREST %)

CBS	24%	TMTM	46%
MBT	38%	TMTD	53%
MBTS	40%	TETD	43%
MOR	25%	TBTD	25%

See also Table 6.3 (p. 114).

When one has a corrected sulphur content which can be defined as the 'cure sulphur', then its value, together with a knowledge of the cure system obtained from an examination of the extract (Chapter 4), enables one to calculate the probable ratio of the sulphur loading to that of the accelerator. Thus a 'cure sulphur' content of 2·7 pphr can reasonably be assumed to be derived from an elemental sulphur addition of 2·5 pphr with the remaining 0·2 pphr being from about 0·5 pphr accelerator. At the other extreme a 'cure sulphur' of 1 pphr would be due to 0·5 pphr elemental sulphur and 2 pphr accelerator, thus an efficient (EV) and conventional cure system may easily be distinguished. A word of warning should be added: if a high sulphur loading is obtained the presence of factice should be considered a possibility.

Formulation Calculation

The two following tables, 12.11 and 12.12 (after Leyden and Rabb, 1979) illustrate data from two vulcanizates which they analysed and for which

TABLE 12.11

CALCULATIONS OF FORMULATION OF CONVENTIONALLY CURED VULCANIZATE

Sample A	*As compounded*	*Derived Leyden and Rabb (1979)*	*Derived from published data (authors)*[a]
NR	50	50	50
SBR	25	25	25
BR	25	25	25
Black	50	49	50
Oil	20	24	22
ZnO	3·0	3	3
Stearic acid	2·0	2	2
Santoflex 13	3·0	2	2
Sulphur	2·25	2·7[b]	2.6 (2·2–2·4)
Vulkacit DZ	1·14	2	(1·6–0·8)

[a]Authors' corrections to the published data: (i) 2·5% of polymer extracted by acetone, (ii) 0·8% of black loading is S.
[b]Not corrected for sulphur in the accelerator.

TABLE 12.12

CALCULATIONS OF FORMULATION OF EFFICIENTLY CURED VULCANIZATE

Sample B	*As compounded*	*Leyden and Rabb (1979)*	*Authors*[a]
Hycar 1042	100	100	100
Black	60	66	60
DOP	10	13	10
ZnO	5	5	5
Stearic acid	1·0	1	1
Struktol WB 212	2·0		
Aminox	2·0		
Sulphur	0·75	1·5[b]	1·25 (0·5–0·8)
TMTD	2·0		(2·0–1·5)
MBTS	2·0		

[a]Authors' corrections: (i) 2·5% of polymer extracted by acetone, (ii) 0·8% of black is S, (iii) Hycar 1042 is 33% nitrile, so 5·3% carbonaceous residue is black.
[b]No correction for accelerators.

they subsequently calculated the formulations (middle column). The right-hand column in each case was calculated by the authors from the published primary data using the corrections discussed for the polyolefins in the case of Sample A, and adding a further one for the carbonaceous residue of the nitrile–butadiene rubber in Sample B. In each case the authors took the sulphur analysis a stage further in interpretation, the left-hand figure in the authors' column relating to sulphur analysis represents the 'cure sulphur' whilst the right-hand figures, in parentheses, represent the 'best estimate' of the levels of ingredients added. These results illustrate the conclusions reached from many hundreds of such calculations: that most corrections do indeed 'cancel out' and that these mathematical manipulations provide an extremely good indication of the formulation to which the sample was compounded.

Note that the 'total sulphur level' in Sample B was measured as 0·95%. This must indicate either an analytical or compounding error as just from 0·75 pphr S (S = 0·75), 2·0 pphr TMTD (S = 1·06) and 2·0 pphr MBTS (S = 0·8) there should be 1·42%. If one includes a low figure from the black (0·5%) this gives 1·6%. The authors' interpretation would suggest that one of the accelerators could have been omitted (if the analytical result is correct) and recourse should be made to a thin layer chromatographic investigation of the extract.

REFERENCES

Baer, M. (1964) *J. Polym. Sci.* **A2**, 417.

Bauminger, B. B. and Poulton, F. C. J. (1949) *Analyst* **74**, 351.

Brazier, D. W. (1980) *Rubber Chem. Technol.* **53**, 437.

Brazier, D. W. and Nickel, G. H. (1975) *Rubber Chem. Technol.* **48**, 661.

Chambers, W. T. (1958) Unpublished work at MRPRA.

Chiu, J. (1966) *Applied Polymer Symposia* **2**, 25.

Dawson, B., Hopkins, S. and Sewell, P. R. (1970) *J. Appl. Polym. Sci.* **14**, 35.

Frey, H. R. (1960) *Methoden zur Chemischen Analyse von Gummimischungen*, 2nd ed., Springer Verlag, Berlin.

Harris, J. (1977) *Synthetic* **8**, 20.

Higgins, G. M. C. and Loadman, M. J. R. (1970) *NR Technol.* **10**, 1.

Higgins, G. M. C. and Loadman, M. J. R. (1971) *Ind. Gomma.* **15**, 50.

Jaroszynska, D., Kleps, T. and Tulak, D. (1977) *Int. Polym. Sci. Technol.* **4**, T20.

Kolthoff, I. M. and Gutmacher, R. G. (1950) *Analyt. Chem.* **22**, 1002.

Leyden, J. J. and Rabb, J. M. (1979) *Paper presented at 116th Meeting, Rubber Div. Am. Chem. Soc.*, Cleveland.

Loadman, M. J. R. (1970) Unpublished work at MRPRA.

Loadman, M. J. R. (1975) Unpublished work at MRPRA.

Loadman, M. J. R. and McSweeney, G. P. (1975) *Rev. Gen. Caoutch. Plast.* **52**, 805.

Loadman, M. J. R. and Tidd, B. K. (1976) *Paper presented at 5th Conference Européenne des Plastiques et des Caoutchoucs*, Paris.

Maurier, J. J. (1969) *Rubber Chem. Technol.* **42**, 110.

Pautrat, R., Metavier, B. and Marteau, J. (1975) *Rev. Gen. Caoutch. Plast.* **52**, 273.

Putman, J. B., Samples, C. R. and Knowles, T. M. (1979) *Paper presented at 116th Meeting, Rubber Div. Am. Chem. Soc.*, Cleveland.

Sircar, A. K. and Lamond, T. G. (1978) *Rubber Chem. Technol.* **51**, 647.

Swarin, S. J. and Wims, A. M. (1974) *Rubber Chem. Technol.* **47**, 1193.

Wake, W. C. (1969) *The Analysis of Rubber and Rubber-like Polymers*, 2nd ed., Maclaren, London.

Chapter 13

Blooms and Visually Similar Phenomena

There is no doubt that every rubber-technologist and analyst knows what is meant by the word bloom—occasionally called 'frosting'—but it is quite apparent from investigations carried out at MRPRA over a number of years that these words include a wide variety of very different effects. These can be divided conveniently into true blooms, modified blooms, pseudo blooms and surface contamination. It is also worth including here a consideration of stains and discolorations, as these are sometimes confused with blooms. Even when this is not the case, there are good reasons for treating them together. They are all visually offensive effects, and the appearance of a surface deposit or a colour change is quite sufficient to cause rejection of the product during manufacture, storage or service. The consequences of this are just as commercially harmful as mechanical damage or a more catastrophic physical failure.

TRUE BLOOMS

The mechanism of the blooming of crystalline materials is simple in broad theoretical outline (Nah and Thomas, 1980); the substance which blooms must have a limited but appreciable solubility in the rubber and be present in excess of this solubility. This excess will exist as discrete particles throughout the mass of the rubber either because it has never dissolved or because, having dissolved at the temperature of vulcanization, it has crystallized out on cooling. These discrete particles can easily be seen in sections cut from pure gum rubber and examined by transmitted light under the microscope. Wax is particularly notable as its crystals are anisotropic and crossed Nicols enable them to be identified easily. In thus crystallizing it must be assumed that local strain is set up in the rubber displaced by the formation of the crystal. This strain results in pressure on the crystal, the solubility of which is increased thereby. At the free surface crystals of the material can form without distortion of the rubber and the solubility will be unaffected. The free energy of crystallization will therefore be less at the surface than in the bulk of the rubber; the solubility of the substance will also be slightly less. There will therefore be a concentration gradient of dissolved material which will cause diffusion from the inside towards the surface and this will persist until all the material crystallized in the bulk has dissolved under the influence of pressure and diffused outwards. The magnitude of the increased solubility due to pressure will, of course, be minute as also will the concentration gradient within the rubber, but large forces are not necessary to account for the observed phenomena.

Free sulphur is probably the most common substance to give a true bloom, and in a vulcanized product such a bloom is due to undercure. This, in itself, could result from a number of factors and the relationship between the time of cure, temperature of cure and suitability of formulation should be considered first of all.

Zinc dithiocarbamates are also known to give blooms and of the three common ones, the dimethyl-, diethyl- and dibutyl-dithiocarbamates, it is the middle one which shows the most rapid and, over a period of time, the densest bloom. The order of solubility is ZDMC < ZDEC < ZDBC and it is therefore concluded that the soluble fraction of ZDMC is relatively low, resulting in a low rate of migration, whilst ZDBC is sufficiently soluble for the solubility limit not normally to be exceeded and thus for there to be no bloom. It is unfortunate that ZDBC gives a slower rate of cure than the methyl or ethyl homologues and that it is therefore not always practicable to use it.

Of the other commonly used accelerators, mercaptobenzothiazole and zinc mercaptobenzimidazole have also been observed to bloom.

Many instances have been recorded of protective waxes being the cause of complaints concerning blooms. This appears to suggest a lack of knowledge of the function of wax added to protect against ozone as the presence of a surface layer (bloom) of wax is the object of its addition, and the reason protection is afforded to the rubber. It must be noted, however, that the extent of a wax bloom is not only a function of its loading, but also of its melting point, and these two parameters can be ‘played off’ against each other for different applications.

MODIFIED BLOOMS

Certain chemicals present within the matrix of a rubber vulcanizate react, either deliberately or not, with constituents of the environment and this results in a significantly different mechanism of blooming. Typical examples are the *para*phenylenediamine (PPD) antiozonants, which protect the rubber by reacting with ozone to form an insoluble protective skin on the surface. This results in a deficiency of PPD in the layer of rubber nearest the atmosphere and there is a migration of the PPD from the bulk rubber to eliminate the concentration gradient. As more PPD reaches the surface layer it reacts with ozone and the process of migration continues until the 'skin' of oxidized PPD prevents further ozone penetration, and enables a constant concentration to be established throughout the bulk of the rubber.

*Para*phenylenediamines may also bloom by the 'true bloom' mechanism and it is therefore important that they should only be added at levels up to their solubility. Although this is usually the case when a formulation is originally devised, subsequent modifications without a full realization of their significance have been known to take formulations 'over the limit'.

Zinc salts of carboxylic acids (in particular zinc stearate) constitute further examples of both true and modified blooms. Zinc stearate has a known solubility in *cis*polyisoprene of about 0·3% and thus the addition of 1 pphr stearic acid and 2–5 pphr zinc oxide should inevitably produce a bloom. However, it is also known that the solubility of zinc stearate is greatly increased when it complexes with amines and, since these are usually present as accelerator decomposition products, or in natural rubber as supplied, the problem is less acute than it would at first glance appear. Zinc stearate itself is oily and thus gives an oily bloom but, in moist atmospheres, this reacts with water vapour at the surface of the rubber to produce 'basic zinc stearate' which forms on the surface as a solid layer, visually indistinguishable from a bloom, and is completely insoluble in the rubber. A true zinc stearate bloom can be dissolved back into the rubber by heating, but this is not the case with the basic salt.

PSEUDO BLOOMS

On a surprisingly large number of occasions it has been found that the matt effect on an initially smooth shiny surface has not been due to the blooming of a particular compound, or to deposition of a contaminant, but to the degradation of the rubber surface itself. The pitted surface which develops results in sufficient light scattering to give the impression of a bloom. This is particularly significant in view of the data of Moakes (1950) who noted blooms of calcium and zinc carbonate. There is no doubt that because of their complete insolubility these inorganic materials *cannot* migrate to give blooms. This regularly observed phenomenon now has been found to be due to the degradation of rubber surrounding the filler particles resulting in the exposure of these particles in a 'crater' of rubber (see SEM examination below).

SURFACE CONTAMINATION

It is always difficult to decide by visual inspection whether a surface deposit is a bloom or contamination and it then rests with the analyst to identify the surface material to such an extent that this can be resolved. One of the most obvious causes of surface contamination is silicone oil, used as a mould release agent. Not only does it impart an oily film to the surface but it also gives a base to which dirt and dusting powders may adhere.

The washing of rubber products also gives rise to contamination if rinsing is inadequate and both inorganic salts and organic materials have found their way on to the surface of rubber articles by this route. Inorganic fillers, used as dusting agents, tend to be present in the air of most factories and can adhere to freshly moulded rubber surfaces, giving the appearance of a bloom.

HAZING (OF TRANSPARENT RUBBERS)

Haze is defined as a cloudy appearance within or on the surface of a transparent article, and from a visual inspection it is often difficult to distinguish between it and a bloom. Blooms and surface breakdown have already been discussed so we must now consider opacity within the bulk of the rubber itself. This will result from the presence of insoluble particles, micelles or droplets (in the case of liquids) having a different refractive index from rubber and so able to cause light scattering. One of the commonest causes of this is the use of zinc oxide either of the wrong grade or in excessive amounts, so the problem can be eliminated by the use of special fine-particle grades at levels not exceeding 1 pphr.

On the other hand, calcium oxide can cause this effect even at the low levels required for desiccant purposes whilst the maldispersion of otherwise suitable compounding ingredients is a further threat to transparency.

STAINING/DISCOLORATION

Although these terms tend to be used interchangeably it is probably better to consider *discoloration* as applying to the rubber article itself, and *stain* as describing the effect produced on a material in contact with the compounded or cured rubber. In the vast majority of cases these effects are brought about by free sulphur or dithiocarbamates in contact with copper, as both copper sulphide and copper dithiocarbamate are very dark coloured, and give a visible stain even at the parts-per-million level. Trace metals such as iron and copper in the rubber itself can also give rise to discoloration, as too can the use of zinc oxide with an over-high level of lead. In these cases the problem is in identifying the source of the excessive levels of trace metal, although the usual culprits are fillers and dyes. Perhaps one of the hardest problems is in defining at what level these elements become effective discolorants. There is no doubt that the form in which the trace metals exist substantially affects the amount which can be tolerated in a vulcanizate but, broadly speaking, if one has discoloration problems and the copper level exceeds 20 ppm, or the iron 150 ppm, then this is probably the root cause of the effect.

Certain phenolic antioxidants are known to give a pink colour to rubber products although this is generally faint and thus only noticed with light-coloured or transparent formulations. Less well known is the phenomenon of *pinking* in latex or goods produced therefrom. This is a light-induced weak coloration which is reported by Sin Siew Weng (1982) only to occur when zinc *diethyl* dithiocarbamate has been added, and which may be removed by washing with dilute potassium hydroxide solution.

The effects of staining antioxidants—particularly *para*phenylenediamines—are, of course, well known and a test for their presence would be the first thing to carry out if a purple, blue or brown stain or discoloration is observed.

PRE-ANALYTICAL CHECK-LIST*

Before discussing in detail the methods available for the analysis of blooms and the other effects described, it is worth while working through a series of questions, the answers to which could assist in finding one's way through this maze:

*We are indebted to our colleague, Mr P. M. Lewis, for permission to quote his question-and-answer scheme.

Q1 Has the bloom increased during storage?
If not, a dusting powder may be responsible.

Q2 Does the bloom disappear on heating?
If not, basic zinc stearate, an insoluble dusting powder such as talc, or surface degradation may be responsible.

Q3 Is light or exposure in the open required for the bloom to form?
If yes, surface degradation or certain antidegradants may be responsible.

Q4 Can the bloom be removed by a solvent wipe?
If not, try other solvents (a complexing agent such as acetyl acetone or lactic acid will remove basic zinc stearate). If these also fail, embedded dusting agent or insolubles may be responsible.

Q5 Is the surface of the rubber pitted or roughened after a solvent wipe?
If yes, surface degradation may be responsible.

If the answers to Q1, Q2 and Q4 are positive and those to Q3 and Q5 negative there is probably a true bloom and the presence of an excess of a compounding ingredient or residue or a protective agent should be suspected.

If the answers to Q4 and Q5 are negative, the haze may be due not to a bloom but to an opaque material within the rubber. If the answers to Q1 and Q2 are also negative the cause may be an insoluble filler such as zinc oxide. Check whether the level of zinc oxide exceeds 1 pphr.

If the answer to Q2 is positive, a vulcanization residue or absorbed water may be responsible. If opaque specks are observed on holding the rubber up to the light, suspect the presence of zinc dimethyldithiocarbamate (or TMTD, TMTM). If water is responsible, the problem should not recur after leaching and drying.

Colour Changes

Q1 Does the discoloration/stain appear to be light-induced?
Check by comparing with a sample kept in the dark or with a surface hidden from light. If ‘yes’, an antidegradant may be responsible. A greyish or brownish discoloration may be indicative of an amine antioxidant, a pinkish discoloration may be indicative of certain phenolic antioxidants.

Q2 Does the discoloration/stain appear to be heat-induced?
If ‘yes’ and there are no signs of ageing (e.g. stickiness, embrittlement, etc.), an amine antioxidant may be responsible.

Q3 Is the discoloration uniform or patchy?

If patchy, external contamination is likely, although an additive in the rubber may still be involved.

Q4 Does the discoloration/stain appear after laundering, contact with metal parts, or only when the rubber is in contact with fibres and textiles?
If 'yes' to the first part, copper or iron contamination may be responsible; see whether there is a brown stain. Check whether dithiocarbamates are present in the formulation.

Q5 Is the discoloration/stain accompanied by poor ageing?
If 'yes', this is further evidence that copper or iron may be responsible.

ANALYTICAL METHODS

It must first be decided whether to examine the surface *as it stands* or to attempt to remove any bloom present and carry out a subsequent analysis on the separated bloom. Some idea of the best approach to adopt will have been gleaned from the tests above, so let us consider first the examination of the surface without attempting to remove anything. To do this there are three basic techniques in addition to the use of a lens or microscope: spot tests, multiple internal reflectance infrared spectroscopy (MIR) (see Fig. 8.1) and scanning electron microscopy with an X-ray analyser (see Chapter 8).

Spot Tests

Spot tests appear to be restricted to the detection of free sulphur. Two extremely sensitive and specific tests are as follows.

Behaviour with Carbon Disulphide

If one small drop of carbon disulphide is spotted on to the bloom the drop spreads out and then dries off, leaving a clean dull circular area surrounded by a line of recrystallized sulphur. The yellow crystalline appearance of this ring seems quite characteristic and is not obtained with accelerators and antioxidants.

Kirchhof's Piperidine Test

The surface of a white or brightly coloured rubber is spotted with piperidine. In the presence of free sulphur a yellow or deep orange–red coloration occurs presumably due to the formation of polysulphide piperidine compounds (Kirchhof, 1925).

Multiple Internal Reflectance

Multiple internal reflectance techniques have been described in detail in Chapter 8 and it will be apparent that if one obtains a spectrum of the surface 2–10 μm of a sample with a bloom on the surface, much of the spectrum will be due to that bloom. Although this is true in principle, the sample requirements of a flat piece some 2″ × 1″ in area, together with the virtual absence of carbon black, which absorbs all infrared radiation, reduces its usefulness appreciably, and eliminates many samples immediately. Nevertheless blooms have been identified using this technique, Petit and Carter (1964) identifying zinc stearate on the surface of a soling material. Excess dusting agents or filler pseudo blooms have similarly been identified although care must be taken when examining a white filled vulcanizate by this technique as there will inevitably be the spectrum of the filler superimposed on that of the elastomer even in the absence of bloom. If the presence of a bloom is suspected on a black filled or other product but no useful spectrum is obtainable, it is frequently worth while to remove the product from the MIR plate and inspect the surface of the plate to see whether any transfer of bloom has occurred. A spectroscopic examination of this coated plate could then provide a spectrum free from the rubber background and thus make it more easily recognizable. This is particularly true of silicone oils, zinc stearate and other 'sticky' materials, including degraded rubber which may itself have been thought to be a bloom.

Scanning Electron Microscopy

The use of the scanning electron microscope has been discussed in detail previously (Chapter 8) and from this the application to bloom analysis will be self evident. Photographs P5–P8 (see pp. 168–9) (Loadman and Brown, 1982) clearly illustrate the type of information which is quickly available by this technique. P5 shows characteristic sulphur crystals bloomed to the surface of a badly cured product, X-ray elemental analysis of an individual crystal showing only sulphur present, whilst P6 again shows a sulphur bloom together with a much larger crystal which, since it was shown to contain large amounts of both zinc and sulphur, is probably due to accelerator. P7 on the other hand shows a sample which was suspected of having a bloom but which was found to have a surface fungal growth. P8 illustrates an example of stress-induced oxidative degradation of an injection moulded sample which had been allowed to cure partly prior to moulding. The matt rather than shiny surface finish led both the manufacturer and customer independently to suspect that blooming had occurred.

REMOVAL OF BLOOM FOR ANALYSIS

When it is necessary to remove the bloom from the surface of the rubber for subsequent examination there is a problem as to whether it is better to use a dry or a wet method and, if wet, which solvent to use. A 'dry wipe' has certain advantages in that one is unlikely to extract chemicals close to, but under, the rubber surface. As early as 1950 Galloway and Foxton described a procedure for identifying a bloom of free sulphur.

> 'To detect sulphur in a bloom on an article, fold in four a piece of filter paper (preferably the slow-absorbing, alkali-resistant type) and rub one of the fold edges over the surface; if the bloom is light, first treat the surface with a drop of carbon disulphide, and then rub the paper round the outline left when the solvent evaporates. Unfold the paper and add 1 drop (about 0·05 cm^3) of strong (20–30%) sodium hydroxide solution, followed by 1 drop of pyridine. A blue–green colour in the pyridine, rapidly succeeded by an orange or brown stain on the paper at the fold, indicates the presence of sulphur.'

This procedure is quite general for other surface blooms in that the spraying of the filter paper with the more specific of the thin layer chromatographic visualizing sprays described in Chapter 4 will often identify the bloom components. The paper can alternatively be extracted with a few drops of a suitable solvent, and the extract examined by TLC, LC, IR, etc. in the same way as a complete rubber extract.

On many occasions it is possible to scrape the bloomed surface with a new clean razor blade and to remove the bloom. This is again then available for examination by the previously described techniques, including IR as a potassium bromide disc, DSC or melting point apparatus, the last being particularly useful for identifying paraffin waxes.

Amos (1967) described in detail a procedure by which he removed blooms from the surface of ethylene propylene rubbers with cotton wool swabs moistened with chloroform and then extracted the swabs to obtain solutions of the removed chemicals for thin layer chromatographic analysis. The great difficulty of wet swabbing is in being certain that there is effectively no penetration of the solvent into the rubber surface, with consequent leaching of non-blooming components. Preferred solvents would thus be those such as acetone or methanol which, whilst being good solvents for most of the chemicals known to bloom, do not swell the polymers appreciably. A reduction in temperature of the rubber and solvent will help to reduce diffusion and so afford clean bloom. A reduction from +20 to −70 °C could reduce the effective penetration one-hundredfold. One particular application of this technique has been reported by Edwards *et al.* (1976) who used a mixture of acetyl acetone and 2-propanol (10:90) to complex and thus render soluble the normally insoluble *basic* zinc stearate/palmitate bloom by soaking for up to 93 hours at −26 °C. Identification and quantification of the zinc and fatty acids were by atomic absorption spectrometry and gas chromatography respectively.

It is recommended that in the analysis of a bloom involving wet swabbing or soaking, a clean control area should be identically treated. This area should ideally be obtained by cutting away the surface and certainly not by washing an area clean! Only those components which are found to be present in the bloom at a level significantly higher than in the 'control area' should be considered to be components of the bloom.

The analysis of stains and discolorations is often extremely difficult because of the very small quantities of materials involved. Staining generally results from the diffusion of a chemical in the rubber product to its surface and then its migration into the material in contact with the rubber. The chemical may be coloured in its own right, in which case it will be removed by extraction or swabbing and identified by comparing an extract of the swab with those of the rubber and the material, or it may be colourless but react with another chemical in the material to produce the staining colour. In the latter case it is quite likely that it will not be possible to remove the stain with the normal range of solvents. It is usual for this situation to arise when the material contains trace metals, such as copper, in its dyes, or perhaps its fillers, which react with dithiocarbamates in the rubber. With this type of stain, an analysis of the metal constituents of the dyes will often give useful information.

In many cases it will be impossible to identify absolutely the cause of the problem but, nevertheless, with sufficient analyses of good and bad regions, slight but constant differences may become apparent which make constructive comments possible.

REFERENCES

Amos, R. (1967) *J. Chromatog.* **31**, 263.

Edwards, A. D., McSweeney, G. P., Roberts, A. D. and Tidd, B. K. (1976) Unpublished work at MRPRA.

Galloway, P. D. and Foxton, R. N. (1950) *J. Rubber. Res.* **19**, 74.

Kirchhof, F. (1925) *Gummi Ztg.* **39**, 849; (1955) *Chem. Ztg.* **79**, 434.

Loadman, M. J. R. and Brown, J. (1982) Unpublished work at MRPRA.

Moakes, R. C. W. (1950) *RAPRA Bull.* **4**, 9 (circulated to members only).
Nah, S. H. and Thomas, A. G. (1980) *J. Polym. Sci. Polym. Phys. Edn.* **18**, 511.
Petit, D. and Carter, A. R. (1964) *Adhesion of translucent rubbers: Application of infrared spectrometry to the problem*, British Boot, Shoe and Allied Trades Research Association, Kettering.
Sin Siew Weng (1982) Private communication.

Chapter 14

Validity of Results

INTRODUCTION

Throughout the previous chapters we have been concerned with the acquisition of data, much of it numerical, and its subsequent manipulation to provide information in whatever form it may be required. At all times we have assumed that the analytical value is a true and absolute result but this of course is not correct, as any analyst knows only too well. This makes it essential for the analyst to understand what limits must be placed on the primary data so that any values passed to a third party may be properly explained to prevent misinterpretation or misrepresentation.

In order to do this it is necessary to examine the ways by which, and reasons for which, a result can deviate from the 'truth', and to define some terms rather more rigorously than usual. The three terms most commonly used to describe these deviations are *error*, *accuracy* (also described as *bias*) and *precision*, so let us start with these.

Error

Error is used to indicate the difference between the true value and that obtained experimentally, be it from a one-off measurement or from the mean of a series of results. As the true value is not normally known, the error has to be based on other information, such as the amount by which the measured value for a reference material, obtained simultaneously, differs from its true value. Correction by this reference error will reduce the sample error but, because the reference error itself contains a degree of uncertainty, we can only quote the likely error of the result based on our previous experience with the equipment, the experimental procedure and the operator him- or herself! An error has two components which we can call the systematic error and the random error, and it is those to which the expressions accuracy (or bias) and precision refer.

Accuracy and Precision

The significance of these terms may be represented diagrammatically as shown in Fig. 14.1.

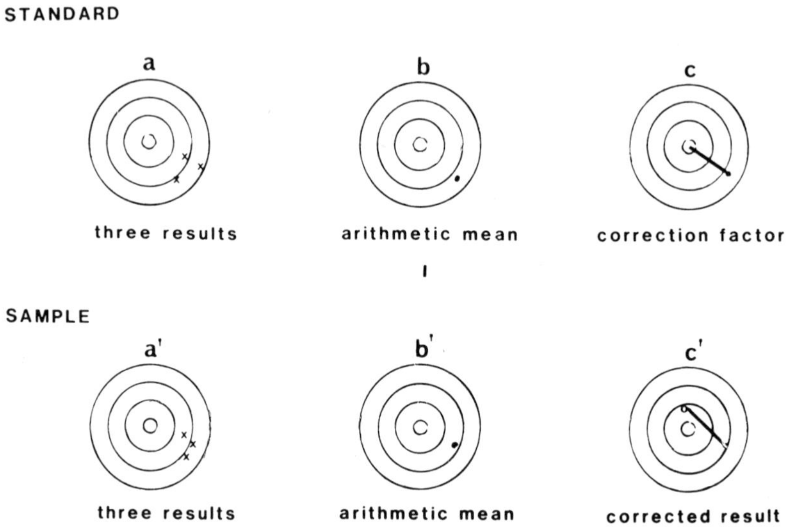

FIG. 14.1. 'Target' representation of precision, accuracy and error.

Targets a and a′ show the scatter (or spread) of three results, the scatter illustrating the *precision* of analysis but in no way indicating how near, or far, it is from the truth. Elimination of the scatter by calculation of each arithmetic mean (b and b′) shows the *accuracy* or *bias* of each result as the distance between the experimental mean and the bullseye for both sample and standard. Calculation of a correction factor using the standard data (from c) and correction of the sample result, using this factor, to point '○' (c′) gives the final *error* as the distance between '○' and the bullseye. Not knowing the 'true' value for the sample, however, we have to define the probable error of the result (○) based on the precision of the standard and sample values.

Precision can be subdivided into *repeatability* (*r*), the spread of results obtained by one individual in one laboratory over a short period of time, and *reproducibility* (*R*), the spread of results obtained on an identical sample by several laboratories.

The use of the word *identical* immediately raises problems. If the material is not perfectly homogeneous, each sample taken for analysis may be considered to possess a different value for the quality being assessed. In this case, however precise the method employed and however careful the analyst to exclude other sources of variation, different samples will show different values. Even if we consider the method and care used to be such that the test portion value is determined with absolute precision, a number of such determinations still enable us to make only an estimation of the mean value for the bulk material, though that estimate will become more and more precise as the number of samples analysed increases. The procedure adopted for taking the samples can significantly influence this process and this is why most specifications lay down fairly precise rules to govern the mode of sampling. Frequently, however, the sampling has already taken place and the analyst is handed the sample from which he weighs, or otherwise measures, the test portion which he requires for a particular analysis.

The significance of sampling therefore has to be considered if a statistical method of evaluating the results is to be used. This is discussed further, on a quantitative basis, later in the chapter.

MEANINGFUL INFORMATION FROM IMPRECISE DATA

Let us now consider what useful information we can obtain. A set of results relating to a specific analysis tends to arrange itself in a regular distribution as illustrated in Fig. 14.2. In some cases this curve is *slewed* or *skew*, but an

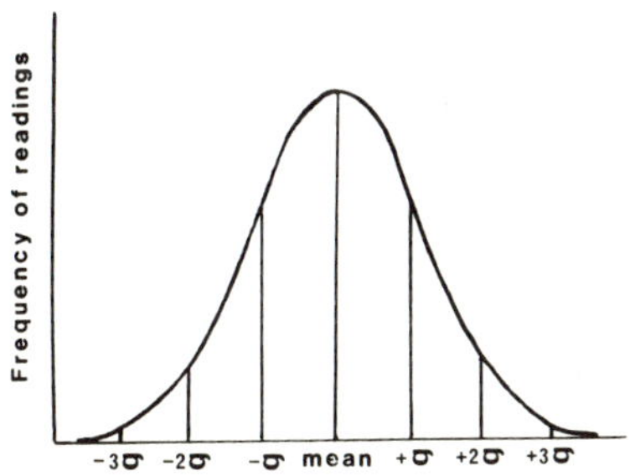

FIG. 14.2. Normal distribution curve.

initial mathematical manipulation of the value being plotted can often improve this. This curve is known as the *normal distribution curve*. The best estimate of the result is known as the *mean*, whilst the scatter of the results about the mean is the *standard deviation* (σ).

Arithmetic Mean

The arithmetic mean is the sum of all the individual readings ($x_1 + x_2 + x_3 + \cdots x_n$) usually written $\Sigma_1^n (x)$, divided by the number of readings (n). This is designated ($\bar{x}$). The arithmetic mean is the best estimate of the result of the test, and the greater the number of readings taken, the more closely will ($\bar{x}$) approach the result which we need.

Standard Deviation

The *standard deviation* (σ) affords a measure of the width of the distribution curve, as shown in Fig. 14.2. It is calculated by taking the difference between each value of x and the true population mean ($\bar{x}$), squaring $(x_1 - \bar{x})^2$, etc., summing, dividing by the number in the population (n) and then taking the square root:

$$\sigma = \sqrt{\frac{\Sigma(x-\bar{x})^2}{n}} \tag{14.1}$$

N.B. $\sum (x-\bar{x})^2 = \sum x^2 - \dfrac{(\Sigma x)^2}{n}$ —an easier expression to calculate!

In practice, since 100% sampling is not carried out one does not have a value for the true population mean, and an estimate of this must be made from the few analyses actually carried out. Under these conditions a more realistic estimate of *sample standard deviation* (s) is obtained by the equation

$$s = \sqrt{\frac{\Sigma(x-\bar{x})^2}{n-1}} \tag{14.2}$$

It will be realized that once n becomes reasonably large s approaches σ in value and thus, in the long run, the standard deviation is independent of the number of results from which it is estimated, whereas it is intuitively accepted that the mean increases in precision as the number of results from which it is derived increases. This increase in precision is quantitatively measured by the *standard error* of the mean. If s is the estimated standard deviation, the standard error of the mean (SE) is given by $s/\sqrt{n}$.

We can, therefore, express analytical results as the mean of n observations ± their standard error, i.e. the required value together with an indication of its reliability. It is sometimes of more interest to examine the relative variability of sets of results which have different mean values. In these instances the concept of *coefficient of variation* (V), defined mathematically as

$$V = \frac{100\sigma}{x} \tag{14.3}$$

is used.

Significance Tests

It would be useful if we could calculate the probability that a result will be obtained which differs from the mean by any given amount. This analytically important piece of information can indeed be calculated quite simply, as too can similar probabilities for many other pieces of statistical information. A few of the more important of such significance tests are considered below, but significance testing in more complex statistical situations is beyond the scope of this book and the interested reader is referred to the specialist books, two of which are referenced here.

Double-sided Significance Tests

In the normal distribution, as illustrated in Fig. 14.2, it is apparent that the frequencies of readings fall off the further one gets from the mean. It can be shown that some 95% of all readings occur in the area within $\pm 2\sigma$ boundaries whilst 99% fall inside the area when the boundaries become $\pm 2{\cdot}6\sigma$. These values are more fully illustrated in Fig. 14.3 and Table 14.1.

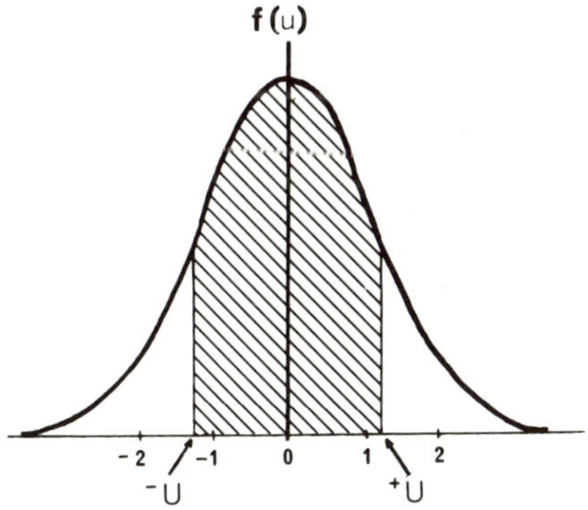

FIG. 14.3. Double-sided normal Gaussian distribution function, $G(U)$= shaded area, = probability ($-U<u<+U$).

TABLE 14.1
RELATIONSHIP BETWEEN U (STANDARD NORMAL DEVIATE) AND PEAK AREA $G(U)$

$G(U)$	U	U	$G(U)$
		0·0	0·000
0·025	0·03	0·1	0·080
0·05	0·06	0·2	0·158
0·075	0·09	0·3	0·236
0·1	0·13	0·4	0·310
0·15	0·19	0·5	0·384
0·20	0·26	0·6	0·452
0·25	0·32	0·7	0·516
0·30	0·38	0·8	0·576
0·35	0·44	0·9	0·632
0·40	0·52	1·0	0·682
0·45	0·59	1·1	0·728
0·50	0·67	1·2	0·770
0·55	0·75	1·3	0·806
0·60	0·84	1·4	0·838
0·65	0·93	1·5	0·866
0·70	1·03	1·6	0·890
0·75	1·15	1·7	0·910
0·80	1·28	1·8	0·928
0·85	1·45	1·9	0·942
0·90	1·64	2·0	0·954
0·925	1·79	2·1	0·964
0·950	1·96	2·2	0·972
0·975	2·24	2·3	0·978
0·990	2·60	2·4	0·984
		2·5	0·988
		2·6	0·990

The shaded area $G(U)$ represents the probability that an individual value will fall between its two boundaries, thus a 95% probability ($G(U)$ = 0·95) is obtained when U equals 1·96 (left-hand pair of columns, Table 14.1) and the value could be anywhere between $(\bar{x} - 1{\cdot}96\sigma)$ and $(\bar{x} + 1{\cdot}96\sigma)$. The area in which there is a certain probability that the results will fall is defined by the *confidence limits* of that result; thus a statement saying that the 95% confidence limits are $\pm 1{\cdot}96\sigma$ means the same as saying that there is a 95% probability that the result lies within the measured values $\pm 1{\cdot}96\sigma$. 95% confidence limits (i.e. 1 in 20 being outside the range indicated) are those generally quoted, but obviously any value required can be calculated. This can also be used to calculate the confidence limits of a population mean ($\bar{x}$). At the 95% level it can be said that the true mean will lie in the range

$$\text{mean} = \bar{x} \pm 1{\cdot}96\frac{\sigma}{\sqrt{n}} \tag{14.4}$$

If n becomes sufficiently large $1{\cdot}96\sigma/\sqrt{n}$ approaches zero and $\bar{x}$ becomes equal to the true population mean.

This relationship is particularly important as it enables us to calculate how many replicate analyses (n) are required to obtain a given precision if the population standard deviation (σ) or the coefficient of variation (V) is known.

Suppose an analysis has a coefficient of variation of 0·3% whilst an answer is required with a 95% probability of its being within 0·2% of the true value. Then

$$\frac{1{\cdot}96 \times 0{\cdot}3}{\sqrt{n}} = 0{\cdot}2$$

$$\sqrt{n} = 2{\cdot}94$$

and nine determinations will be required.

Another question which often confronts the analyst is that of deciding whether the difference between duplicate results is within allowable limits, or whether they are significantly different, in which case the interpretation is that either an error has been made or the samples are inhomogeneous. Given that we are still using 95% confidence limits, two results may be considered sufficiently close for the difference to be acceptable if they are within $2{\cdot}77\sigma$ where σ is already known from a large number of determinations. This value is known as the *least significant difference* and can be expanded to give the equivalent value for two means of two sets of results.

The least significant difference between means then becomes

$$1{\cdot}96\sigma\sqrt{1/n_1 + 1/n_2}$$

It can be seen that if the two sets of results have the same number of analyses (n)

$$\text{the value} = 1{\cdot}96\sigma\sqrt{\frac{2}{n}}$$

(note that for the relationship between two single analyses (2×1) $n = 1$, therefore the least significant difference $= 1{\cdot}96\sqrt{2}\sigma = 2{\cdot}77\sigma$).

Single-sided Significance Tests

All the tests described so far are *double sided*, that is the variation both sides of the mean is of equal significance. In certain cases, however, one may only be concerned that a minimum specified value is being exceeded, that a certain contaminant is not exceeding a given value, or that the uptake of a chemical during ageing (e.g. chloride in seawater soaking) is below specified limits.

If we consider the normal distribution curve, it is apparent that the situation we are describing is as shown in Fig. 14.4 which illustrates a *single-sided function*.

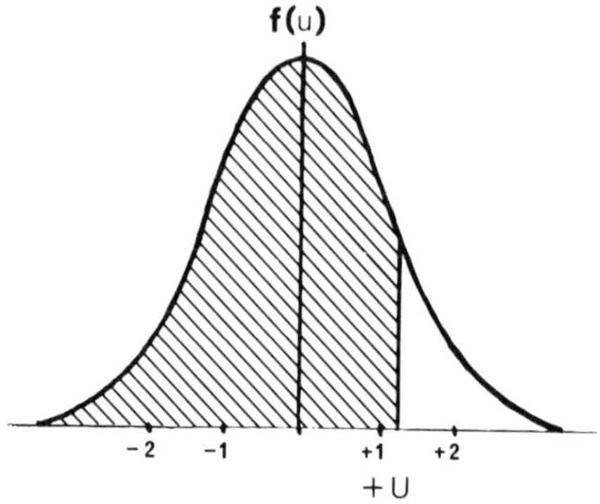

FIG. 14.4. Single-sided normal Gaussian distribution function. $F(U)$ = shaded area, = probability ($u < U$).

Although separate tables exist to enable the determination of U it is as simple, and offers a better understanding of the logic, to consider Table 14.1 again.

In the double-sided situation (Table 14.1) we can see that there is a 90% probability of all values lying within $\pm 1{\cdot}64\sigma$ of the mean. As the normal distribution curve is symmetrical there is 45% probability of a value lying between $(\bar{x} - 1{\cdot}64\sigma)$ and $\bar{x}$ and a 45% probability of a value lying between $\bar{x}$ and $(\bar{x} + 1{\cdot}64\sigma)$. Now in a single-sided function 50% of the values will lie on one side of the mean whilst the probability level for values on the other side is the same as that for *each* side of the double-sided function; thus for a single-sided function:

$$F(U) - 0{\cdot}5 = \frac{G(U)}{2} \qquad (14.5)$$

or

$$2(F(U) - 0{\cdot}5) = G(U)$$

If we consider the situation illustrated in Fig. 14.4 and require the 95% confidence limit for the maximum value,

$$2(0{\cdot}95 - 0{\cdot}5) = 0{\cdot}9 = G(U)$$

which, by reference to Table 14.1, gives a value of $(\bar{x} + 1{\cdot}64\sigma)$ which would increase to (mean $+ 2{\cdot}35\sigma$) if we require 99% confidence limits.

Exactly the same arguments can be applied to the confidence limits of a population mean (see eqn (14.4)); by replacing 1.96 with 1.64 one can say that it is 95% certain that a population mean is greater than

$$\bar{x} - 1{\cdot}64\sigma/\sqrt{n}$$

or conversely that it is 95% certain that the population mean is less than

$$\bar{x} + 1{\cdot}64\sigma/\sqrt{n}$$

Most of the statistical interpretations covered so far require that one has sufficient data to have obtained a meaningful value for the standard deviation (σ) and thus its squared term (σ^2) known as the *variance*. In many practical instances this is not the case so further consideration must be given to the information which is available. Equation (14.2) indicated the means whereby an estimate of the sample standard deviation (s) is obtained when only a limited number of readings is available.

It will be apparent that the lower the number of readings used to calculate s the greater must be the limits between which a particular result will be found for a given probability value. It therefore become invalid to use U (the standardized normal deviate) for calculation with s, and it must be replaced by Student's t, the value of which is dependent upon the number of readings taken and may be obtained from Table 14.2. It should be noted that the values of t are arranged according to the symbol γ which represents the number of *degrees of freedom*. This is a difficult concept to understand, as it indicates the number of independent comparisons which can be made between the individual values in a set of analytical data. If, as is usually the case, the mean ($\bar{x}$) has been calculated from a set of n observations, there are $n - 1$ degrees of freedom (γ). Note that the nth comparison is not independent as a knowledge of x and $n - 1$ values must define the nth value unambiguously. If, on the other hand, the mean is known independently and only the standard deviation needs to be estimated from the results of n samples, then there are n individual comparisons, hence n degrees of freedom.

TABLE 14.2

TABLE OF PERCENTAGE POINTS OF STUDENT'S t-DISTRIBUTION

Degrees of freedom (γ)	*For use in double-sided significance test (or confidence interval)* *Percentage point*			*For use in single-sided significance test (or confidence interval)* *Percentage point*		
	1%	5%	10%	1%	5%	10%
1	63·66	12·71	6·31	31·82	6·31	3·08
2	9·92	4·30	2·92	6·97	2·92	1·89
3	5·84	3·18	2·35	4·54	2·35	1·64
4	4·60	2·78	2·13	3·75	2·13	1·53
5	4·03	2·57	2·02	3·37	2·02	1·48
6	3·71	2·45	1·94	3·14	1·94	1·44
7	3·50	2·36	1·89	3·00	1·89	1·42
8	3·36	2·31	1·86	2·90	1·86	1·40
9	3·25	2·26	1·83	2·82	1·83	1·38
10	3·17	2·23	1·81	2·76	1·81	1·37
11	3·11	2·20	1·80	2·72	1·80	1·36
12	3·05	2·18	1·78	2·68	1·78	1·36
13	3·01	2·16	1·77	2·65	1·77	1·35
14	2·98	2·15	1·76	2·62	1·76	1·35
15	2·95	2·13	1·75	2·60	1·75	1·34
16	2·92	2·12	1·75	2·58	1·75	1·34
17	2·90	2·11	1·74	2·57	1·74	1·33
18	2·88	2·10	1·73	2·55	1·73	1·33
19	2·86	2·09	1·73	2·54	1·73	1·33
20	2·85	2·09	1·72	2·53	1·72	1·33
25	2·79	2·06	1·71	2·49	1·71	1·32
30	2·75	2·04	1·70	2·46	1·70	1·31
40	2·70	2·02	1·68	2·42	1·68	1·30
60	2·66	2·00	1·67	2·39	1·67	1·30
120	2·62	1·98	1·66	2·36	1·66	1·29
	2·58	1·96	1·64	2·33	1·64	1·28

Courtesy Longman Group Ltd.

The calculation of degrees of freedom can become complicated in certain complex statistical analyses and for further information the reader is referred to the work of Davies and Goldsmith (1972).

It has already been pointed out that as the number of readings used to obtain s increases, s tends towards σ. Student's t table illustrates that as the number of readings increases, so t tends towards U.

We can thus compare the 'ideal' and 'practical' equations:

	Ideal	Practical
(a) Confidence limit for a mean (95% probability)	$\bar{x} \pm \dfrac{1{\cdot}96\sigma}{\sqrt{n}}$	$\bar{x} \pm \dfrac{ts}{\sqrt{n}}$
(b) Least significant difference between:		
(i) two means (different n values) 95% probability	$1{\cdot}96\sigma\sqrt{\dfrac{1}{n_1}+\dfrac{1}{n_2}}$	$ts\sqrt{\dfrac{1}{n_1}+\dfrac{1}{n_2}}$
(ii) two means (same n values) 95% probability	$1{\cdot}96\sigma\sqrt{\dfrac{2}{n}}$	$ts\sqrt{\dfrac{2}{n}}$

and so on. Note that for the first example (a) we will obtain t from the 95% column at a level where $\gamma = n-1$ whilst for the latter case (b)(i), $\gamma = (n_1-1)+(n_2-1)$ and (b)(ii), $\gamma = (n-1)+(n-1)$, i.e. $(2n-2)$.

Analysis of variance

Now that we are considering practical standard deviations we must consider means of assessing operator variability—i.e. if two operators carry out replicate analyses, does one show a broader scatter of results than the other? If operator A carries out n_A analyses with a variance of s_A^2 whilst operator B carries out n_B with a variance s_B^2, the value s_A^2/s_B^2 can be determined where $s_A^2 > s_B^2$ and compared with the tabulated 5% points of the F distribution (Table 14.3). It should be noted that the number of degrees of freedom for both operators γ_A and γ_B must be taken into account. If the calculated value of F is larger than the tabulated value, this indicates a 95% probability that the standard deviation for operator A (s_A) is actually larger than that of operator B (s_B); that is, the results from one operator show more spread (i.e. lie on a broader distribution curve) than those of the other. This procedure can also be used to compare the precision of different techniques.

Perhaps the culmination of all that has been said in this section so far is the use of the statistical relationships in the analysis of variance, so that specific sources of variation in a set of analytical results may be identified. We have already identified sources such as homogenizing, sampling,

TABLE 14.3

5% POINTS OF F DISTRIBUTION

γ_B \ γ_A	1	3	5	7	8	10	12	24	*Large s* ∞
1	161·4	215·7	230·2	236·8	238·9	241·9	243·9	249·0	254·3
3	10·13	9·28	9·01	8·89	8·85	8·79	8·74	8·64	8·53
5	6·61	5·41	5·05	4·88	4·82	4·74	4·68	4·53	4·36
7	5·59	4·35	3·97	3·79	3·73	3·64	3·57	3·41	3·23
8	5·32	4·07	3·69	3·50	3·44	3·35	3·28	3·12	2·93
10	4·96	3·71	3·33	3·14	3·07	2·98	2·91	2·74	2·54
12	4·75	3·49	3·11	2·91	2·85	2·75	2·69	2·51	2·30
24	4·26	3·01	2·62	2·42	2·36	2·25	2·18	1·98	1·73
∞ Small (s)	3·84	2·60	2·21	2·01	1·94	1·83	1·75	1·52	1·00

Courtesy Longman Group Ltd.

repeatability and reproducibility, and if one considers compounds or vulcanized products we can extend the list much further!

In order to illustrate the analysis of variance we shall consider the analysis of a sample of raw rubber, homogenized to produce four test portions, each of which was analysed in triplicate. The results obtained are illustrated in Table 14.4.

We require to know the 'true' ash content and also whether there is significant sample–sample variation. It is a simple matter to calculate the mean ($\bar{x}$) for the four samples:

$\bar{x}_1$	$\bar{x}_2$	$\bar{x}_3$	$\bar{x}_4$	$\bar{x}_{total}$
0·3167	0·3033	0·3633	0·3700	0·3383

and we must now decide whether this variation is likely to occur with truly homogeneous samples or not. This is done by constructing an analysis-of-

TABLE 14.4

TRIPLICATE ANALYSES OF FOUR RUBBER SAMPLES FOR ASH CONTENT

Sample 1	*Sample 2*	*Sample 3*	*Sample 4*
0·32	0·30	0·38	0·35
0·31	0·29	0·36	0·39
0·32	0·32	0·35	0·37

variance table which separates the total variation into identifiable parts. In this instance we are considering sample–sample variation and replicate analysis variation.

A measure of the total variation is given by the sum of the squared deviations from the mean of all results, i.e $\Sigma(x - \bar{x}_{\text{total}})^2$.

A measure of the replicate analysis variation is given by the sum of the squares of the deviations of each result from its own set mean, i.e. $(x_{1A} - \bar{x}_1)^2 + (x_{1B} + \bar{x}_1)^2 + \ldots + (x_{4C} - \bar{x}_4)^2$.

A measure of the sample–sample variation is obtained by measuring the differences between the total mean and the individual sample means, squaring, weighting by multiplying each value thus obtained by the number of analyses used to obtain the mean, and summing:

$$3(\bar{x}_1 - \bar{x}_{\text{total}})^2 + \ldots + \ldots + 3(\bar{x}_4 - \bar{x}_{\text{total}})^2$$

We can thus calculate a total sum of squares (SS_{T}):

$$(0{\cdot}32 - 0{\cdot}3383)^2 + \ldots + (0{\cdot}37 - 0{\cdot}3383)^2$$

$$SS_{\text{T}} = 0{\cdot}01177$$

The replicate analysis sum of squares (SS_{a}) is given by

$$(0{\cdot}32 - 0{\cdot}3167)^2 + \ldots + (0{\cdot}37 - 0{\cdot}37)^2$$

$$= 0{\cdot}0000667 + 0{\cdot}000467 + 0{\cdot}000467 + 0{\cdot}000800$$

$$SS_{\text{a}} = 0{\cdot}001801$$

The between sample sum of squares (SS_{s})

$$= 3 \times 0{\cdot}0004666 + 3 \times 0{\cdot}001225 + 3 \times 0{\cdot}000625 + 3 \times 0{\cdot}001005$$

$$SS_{\text{s}} = 0{\cdot}009965$$

As a crosscheck note that $SS_{\text{T}} = SS_{\text{a}} + SS_{\text{s}}$.

The analysis-of-variance table can now be constructed as shown in Table 14.5.

From the figures given in Table 14.5 one can now calculate the ratio

$$\frac{\text{between samples mean square}}{\text{replicate analyses mean square}} = \frac{0{\cdot}003321}{0{\cdot}000225} = 14{\cdot}76$$

If this ratio is greater than the appropriate value from the *F* distribution table, it suggests that the between-samples and between-analyses mean squares are significantly different; thus it is 95% certain that there is

TABLE 14.5

ANALYSIS OF VARIANCE

Source of variation	*Sum of squares*	*Degrees of freedom*	*Mean square*
Replicate analyses	0·001801	8[a]	0·000225
Between samples	0·009965	3[b]	0·003321
Total	0·01177	11	0·0011

[a]2 degrees of freedom per sample since 3 analyses performed. 4 samples hence $2 \times 4 = 8$ degrees of freedom.
[b]4 samples hence 3 degrees of freedom.

sample–sample variation. Using the F distribution table at $\gamma_1 = 3$ and $\gamma_2 = 8$ we have a value of 4.07. Our found value exceeds this substantially so we can say that it is over 95% certain that the samples are not the same and exhibit intersample variation. We can, however, still use these data to put various limits on the analyses; the analysing variance (s_a^2) is best represented by $0{\cdot}001801/8 = 0{\cdot}000225$ whilst the sampling variance is best represented by

$$s_s^2 = \frac{\text{between samples mean square} - \text{replicate analyses mean square}}{\text{number of analyses on each sample}}$$

$$= \frac{0{\cdot}003321 - 0{\cdot}000225}{3}$$

$$= 0{\cdot}00103$$

From these values we can calculate the approximate 95% confidence limits for the true ash levels as

$$\bar{x} \pm 2\sqrt{\frac{s_a^2}{np} + \frac{s_s^2}{n}}$$

where

$$\left(\frac{s_a^2}{p} + s_s^2\right) = s_{\text{total}}^2 \quad \text{(see p. 281 (a))}$$

where n is the number of samples, p is the number of tests on each sample, which is

$$0{\cdot}3383 \pm 2\sqrt{\frac{0{\cdot}000225}{4 \times 3} + \frac{0{\cdot}00103}{4}}$$

$$= 0{\cdot}3383 \pm 0{\cdot}0332$$

or

$$0{\cdot}338 \pm 0{\cdot}033$$

Of necessity this represents an extremely simple example of the analysis of variance, but the basic principles of identifying the causes of variation, dividing the total variation into sections attributable to each cause, defining the degrees of freedom for each cause, constructing the analysis-of-variance table and comparing the correct F distribution value with the calculated ratio, remain unchanged.

It may not be quite so obvious that this technique has further ramifications in that where information on the sampling error as well as the experimental error is available, it is possible to decide on rational grounds what size of sample should be homogenized before the test portion is taken. The following argument applies to the separation of a single test portion, not a pair of duplicate test portions.

If the standard deviation of a single determination on a test portion from a completely homogeneous sample is z and the standard deviation of a corresponding determination on an ordinary (assumed heterogeneous) sample is z_1 then the standard deviation attributable to the sampling variation is given by $\sqrt{z_1^2 - z^2}$ which we will designate y (i.e. $y^2 + z^2 = z_1^2$).

Unless the particular point of making the determination is an investigation of sampling error, the analyst will wish to take the test portion from a sample which has been homogenized and which is sufficiently large so that the homogenization has reduced the sampling error to negligible proportions compared with the experimental error. If the piece for homogenization is n times the size required for the test portion, we may regard this piece as equivalent to n test portions and by determination of the value appropriate to it we may regard this value as approximating to that of the mean of n test portions. The standard error (due to the sampling error) of the test portion will be $y/\sqrt{n}$. The condition for this error to be ignored is that the variance ratio, given by $z^2/(y/\sqrt{n})^2$, should be equal to or exceed the variance ratio required for significance at the desired level based on 1 and $(n-1)$ degrees of freedom. This is algebraically equivalent to $z^2/y^2 > F/n$.

In the usual situation of y^2 being greater than z^2 this condition is met when n is greater than F. As can be seen from Table 14.3 (at a probability level of 0·5 and where γ_1 and γ_2 refer to the degrees of freedom pertaining to z_1 and y respectively), this occurs when n is not less than 7; for the more rare cases when y^2 is less than z^2 the sampling error is reduced to negligible proportions if homogenization extends only over 5 times the size of the test portion, or even less depending upon the actual magnitude of y with respect to z.

In the case of acetone extraction, where the micro-scale determination requires a sample of 20 mg, it is probable that $y^2 < z^2$ and it would be safe to take the test portion from a sample of 0·1 g. This is in line with the recommendation made, except that the choosing of the 0·1 g piece is itself a sampling operation made after homogenization. Thus double homogenization is a precaution which is always advisable when mg quantities form the test piece and does ensure adequate safety when the sampling error is unknown.

To the analyst who despairs of statistics we would suggest that a graphical presentation of a mass of data provides a rapid way of carrying out an initial assessment of their validity and significance. This is extremely well illustrated in the data given in Table 14.6, the results of an interlaboratory check on the ash contents of natural rubber samples; 24 laboratories, two samples of rubber, each divided into two test portions, one test portion of each rubber being analysed in triplicate on one day, and the procedure repeated on the next.

So far so good; the pairs of means are in good qualitative agreement and, as inspection of Table 14.6 indicates, sample 2 shows a higher mean than sample 1. The values for sample 2, however, are grouped around the 0·4% level, whereas those for 1 show rather more spread, with a number of values in the 0·2–0·3% region. It is also noticeable that the analytical variation appears relatively constant, with the low values holding for all six analyses of a particular sample. Before 'number-crunching' the 200-odd results one could plot a bar graph as illustrated in Fig. 14.5 and it immediately becomes clear that sample 1 is not going to give a normal distribution—indeed it is developing a clear bimodal distribution, one mean of which looks similar to that of sample 2 whilst the other is appreciably lower. The most obvious conclusion from these data is that two quite different samples are involved. One accounts for all 24 test portions of sample 2 and 13 of the 24 in the case of sample 1, whilst the other accounts for the remaining 11 test portions of sample 1. These data can still be analysed statistically rather than discarding the total experiment, but *in a*

TABLE 14.6

INTERLABORATORY ASH ANALYSIS (NR)

	Sample 1 Day 1			*Sample 2 Day 2*			*Sample 2 Day 1*			*Sample 2 Day 2*		
Laboratory	1	2	3	1	2	3	1	2	3	1	2	3
1	0·30	0·29	0·31	0·28	0·30	0·30	0·40	0·39	0·43	0·42	0·40	0·40
2	0·34	0·39	0·30	0·44	0·41	0·42	0·31	0·35	0·30	0·41	0·41	0·43
3	0·29	0·28	0·29	0·26	0·26	0·26	0·46	0·48	0·46	0·42	0·43	0·43
4	0·42	0·42	0·43	0·44	0·42	0·43	0·42	0·43	0·42	0·40	0·41	0·40
5	0·39	0·41	0·39	0·41	0·45	0·43	0·37	0·37	0·36	0·42	0·41	0·41
6	0·45	0·47	0·45	0·45	0·41	0·43	0·47	0·47	0·47	0·42	0·45	0·42
7	0·40	0·42	0·42	0·44	0·43	0·46	0·38	0·36	0·35	0·35	0·36	0·36
8	0·28	0·29	0·28	0·31	0·29	0·31	0·40	0·42	0·40	0·39	0·37	0·39
9	0·31	0·29	0·30	0·30	0·28	0·33	0·41	0·41	0·39	0·39	0·39	0·39
10	0·43	0·41	0·41	0·40	0·42	0·41	0·40	0·42	0·42	0·42	0·42	0·41
11	0·41	0·40	0·42	0·45	0·42	0·44	0·44	0·44	0·45	0·45	0·44	0·43
12	0·32	0·31	0·32	0·31	0·31	0·31	0·41	0·40	0·41	0·40	0·40	0·40
13	0·38	0·44	0·47	0·40	0·42	0·43	0·42	0·45	0·46	0·42	0·46	0·41
14	0·39	0·38	0·40	0·42	0·45	0·44	0·38	0·39	0·39	0·44	0·42	0·43
15	0·38	0·39	0·39	0·37	0·39	0·37	0·42	0·41	0·42	0·43	0·42	0·42
16	0·25	0·24	0·27	0·28	0·24	0·25	0·39	0·33	0·37	0·35	0·39	0·37
17	0·46	0·46	0·46	0·42	0·42	0·42	0·44	0·44	0·44	0·40	0·40	0·40
18	0·30	0·28	0·29	0·29	0·29	0·29	0·43	0·44	0·41	0·42	0·42	0·41
19	0·31	0·28	0·28	0·26	0·28	0·30	0·43	0·46	0·43	0·36	0·39	0·46
20	0·27	0·25	0·27	0·28	0·26	0·27	0·41	0·39	0·40	0·40	0·42	0·40
21	0·29	0·28	0·30	0·30	0·29	0·32	0·40	0·43	0·43	0·41	0·41	0·45
22	0·42	0·39	0·41	0·39	0·44	0·44	0·45	0·38	0·38	0·46	0·40	0·37
23	0·43	0·40	0·41	0·43	0·45	0·44	0·39	0·40	0·38	0·42	0·41	0·44
24	0·30	0·30	0·29	0·30	0·33	0·29	0·43	0·45	0·45	0·45	0·41	0·40
	Sample 1 Day 1			*Sample 1 Day 2*			*Sample 2 Day 1*			*Sample 2 Day 2*		
	$\bar{x}$ = 0·358			0·362			0·411			0·410		

Table reproduced with permission of the Rubber Research Institute of Malaysia.

different way from the obvious one which assumes a normal distribution pattern. One thus identifies six sets of results: 1.1 low, 1.1 high, 1.2 low, 1.2 high, 2.1 and 2.2. Whilst the number of samples in the four '1 sets' are relatively small, one can obtain means, standard deviations, standard errors etc. as a check on the laboratory and replication precisions. Some results are illustrated in Tables 14.7 and 14.8.

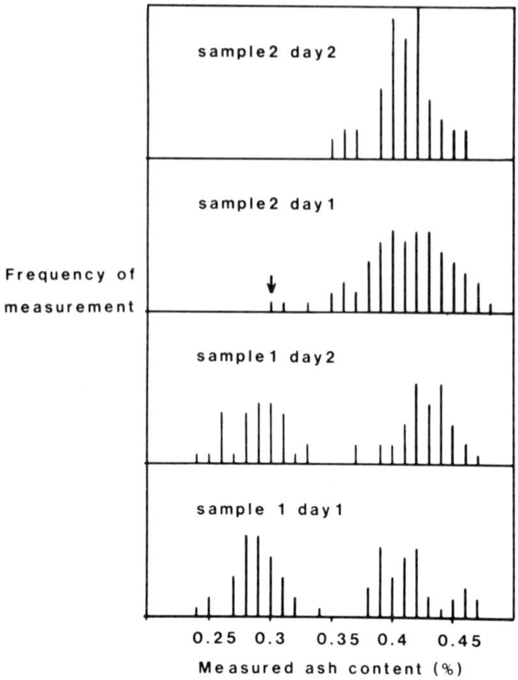

FIG. 14.5. Bar graph of interlaboratory ash determinations.

The values in Table 14.8 afford 'least significance difference' values for several pairs of means in the region of 0·025 which would statistically confirm the observation that the four samples are probably from the same piece of rubber.

TABLE 14.7

'LOW VALUE' COMPONENTS IDENTIFIED IN SAMPLE 1

		1·1 Low			*1·2 Low*	
	1	2	3	1	2	3
$\bar{x}$	0·293	0·281	0·291	0·288	0·285	0·294
s	0·020	0·021	0·16	0·18	0·025	0·025

TABLE 14.8

'HIGH VALUE' COMPONENTS OF SAMPLE 1 AND ALL COMPONENTS OF SAMPLE 2

	1·1 High (n = 13)			*1·2 High* (n = 13)			*2·1* (n = 24)			*2·2* (n = 24)		
	1	2	3	1	2	3	1	2	3	1	2	3
$\bar{x}$	0·408	0·418	0·412	0·420	0·425	0·428	0·418	0·414	0·409	0·410	0·410	0·410
s	0·032	0·032	0·041	0·025	0·019	0·022	0·034	0·039	0·039	0·028	0·023	0·025

Outliers

Reference to Fig. 14.5, sample 2, day 1, shows some analytical results which are appreciably different from the main set, and occasions arise when one must decide whether such values should be included in, or rejected from, the bulk of the data. In principle a reading should never be rejected unless there is a valid reason for so doing. Possible reasons can be summarized:

Operator error
Instrument malfunction
Damaged sample
Statistical reasons

The first two are self evident, although sometimes difficult to prove. Rejection on the grounds of a damaged sample is critically dependent upon the reason for the test. If a sheet of rubber is being used to measure gas permeability, the presence of a pinhole flaw will give a value which will be rejected. If, on the other hand, the gas permeability is being used to check the quality of the rubber sheet or fabric, the occasional pinhole will be extremely important. Each sample must be judged on its merits.

There are statistical methods for deciding which outlying values should be rejected and again it should be pointed out that these assume a normal distribution. If the data do not, approximately at least, fit this pattern, the statistical treatment will not be strictly valid and the conclusions should not be interpreted too rigidly. It is also the case that for statistical tests for outliers, low significance limits are used such as 1%, with levels greater than 5% being avoided. Tests exist to detect one or more outliers on one or both sides of the normal distribution curve, but here we will initially describe two tests to check the most common situation of there being an outlier at one side of the distribution. Both tests are simplified by tabulating the results in order of increasing value, with x_1 being the smallest, x_2 the next smallest, and eventually, x_n the largest.

T Test

The first test, the *T* test, measures the distance the extreme reading (either x_1 if we suspect a low outlier, or x_n if we suspect a high one) lies from the mean value of the readings ($\bar{x}$). This value is then divided by the sample standard deviation (s) and the quotient (T) is noted on the graph of T *vs*.

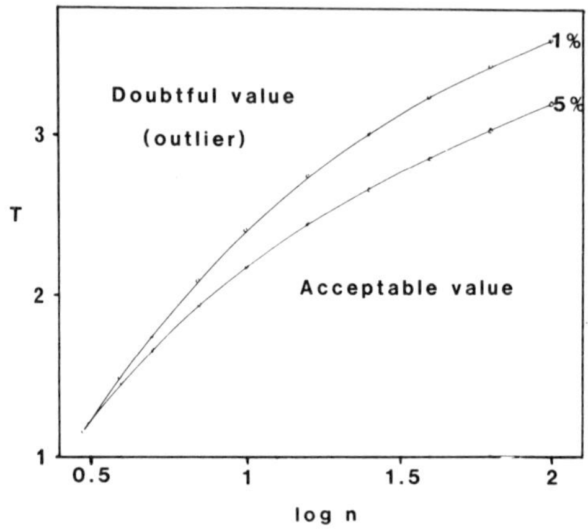

FIG. 14.6. *T* test for outlier (using data of Grubbs and Beck, 1972).

log *n*. This is illustrated for two probabilities, indicating 1 in 20 (5%) and 1 in 100 (1%) boundaries for acceptability, in Fig. 14.6. If we do not specify on which side the outlier is being tested for, the probability of its being on either side is twice that of its being on a specified one. The 1% probability boundary would then become the 2% (1 chance in 50).

Dixon's Test

The second test, Dixon's test, compares the interval between the suspected outlier and one of its neighbours, with the spread of the results between a large number of readings, the exact intervals measured depending upon the number of readings taken. These are listed under *criterion 1* to *criterion 4* and again the resulting value is compared with the boundary lines shown in Fig. 14.7 to see whether it falls in the *acceptable value* or *possible outlier* region.

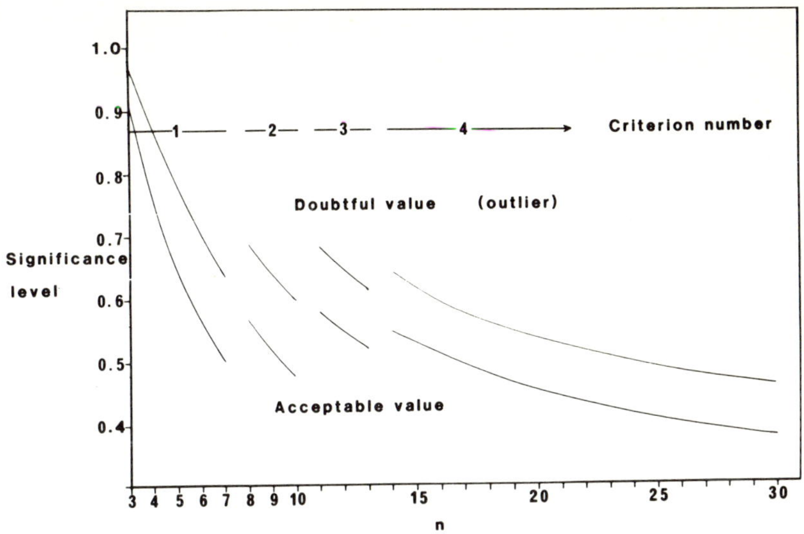

FIG. 14.7. Dixon's test for outlier (using data of Dixon, 1953).

Criterion 1: 3–7 readings ($n = 3–7$)

If the suspected outlier is on the low side, the interval between it (x_1) and its neighbour (x_2) is ($x_2 - x_1$). The interval between a large number of readings (all for this low number of readings, 3–7) is ($x_n - x_1$), thus:

$$\text{Dixon's significance level } (d_s) = \frac{(x_2 - x_1)}{(x_n - x_1)}$$

Similarly if the outlier is suspected of being on the high side of the distribution:

$$d_s = \frac{(x_n - x_{n-1})}{(x_n - x_1)}$$

Criterion 2: $n = 8–10$

For low outlier (x_1)

$$d_s = \frac{(x_2 - x_1)}{(x_{n-1} - x_1)}$$

For high outlier (x_n)

$$d_s = \frac{(x_n - x_{n-1})}{(x_n - x_2)}$$

Criterion 3: $n = 11–13$

For low outlier (x_1)

$$d_s = \frac{(x_3 - x_1)}{(x_{n-1} - x_1)}$$

For high outlier (x_n)

$$d_s = \frac{(x_n - x_{n-2})}{(x_n - x_2)}$$

Criterion 4: $n > 14$

For low outlier (x_1)

$$d_s = \frac{(x_3 - x_1)}{(x_{n-2} - x_1)}$$

For high outlier (x_n)

$$d_s = \frac{(x_n - x_{n-2})}{(x_n - x_3)}$$

Inevitably with borderline cases the two tests sometimes show differences of opinion. As a general rule the Dixon's test is the quicker and can be used to screen many results rapidly but Ferguson (1961) shows the T test to be the better one to use in the event of the presence of a single outlier being suspected.

In the case of two outliers, one on each side, one can again use the T test on the reading which is further from the mean and if this proves likely to be an outlier reject the value and recalculate $\bar{x}$, s and the T value for the other possible outlier. Alternatively one may prefer to use the test of Teitjen and Moore (1972). If it is suspected that two readings on the same side are outliers it is probably simplest to ignore the outer one and use the T test on the remainder. If the remaining suspect one proves to be probably an outlier it is apparent that the rejected one must also be. One can also refer to the more sophisticated method of Grubbs (1969).

The two tests are illustrated on the low reading shown in Fig. 14.5, Sample 2, day 1 (marked ↓).

$$n = 72 \qquad \bar{x} = 0{\cdot}411 \qquad s = 0{\cdot}037$$

T test:

$$\frac{\bar{x} - x}{s} = T$$

$$\frac{0{\cdot}411 - 0{\cdot}30}{0{\cdot}037} = 3{\cdot}00$$

The maximum value, above which T indicates a possible outlier, is 3·09 (at 5%) for $n = 72$; therefore the reading of 0·30 is just acceptable.

Dixon's test:

$$d_s = \frac{x_3 - x_1}{x_{n-2} - x_1}$$

$$d_s = \frac{0{\cdot}33 - 0{\cdot}30}{0{\cdot}46 - 0{\cdot}30} = 0{\cdot}1875$$

The graph (Fig. 14.7) does not continue to 72 readings but the value of d_s is obviously very low which suggests acceptability. One may therefore conclude that it is reasonable to expect a value as low as 0·3 in this set and hence that this sample is probably from the same set as the others.

Of necessity this constitutes an extremely brief journey into the field of statistics related to the analysis of rubbers and rubber-like materials. The interested reader is referred to the many books on statistics such as those by Brooks and Dick (1955) and by Davies and Goldsmith (1972), whilst certain relevant standards are included in Appendix I.

REFERENCES

Brooks, B. C. and Dick, W. F. L. (1955) *Introduction to Statistical Method*, Heinemann, London.

Davies, O. L. and Goldsmith, P. L. (1972) *Statistical Methods in Research and Production*, 4th ed., Oliver & Boyd, Edinburgh.

Dixon, W. J. (1953) *Biometrics* **9**, 74.

Ferguson, T. S. (1961) *Rev. Inst. Int. de Stat.* **3**, 29.

Grubbs, F. E. (1969) *Technometrics* **11**, 1.

Grubbs, F. E. and Beck, G. (1972) *Technometrics* **14**, 847.

Teitjen, G. L. and Moore, R. H. (1972) *Technometrics* **14**, 583.

Appendix I

Table of Official National and International Standards

Many countries have standard test procedures and methods of analysis which are normally accepted in any dispute as being reliable and accurate. In the following table are listed many standards concerning the chemical analysis of rubbers (latex, raw, compounded and vulcanized) and carbon black. A few additional ones are included which are particularly significant to the rubber-analyst such as nomenclature, sampling, statistical interpretation, etc.

The table is based on the standards issued by the *International Organization for Standardization* (ISO), whilst listed with the appropriate ISO number are the numbers to the *nearest equivalent* British (BS), American (American Society for Testing and Materials, ASTM) and German (Deutsches Institut für Normung e. V, DIN) Standards. In some cases these are identical—the trend with both BS and DIN is to adopt ISO Standards and 'dual number' them—whilst in other cases the standards are closely similar although not identical. Obviously if it is necessary to test to a specific standard then it *must* be used, but if a reliable procedure is required, any appropriate standard may be used although this must be specified with the result.

It is hoped that the list is exhaustive (in respect of ISO and 'equivalent') at the date of publication, but at the end of the Appendix are listed two extra sections, ISO *Draft Proposals* and ISO *Draft International Standards* (ISO/DP and ISO/DIS respectively). Copies of these are available from the National Standards Institutions but they are not *yet* standards. Should they become so the reference numbers will usually remain the same although a few exceptions are noted in the following table. The BS/ASTM/DIN numbers listed alongside these are *existing standards* in their respective countries.

Description of sample	*ISO No.*	*BS No.*	*ASTM No.*	*DIN No.*
General				
Rubber, vocabulary	1382	3558	D 1566	53501
Rubber/latex nomenclature	1629	3502/3	D 1418	ISO1629(E)
Test sieves, specn. and use	2395 2591	410 1796	E11	4188
Glass viscometers	3105	188	D 746	
Rubber, preparation of test pieces	4661	1673/1.2	D 3183	53502
Latex				
Sampling	123	1672 3397	D 1417 D 1076	53562
Total solids	124	6057/3, 3.2	D 1417 D 1076	53563
NR, alkalinity	125	6057/3, 3.3	D 1076	53565 (E)
NR, dry rubber content	126	1672	D 1076	53564
NR, KOH value	127	1672	D 1076	53566
NR, density	705	1672	D 1076	53597
NR, coagulum	706	1672	D 1076	53594
pH	976	6057/3, 3.9	D 1417 D 1076	53606
Viscosity	1652	5858	D 1417 D 1076	
Copper (diethyldithiocarbamate)	1654	1673/2	D 1278	53569
Manganese (periodate)	1655	1673/2	D 3533	
Iron (1,10 phenanthroline)	1657	1673/2	D 1278	53620
NR, boric acid	1802	1672	D 1076	53605
NR, centrifuged/creamed-ammonia preserved: spec.	2004	6057/1	D 1076	
NR, sludge	2005	1672	D 1076	53592
SBR, volatile unsaturates	2008	3397	D 1417	53675(E)
NR, evaporated-preserved: spec.	2027	6057/1.2		53595(E)
BR, mono/copolymer: dry film prepn.	2028	3397	D 1417	
SBR, volatiles	2058	3397	D 1416	53526(E)
Synthetic, codification	2438	5123		53549
SBR, bound styrene	3136	3397	D 1417	
NBR, residual acrylonitrile	3899	3397	D 1417	
NBR, bound acrylonitrile	3900	3397		
SBR, reinforced bound styrene	4655		D 1416/ 1417 D 297	

Description of sample	*ISO No.*	*BS No.*	*ASTM No.*	*DIN No.*
Raw rubbers				
Ash	247	5923/1	D 1416 D 1278	 53568
Volatile matter	248	5923/3	D 1416 D 1278	53526
NR, dirt	249	1673/2	D 1278	53527
Viscosity (shearing disc)	289	1673/3		53523
Solvent extraction	1407	1673/2	D 1416 D 1278	53553
NR, bale coating	1434		D 2449	
Copper (diethyldithiocarbamate)	1654	1673/2	D 1278	53569
Manganese (periodate)	1655	1673/2	D 1278	53589
NR, nitrogen	1656	5923	D 3533	
Iron (1,10 phenanthroline)	1657	1673/2	D 1278	
Bale sampling	1795	1673/1	D 1485	53525
Sample preparation	1796	1673/1	D 1485	53525
NR, specification	2000	4396	D 2227	
SBR, organic acids soaps	2002 2003	3397 1673/5.5	D 1416 D 1416	
SBR, volatiles	2058	1673/5, 5.2–4	D 1416	535 (E)
SBR, copolymers: bound styrene	2433	1673/5, 5.1	D 1416	
NR, colour index	4660	4396	D 3157	
Single polymer, identification by pyrolysis gas chromatography	5475		D 1417	
Isoprene content	5945		D 1278	
SBR, block styrene content	6235		D 3314	
See also:				
Elastomers, detection of factice			D 297	53588
NR, chemical analysis		1673/2		
SBR, chemical analysis		1673/5, 5.2–4		
NBR, chemical analysis		1673/7		
CR, chemical analysis		1673/8		
BR, chemical analysis		1673/9		
Oil				
Extender and processing oils, chromatography			D 2007	
Compounded/vulcanized rubbers				
Ash	247	5923/1	D 297	53568
Zinc (diethyldithiocarbamate)	1396	903/B17	D 1278	
Manganese (periodate)	1397	903/B15	D 1278	

Description of sample	ISO No.	BS No.	ASTM No.	DIN No.
Solvent extract	1407	1673	D 297	53553
Carbon black (pyrolytic)	1408	903/B14	D 297	53585
Vulcanized rubber:storage	2230	3574		
Zinc (EDTA)	2454	5923/2	D 297	53581
Density	2781	903/A1	D 297	
Test for staining with organic materials	2865	903/A33	D 297	53540
Test piece, dimensions	4648	903/A38		
Prepn. of test pieces for chemical analysis	4661	903/A36	D 297 D 3183	
Isoprene content	5945	903/B12		53621
Zinc (atomic absorption spectrophotometry)	6101/1		D 4004	
Carbon black				
Bulk/bin delivery, sampling	1124	5293	D 1900	53602
Ash	1125	5293	D 1506	53586
Loss on heating	1126	5293	D 1509	53552
Total sulphur	1138	5293	D 1619	53584
Iodine No.	1304	5293	D 1510	53582
Package shipments, sampling	1310	5293	D 1799	
Sieve residue, determination	1437	5293	D 1514	
Sieve residue, specification	1867		D 1765	
Volatiles	1868		D 1765	
Toluene extract: light transmission	3858	5293	D 1618	
Nitrogen surface area	4652		D 3037	
Dibutyl phthalate absorption No.	4656	5293	D 9414	53601
Tinting strength	5435		D 3265	
Solvent extractables	6209			53553
See also:				
Classification for carbon black			D 1765	
Visual inspection of dispersion			D 2663	
Agglomerate counts			D 2663	
Standard test/formulations for evaluation of rubber				
NR	1658	1767	D 3184	53670
IIR	2302	4470	D 3188	53670(E)
SBR	2322	3472 3650	D 3185	53670
CR	2475	5375	D 3190	53670(E)
BR	2476	5047	D 3189	53670(E)
EPDM	4097	6063	D 3568	
SBR (black/oil masterbatches)	4659	5563	D 3186	

Description of sample	*ISO No.*	*BS No.*	*ASTM No.*	*DIN No.*
Standard test formulations for carbon black in:				
NR			D 3192	
SBR	3257		D 3191	
Relevant statistics				
Statistical interpretation of test results, estimation of mean, confidence limits	2602	2846/2	[a]	55303(E)
Statistical interpretation of data techniques and tests relating to means and variances	2854	2846/4	[a]	55303(E)
Sampling procedures and tables for inspection by attributes	2859	6000 6001	[a]	
Statistical interpretation of data, determination of a statistical tolerance level	3207	2846/3	[a]	
Statistical interpretation of data, comparison of two means for paired observations	3301	2846/6	[a]	55303(E)
Statistical interpretation of data, power of tests relating to mean and variances	3494	2846/5	[a]	55303(E)
Statistics—vocabulary and symbols	3534	5532	E 456	13303(E)
Precision of test methods, determination of repeatability and reproducibility	5725			ISO 5725

Note: These ISO Standards are available in one publication: ISO Handbook No. 3 (1979)

Draft International Standards (DIS)				
NBR, residual acrylonitrile	4581			
Antidegradants by thin layer chromatography	4645/2		D 3156	53622
Rubber, identification by infrared spectroscopy	4650	4181	D 3677	
Castor oil content by thin layer chromatography	6225/1.2			
Castor oil content by gas chromatography	6225/2.2			
Rubber compounding materials, abbreviations	6472			

Description of sample	*ISO No.*	*BS No.*	*ASTM No.*	*DIN No.*
Rubber, total sulphur content, 3 methods	6528/1–3	903/B6	D 297	53561
Carbon blacks, ref. grades	6809			
Carbon black, surface area by surfactant absorption No.	6810			
Carbon black, dibutylphthalate absorption No.	6894		D 3493	
Rubber, vulcanized, free sulphur	7269	903/B7	D 297	
Rubber, identification by pyrolysis gas chromatography	7270[b]			
Manganese (periodate)	7780[c]			
Draft Proposals	(DP)			
Rubber, vulcanized, carbon black content	1408		D 297 D 1416	
NBR latex, residual acrylonitrile	6100		D 1417	
Lead } atomic absorption spectrophotometry	6101/2		D 4004	53599
Copper } atomic absorption spectrophotometry	6101/3			53569
Manganese } atomic absorption spectrophotometry	6101/4			53589
Rubber, binary blends by pyrolysis–gas chromatography	6226[b]			
Chemical tests on water extractables	6913			
by acid digestion	7234/1			
by ozonolysis	7234/2			
Rubber, Cl/Br by oxygen flask	7725/1			3566
sodium peroxide	7225/2			3566
Rubber, copper (photometric)	8053			
Rubber, sulphide sulphur	8054	903/B10	D 297	
See also:				
Sampling, No. of units to obtain a specified precision			D 122	
Use of the terms precision and accuracy			D 177	
Dealing with outliers			D 178	
Conducting an interlaboratory test programme			D 691	

[a]Much of this covered in: 'Preparing precision statements for standards related to rubber and rubber testing'. D 3080

[b]ISO/DIS 7270 and ISO/DP 6226 will be combined to replace ISO 5475.

[c]ISO/DIS 7780 will replace ISO 1655 and ISO 1397 if adopted as full standard.

(E) following a DIN number indicates a Draft German Standard.

Appendix II

Elastomers: Nomenclature, Description and Properties

The classification system used is based on that described in ISO 1629-1976. The last letter of the identification code defines the basic group to which the polymer belongs whilst the earlier ones provide more specific information and in many cases define the polymer absolutely.

It is unrealistic to attempt a comprehensive catalogue of all elastomers in an appendix of this nature. It is hoped that most common ones, and those which could become common in the next decade, are included. T_gs and density values are given where available but it should be noted that, except in the case of a stereospecific homopolymer, variations in structure and blend composition will have an effect whilst there is a further dependence on the molar mass and method of measurement used. The figures quoted are all DTA/DSC values and the samples are of such a molar mass that variation in this would have an insignificant effect on the T_g.

'M' Group: Rubbers Having a Saturated —C—C— Main Chain

IM: Polyisobutylene (e.g. VISTANEX), a soft inert plastic; low molecular weight material used as a plasticizer and adhesive.

$d = 0{\cdot}91$ T_g $-70 \rightarrow -73$ °C

EPM: Copolymer of ethylene and propylene; the rubber-like materials have a wt/wt composition between 70–30 and 30–70.

$d = 0{\cdot}87$ $T_{g\ 50/50}$ -60 °C

EPDM: A terpolymer of ethylene, propylene and a di- or polyene giving pendent olefin groups as crosslinking sites (e.g. NORDEL). An ozone- and oxidation-resistant rubber.

$d = 0{\cdot}85$ T_g (NORDEL 1470, 1660) -66 °C

CSM: Chlorosulphonated polyethylene (e.g. HYPALON), containing both C—Cl and C—SO_2Cl groups. Cl content 20–45%; S content 0·5–2·5%. Optimum properties 30% Cl, 1·5% S; ozone-resistant rubber and used in varnishes.

d = 1·27 T_g (HYPALON 30) +9 °C

FPM: Fluoro/fluoroalkyl groups on C—C backbone (e.g. VITON, FLUOREL)—copolymers of hexafluoropropylene and vinylidene fluoride.
(e.g. TECHNOFLON)—copolymer of vinylidene fluoride and 1-hydropentafluoropropylene.

d = 1·85 T_g (VITON B) −18 °C

CFM: As above, but containing Cl as well as F; vinylidene fluoride (VF): chlorotrifluoroethylene (CTFE) copolymer (e.g. VOLTALEF, KEL F).

d = 1·85 T_g VF −45 °C
CTFE +52 °C

All the fluoropolymers are thermally stable and relatively inert. Various copolymers show a linear relationship between CTFE wt % and T_g.

'O' Group: Rubbers Having Carbon and Oxygen in the Main Chain

CO: Poly(epichlorohydrin) (HERCLOR H)—the parent material from which came:

ECO: Copolymer of epichlorohydrin and ethylene oxide (HERCLOR C)

d = 1·27 T_g −47 °C

GPO: Copolymer of propylene oxide and allyl glycidyl ether (PAREL)

d = 1·01 T_g (PAREL 58) −73 °C

All these materials have good heat resistance and excellent low temperature properties.

'Q' Group: Silicone Rubbers

MQ: Polydimethylsiloxane; depending upon molar mass can be oil, wax or rubber.

T_g −127 °C (unknown sample)

MPQ: As MQ with the addition of phenylmethylsiloxane.

Typical T_g −86 °C

MPVQ: As above but with vinyl groups.

MFQ: As MQ but fluorinated.

These are all relatively stable thermally and because of their cold cure characteristics may be used as electrical insulants, seals, moulds, etc.

'R' Group: Rubbers Having an Unsaturated Carbon Backbone

ABR: Refers to copolymers of butadiene and methyl methacrylate (e.g. BUTAKON ML)—used to impregnate paper

T_g −57 °C at 25% PMMA

but also includes the terpolymer with acrylonitrile (primer, before adhesive layer applied) and tetrapolymer with styrene (used as a synthetic rubber).

BR: Poly(butadiene)—available as high *cis* (98% +), high *trans* (98% +) and anywhere in between. Can also have vinyl groups present at any level.

d = 0·91–0·93 T_g −107 °C (100 *cis* and 100 *trans*) → −15 °C (100 vinyl).

A linear relationship exists between these ranges depending upon the vinyl content *not* affected by *cis/trans* ratio. General-purpose rubbers usually 90 +% *cis* or about 45% *cis*, 45% *trans*, 10% vinyl. High vinyls have some specialist uses.

CR: Poly(β-chlorobutadiene) (e.g. CHLOROPRENE, NEOPRENE). Two main types, 'G', amber in colour with large molar mass range centred at about 100 000; 'W', white, molar mass of narrower range and centred about 200 000. Used as an adhesive or where oil or ozone resistance required; gaskets, subaqua suits, etc.

d = 1·23 T_g −45 °C

IIR: Copolymer of isobutylene and isoprene (BUTYL). Only a small amount of diene added (~5) to give crosslinkable sites. Has a low gas permeability, hence uses in inflatable products, and as general-purpose rubber.

d = 0·92 T_g −67 °C → −75 °C (varies with isoprene loading)

BIIR: Brominated IIR
CIIR: Chlorinated IIR } with 2–3% w/w halogen to decrease gas permeability and improve self-adhesion on building (e.g. HYCAR 2202, BUTYL HT 1066, 1088). Uses as for IIR.

$d = 0{\cdot}95$ T_g (HT 1066) −70 °C

IR: Synthetic *cis*-poly(isoprene) (e.g. CARIFLEX, NATSYN, SKI3) *cis* level 90–99%, remainder *trans* and *vinyl.* General-purpose rubber.

$d = 0{\cdot}91$ T_g −68 → −70 °C

NBR: Copolymer; acrylonitrile and butadiene (e.g. KRYNAC, NITRILE) available with a wide range of ACN loadings to alter hardness; oil-resistant applications.

$d = 0{\cdot}95$–$1{\cdot}05$ T_g, linear dependency upon ACN loading; −65 °C at 15% → 0 °C at 50% for the random copolymer. Also available is terpolymer (see ABR) and tetrapolymer with styrene.

NR: *cis*-poly(isoprene) natural rubber, 100% *cis*, no *trans* or *vinyl* species detected (<0·1%). Contains *ca.* 95% polyisoprene. Various grades available—RSS, SMR, SIR, SLR, NIG, and modified NRs: PA, SP, OENR, ENR, DPNR. The original general purpose rubber.

$d = 0{\cdot}92$ T_g −72 °C

SBR: Random copolymer of styrene and butadiene. Styrene level varies from 10 to 80% w/w but the 'usual' level is 23·5%. Many types available and the exact type specified by a numerical code. General-purpose rubbers; like NR, vast amounts used in tyres.

$d = 0{\cdot}93$ T_gs show a lot of variation:
−60 °C for 23·5% styrene
→ −38 °C for 36% styrene

Also available as ter/tetra polymer systems (see ABR and NBR).

'T' Group: Rubbers Having Carbon (Oxygen) and Sulphur in the Main Chain

OT: Polymer of *bis*chloroalkylether (or formal), with sulphur. Most common one uses *bis*-2-chloroethylformal; $CH_2(OCH_2CH_2Cl)_2$

(with a little 1,2,3-trichloropropane for crosslinking) THIOKOL ST.

$d = 1{\cdot}3$ T_g $-59\,°C$

EOT: As above, but co-polymerized with ethylene dichloride. All of these smell strongly of sulphur but are used for oil and solvent seals. The liquid polymers cold cure and find a wide acceptance as sealants in the building trade. Popular ones include:

Poly(ethylene disulphide) $d = 1{\cdot}4$ T_g $-27\,°C$

Poly(butyl ether disulphide) $d = 1{\cdot}1$ T_g $-76\,°C$

'U' Group: Polymer Chain Contains Carbon, Oxygen and Nitrogen

AU: Polyester urethanes } See Chapter 1, Table 1.1
EU: Polyether urethanes } for structural details.

A wide range of materials used as oil-resistant materials, in oxidation-resisting applications and as lightweight shoe soling.

$d = 1{\cdot}2$ T_g AU often $-30\,°C$

T_g EU often $-50\,°C$

but values are very variable.

Although not elastomers, certain other polymeric materials merit inclusion because of their application as rubber-like materials:

Designation

PVC: Polyvinylchloride; hard brittle material ($d = 1{\cdot}4$) often co-polymerized with vinylidene chloride, vinyl acetate, styrene, ABR, ethylene vinyl acetate etc. for a wide range of applications. When plasticized with esters (often phthalates) it becomes quite 'rubbery', used in conveyor belting, and in paints, varnishes, floor coverings, gramophone records, erasers, flexible tubing, waterproof fabrics, Wellington boots.

PE: Polyethylene; a vast range of types available, HDPE (high-density polyethylene) used blended with elastomers such as EPDM to give a thermoplastic elastomer.

PP: Polypropylene; also used to a very appreciable extent in thermo-plastic elastomers.

PS: Polystyrene; only occasionally met as a reinforcing plastic, within an elastomeric continuous phase (e.g. in shoe soling) but can be considered to be present in some thermoplastic elastomers such as the block copolymers:

SIS styrene–isoprene–styrene

SBS styrene–butadiene–styrene

Spectral and some thermal data show the styrene as 'polystyrene' rather than randomly dispersed styrene as in SBR.

TPR: Thermoplastic rubber; for full classification, see Chapter 1, Table 1.1.

Chlorinated rubber; refers specifically to chlorinated natural rubber, used for paints and adhesives. The theoretical level for $(C_5H_8Cl_2)_n$ is 51% but commercial chlorinated rubber contains ~65% Cl which is ascribed to the structure

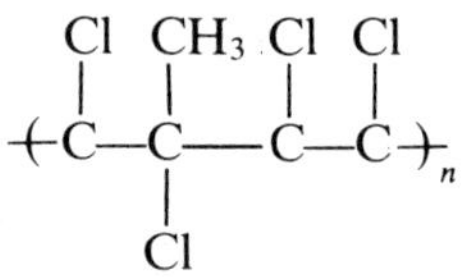

which requires 68·3% Cl.

Rubber hydrochloride; again refers specifically to hydrochlorinated natural rubber—usually with about 90% of the double bonds hydrochlorinated (30% Cl). Plasticized material produced as film (e.g. PLIOFILM) for packaging.

M.G. Rubber; natural rubber to which methyl methacrylate has been grafted, commercial materials generally contain 30% or 49% w/w methacrylate.

Gutta Percha; *Balata*; } Polyisoprene, with 100% of the units *trans*; pure material not unlike PVC in feel and, when plasticized, can have similar uses.

Chicle; a naturally occurring mixture of *cis* and *trans* polyisoprene (25:75), with resins, used in chewing-gum.

Guayule; natural *cis*-polyisoprene isolated from the shrub *Parthenium argentatum* by solvent extraction. Uses and properties as for NR, but smell reminiscent of gin. Availability likely to increase substantially over the next few years.

For comprehensive details of manufacturers, trade names, detailed technical data, and application of the whole range of elastomers, together with similar details for chemicals used in the rubber industry, the interested reader is referred to the two annual publications:

Rubber Red Book, Communication Channels Inc., 6285 Barfield Road, Atlanta, Ga. 30328, USA; and *The Blue Book*, Bill Communications Inc., 633 Third Avenue, New York, USA.

Appendix III

Intercorrelation of Analytical Techniques

At the very beginning of this book the problems of devising a comprehensive analytical scheme, without some prior knowledge of the sample, were discussed. Throughout subsequent chapters specific techniques and analytical schemes have been described which led to the derivation of the formulations of two samples of vulcanized rubber from a wide variety of primary analytical data.

Although a single ideal scheme is not practicable, an attempt has been made, in Fig. A1, to illustrate the interrelationships which do exist between many of the analytical procedures which could be available to the rubber-analyst. It is not so much intended as a comprehensive flow chart for routine analysis but more as a stimulus, to be inspected when an unusual problem arises, in the hope that its breadth suggests a possible new line of attack.

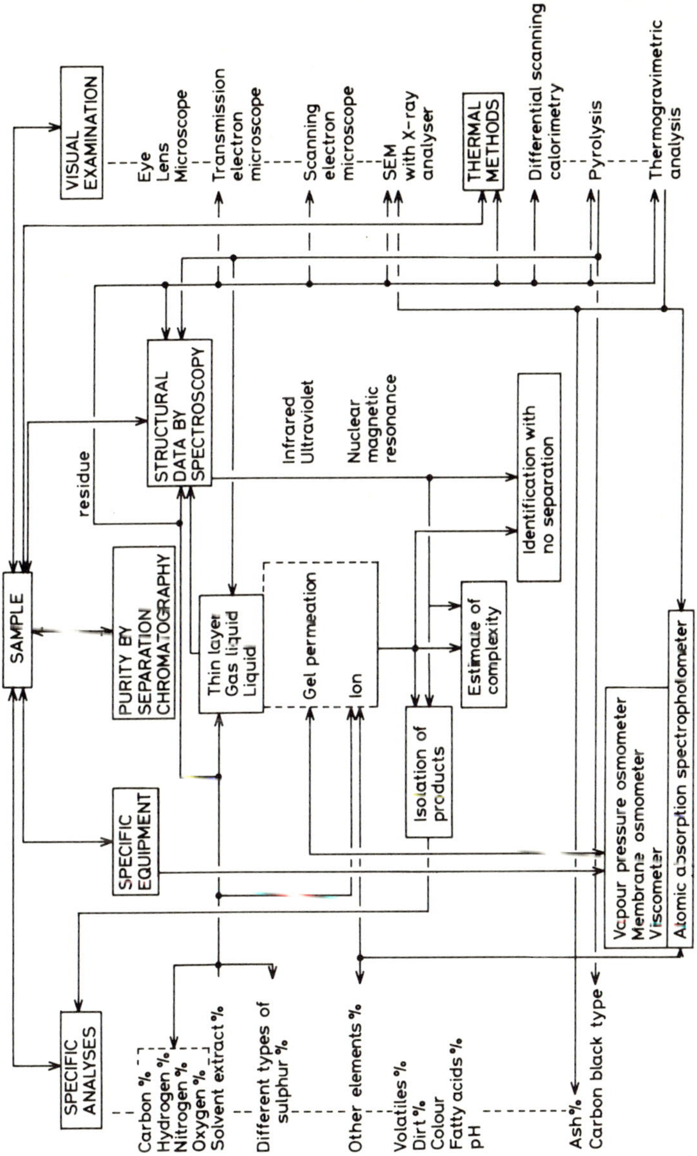

FIG. A1. Interrelationships between the analytical procedures available to the rubber-analyst.

Author Index

Numbers in italic type indicate those pages on which references are given in full.

Index